Queueing Theory 1

SCIENCES

Mathematics, Field Director – Nikolaos Limnios

Queuing Theory and Applications, Subject Head – Vladimir Anisimov

Queueing Theory 1

Advanced Trends

Coordinated by
Vladimir Anisimov
Nikolaos Limnios

iSTE

WILEY

First published 2020 in Great Britain and the United States by ISTE Ltd and John Wiley & Sons, Inc.

ISTE Ltd
27-37 St George's Road
London SW19 4EU
UK

www.iste.co.uk

John Wiley & Sons, Inc.
111 River Street
Hoboken, NJ 07030
USA

www.wiley.com

Library of Congress Control Number: 2019955378

British Library Cataloguing-in-Publication Data
A CIP record for this book is available from the British Library
ISBN 978-1-78945-001-9

ERC code:
PE1 Mathematics
PE1_21 Application of mathematics in industry and society

MIX
Paper from responsible sources
FSC® C013604

Contents

Preface

Vladimir ANISIMOV[1] and Nikolaos LIMNIOS[2]

[1] *Amgen Inc., London, United Kingdom*

[2] *University of Technology of Compiègne, France*

Queueing theory is a huge and very rapidly developing branch of science that originated a long time ago from the pioneering works by Erlang (1909) on the analysis of the models for telephone communication.

Now, it is growing in various directions including a theoretical analysis of queueing models and networks of rather complicated structure using rather sophisticated mathematical models and various types of stochastic processes. It also includes very wide areas of applications: computing and telecommunication networks, traffic engineering, mobile telecommunications, etc.

The aim of this book is to reflect the current state-of-the-art and some contemporary directions of the analysis of queueing models and networks including some applications.

The first volume of the book consists of 10 chapters written by world-known experts in these areas. These chapters cover a large spectrum of theoretical and asymptotic results for various types of queueing models, including different applications.

Chapter 1 is devoted to the investigation of some theoretical problems for non-classical queueing models including the analysis of queues with inter-dependent arrival and service times.

Chapter 2 deals with the analysis of some characteristics of fluid queues including busy period, congestion analysis and loss probability.

Some contemporary tendencies in the asymptotic analysis of queues are reflected in the following three survey Chapters 3, 7 and 10.

Chapter 3 includes the results on the average and diffusion approximation of Markov queueing systems and networks with a small series parameter ε including applications to some Markov state-dependent queueing models and some other type of models, in particular, repairman problem, superposition of Markov processes and semi-Markov type queueing systems.

Diffusion and Gaussian limits for multi-channel queueing networks with rather general time-dependent input flow and under heavy traffic conditions including some applications to networks with semi-Markov or renewal type input and Markov service are considered in Chapter 7.

Chapter 10 is devoted to the asymptotic analysis of time-varying queues using the large deviations principle for two-time-scale non-homogeneous Markov chains including the analysis of the queue length process and some characterizations of the quality and the efficiency of the system.

The analysis of so-called retrial queueing models is reflected in two Chapters – 4 and 8.

In Chapter 4, two models that provide some modifications of "First-Come First-Served" retrial queueing system introduced by Laszlo Lakatos are investigated.

Chapter 8 gives a review of recent results on single server finite-source retrial queueing systems with random breakdowns and repairs and collisions of the customers.

The analysis of transient behavior of the infinite-server queueing models with a mixed arrival process and Coxian service times and of the Markov-modulated infinite-server queue with general service times is considered in Chapter 5.

Chapter 6 deals with the applications of fast simulation methods used in queueing theory to solve some high-dimension combinatorial problems in case the other approaches fail.

A survey on the analysis of a strong stability method and its applications to queueing systems and networks and some perspectives are considered in Chapter 9.

The second volume of the book will include the additional chapters devoted to some other contemporary directions of the analysis of queueing models.

The volumes will be useful for graduate and PhD students, lecturers, and also the researchers and developers working in mathematical and stochastic modelling and various applications in computer and communication networks, science and engineering in the departments of Mathematics & Applied Mathematics, Statistics, or Operations Research at universities and in various research and applied centres.

Dedication to Igor Mykolayovych Kovalenko who died shortly after the writing of this book

Ukrainian and world science is mourning the loss of a brilliant scientist, Professor Igor Mykolayovych Kovalenko, who died on October 19, 2019, after a difficult fight with heart disease.

Prof. Igor Kovalenko was a prominent Ukrainian mathematician in the field of probability theory and its practical applications, a disciple and associate of Boris Gnedenko and Vladimir Korolyuk. He became famous worldwide for his book *Introduction to Queueing Theory*, written together with Gnedenko. He founded a scientific school in the theory of reliability, queueing theory and cryptography, well known in Ukraine and all over the world.

Igor Kovalenko was born on March 16, 1935 in Kyiv, Ukraine. After graduating from the Faculty of Mechanics and Mathematics of Kyiv Taras Shevchenko University, he worked at the Computing Centre of the Academy of Sciences of Ukraine. From 1962 till 1971, Kovalenko worked in Moscow, where he headed a laboratory at the Moscow Institute of Electronic Engineering, and together with other Gnedenko's disciples, was the head of the seminar on queueing theory at Moscow State University. Many leading scientists of the former Soviet Union and foreign countries attended this seminar.

Based on the model of piecewise linear Markov processes developed by him, Kovalenko built a mathematical model of a complex defence system reliability and developed numerical algorithms for its implementation based on the method of a small parameter.

In 1964, Igor Kovalenko became a Doctor of Technical Sciences. He formulated the principle of monotonous failures, which, while maintaining high accuracy, significantly simplified the calculations of system reliability. In 1970, Kovalenko was awarded the degree of Doctor of Physics and Mathematics for another thesis on the probabilistic theory of systems of random Boolean equations. Being a doctor twice over is a very rare practice in the scientific world.

After returning to Kyiv in 1971, Prof. Kovalenko founded and headed the Department of Mathematical Methods of the Theory of Complex Systems Reliability at the V.M. Glushkov Institute of Cybernetics. Two areas of research formed the mainstream of investigations: approximate combined analytical and statistical methods of reliability analysis, and theoretical and applied cryptography, systems and methods for data protection. Under his guidance, the first national standard in the field of cryptographic information security was developed in Ukraine.

Prof. Kovalenko is the author of 25 monographs and more than 200 articles. He was elected as an Academician of the National Academy of Sciences of Ukraine in 1978 (Corresponding Member since 1972). He was an extremely hard-working, honest and sincere person, a competent manager and, thanks to his human qualities, professional experience and knowledge, highly respected among his colleagues.

Prof. Igor Kovalenko left many disciples, among them there are many professors and associate professors. All of them preserve in their memory the unforgettable days of joining the science and independent creativity under the guidance of a Great Scientist and Teacher, hours of direct communication with a person of great erudition and high culture.

1

Discrete Time Single-server Queues with Interdependent Interarrival and Service Times

Attahiru Sule ALFA
University of Manitoba, Canada
and
University of Pretoria, South Africa

Discrete time single-server queues in which the interarrival and service times are interdependent are presented. First, we study the simple Geo/Geo/1 system, let the interarrival times depend on the service times and then consider special cases. The idea is then extended to the PH/PH/1 system. We then consider the case where the interarrival times are constructed from a combination of a set of interarrival times driven by the service times distribution. The initiating vector for the resulting combined PH distribution for the interarrival times is constructed as a function of the service times distribution, so that any changes in the service time distributions are reflected in this initiating vector. We capitalize on the structures of discrete phase type distributions in generalizing the resulting interarrival times. Finally, we introduce a general case where there is interdependence between service and interarrival times. We present a generalized matrix version of the bivariate geometric distribution, which can be used to capture some interdependence between the interarrival and service times.

1.1. Introduction

Most of the queueing models in the literature usually assume that the interarrival times and service times are independent of each other. While this may be true in

Queueing Theory 1,
coordinated by Vladimir ANISIMOV, Nikolaos LIMNIOS. © ISTE Editions 2020.

some instances, it was pointed out by Cidon *et al.* (1993) that there are instances where there are dependencies between the two distributions. They mentioned that in communication systems where finite speed of communications links constrain the amount of data that can be received, one finds the interarrival times and service times to be correlated. They studied communication systems in which the interarrival time between nth and $(n+1)st$ customer depends naturally on the service time B_n of the nth customer. In a subsequent paper, Cidon *et al.* (1996) studied queues with interarrival times proportional to service times. Earlier, Fendick *et al.* (1989) did demonstrate that the dependence between successive service times and interarrival times for data packets can be important. Also in another earlier paper, John (1963) considered the case where the service time depends on the interarrival times and gave examples of applications. Benes (1957, 1960) was the first not to assume that interarrival times and service times are both independent of each other. It is, however, surprising that there has been limited amount of work following up on that.

Chao (1965) captured the interdependence between interarrival times and service times by using a bivariate exponential distribution to capture the two processes. By defining the correlation between the interarrival and service times as ρ, Chao (1965) pointed out that a customer's waiting time is monotonically decreasing in ρ in increasing convex ordering sense. While this seems common sense, it was the first confirmation based on rigorous analysis. Later, Muller (2000) studied the waiting times when there is dependency between amount of work (service time) brought by a customer and the subsequent interarrival times. He showed that the stronger the dependency is between interarrival times and service times, the more the waiting time decreases in increasing convex ordering sense (similar to Chao 1995). Hadidi (1981a, 1981b) used bivariate exponential distribution for the joint interarrival and services times to capture the dependency, and carried out simulation and numerical studies to understand the effects of the correlation coefficient on waiting times.

One of the most recent works on this subject is by Iyer and Manjunanth (2006) who also studied the interarrival and service times of queue by using mixtures of bivariates. Panda *et al.* (2017), in a work similar to the one by Iyer and Manunanth (2006), studied queues in which batch interarrival and service times are correlated by representing them with a bivariate mixture of rational (R) distributions.

Generally, most of the reported works focus on proportional relationships between interarrival times and service times. This is quite limiting. In this chapter, this aspect is relaxed, and the interdependence between the interarrival times and services is captured in a one-way dependence of interarrival time on service time and later in a two-way dependence. However, that dependence is allowed to be general. In the one-way case, we start by considering the Geo/Geo/1 system in which the arrival probability depends on service completion probability. Specifically in the general one-way case, we assume that there is a set of several, but finite, number of interarrival times. At the end of a service completion, the next interarrival time of a user is selected based on the service time. Even though the distributions of the interarrival times and service times will be general, we will be using the discrete phase type distribution to represent them based on the fact any discrete general distributions, with finite support, can be represented by a Markov chain (Alfa 2004).

1.2. The Geo/Geo/1 case

Consider the simple Geo/Geo/1 system where the probability of arrival is a and of service completion is b. Let $X_n(a,b),\ n = 0,1,2,\cdots$ be the number in the system at time n for a specified pair (a,b), with $x_i(a,b) = Pr\{X_n(a,b) = i\}|_{n\to\infty}$ for the case of a stable system. Based on the results in Alfa (2016), the associated discrete time Markov chain (DTMC) has the transition matrix P written as

$$P = \begin{bmatrix} 1-a & a & & & \\ (1-a)b & ab+(1-a)(1-b) & a(1-b) & & \\ & (1-a)b & ab+(1-a)(1-b) & a(1-b) & \\ & & \ddots & \ddots & \ddots \end{bmatrix}.$$

For a stable system with $\frac{a(1-b)}{b(1-a)} < 1 \Rightarrow \frac{a}{b} < 1$, we have

$$x_0(a,b) = \frac{b-a}{b},\ x_i(a,b) = (b(1-b))^{-1}(b-a)\theta(a,b)^i,\ i \geq 1,$$

and

$$E(X(a,b)) = \frac{a(1-a)}{b-a},$$

where $\theta(a,b) = (b(1-a))^{-1}(a(1-b))$ and $E(X(a,b))$ is the expected value of $X_n(a,b)|_{n\to\infty}$.

Next we consider the cases of $a = f(b)$. We require that the following condition applies all the time in this section:

$$0 < a = f(b) < b < 1. \qquad [1.1]$$

We consider two main cases: one when $f(\cdot)$ is a decreasing function and the second one when it is an increasing function.

1.2.1. *Arrival probability as a function of service completion probability*

When we consider the case where $a = f(b)$, we have

$$x_0(b) = \frac{b - f(b)}{b}, \text{ and } x_i(b) = (b(1-b))^{-1}(b - f(b))\theta(b)^i,\ i \geq 1,$$

and

$$E(X(b)) = \frac{f(b)(1 - f(b))}{b - f(b)},$$

where $\theta(b) = (b(1 - f(b)))^{-1}((1 - b)f(b))$.

Next, we study the behavior of this system as a function of b.

1.2.1.1. *Consider* $x_0(b)$

By differentiating $x_0(b)$ with respect to b, we have

$$\frac{dx_0(b)}{db} = \frac{f(b) - bf'(b)}{b^2}, \text{ where } f'(b) = \frac{df(b)}{db}.$$

PROPOSITION 1.1.– By studying the derivative of $x_0(b)$ w.r.t. b, we observe

1) if $f'(b)$ is negative, then an increase in b leads to an increase in the probability that the system is empty;

2) if $f'(b)$ is positive, then the probability that the system is empty could be increasing or decreasing, depending on whether $f(b) > bf'(b)$ or $f(b) < bf'(b)$, respectively. Hence, an increase in the arrival rate as a function of the service rate does not necessarily mean a decrease in the emptiness probability.

We obtain additional information by considering the rate of increase in the probability of empty system as a rate of increase in the service completion probability, i.e.

$$\frac{d^2(x_0(b))}{db^2} = \frac{2b^2f'(b) - 2bf(b) - b^3f''(b)}{b^4}, \text{ where } f''(b) = \frac{d^2(f(b))}{db^2}.$$

PROPOSITION 1.2.– If $f(b)$ is convex and $f'(b) < 0$, then $x_0(b)$ is definitely concave. If $f(b)$ is decreasing fast, then $x_0(b)$ is increasing slowly, i.e. $f(b)$ decreasing faster than $x_0(b)$ increases. If, however, $f(b)$ is convex and $2b^2f'(b) - 2bf(b) > b^3f''(b)$, then $x_0(b)$ is convex.

1.2.1.2. *Consider* $E(X(b))$

Next let us consider $E(X(b))$. By differentiating $E(X(b))$ w.r.t. b, we have

$$\frac{dE(X(b))}{db} = \frac{(b-f(b))[f'(b)(1-f(b))-f(b)f'(b)]-f(b)(1-f(b))(1-f'(b))}{(b-f(b))^2}$$

$$= \frac{(b-f(b))f'(b)[1-2f(b)]-f(b)(1-f(b))(1-f'(b))}{(b-f(b))^2}.$$

Define

$$A(b) = f(b)(1-f(b)), \text{ and } B(b) = (b-f(b))(1-2f(b)).$$

For a stable system, we know that $b - f(b)$ and $1 - f(b)$ are always positive, hence both $A(b)$ and $B(b)$ are always positive if $f(b) < \frac{1}{2}$.

We thus have that the behavior of $E(X(b))$ depends entirely on $f'(b)$, $A(b)$ and $B(b)$ and whether $f(b) < \frac{1}{2}$ or not. $E(X(b))$ depends on the sign and values of the first three parameters, and the value of the fourth one.

PROPOSITION 1.3.– For a stable system,

1) if $A(b) > f'(b)(A(b)+B(b))$, then $E(X(b))$ will be decreasing as b increases;

2) if $A(b) < f'(b)(A(b)+B(b))$, then $E(X(b))$ will be increasing as b increases.

The first part of proposition 1.3 is straightforward in that an increase in the service rate reduces the arrival, thereby reducing the traffic intensity and decreasing the expected number in the system. However, an increase to the service rate could also lead to an increase in the mean number in the system as observed in the second part of the proposition, which can be explained by the rate at which the arrival probability increases compared to the service completion probability (b), especially if that rate is high.

1.2.1.3. *Special cases*

CASE 1: Consider the case where $f(b)$ is a decreasing function of b, say for example $a = f(b) = 1 - b^2$, with the strict condition that $f(b) = 1 - b^2 < b$. For this, we have

$$E(X(b)) = \frac{b^2(1-b^2)}{b-1+b^2} = EX_1.$$

In this case, as b increases we see that $E(X(b))$ decreases as well. This is because the arrival probability decreases at a faster rate than service completion increases, i.e. $f'(b) = -2b$.

REMARK 1.1.– For the case where $a = f(b) = 1 - b^2$, the arrival probability is a decreasing concave function of the service completion probability and the mean number in the system is decreasing in a convex form.

CASE 2: Next consider the case where $f(b)$ is an increasing function of b, say for example $a = f(b) = b^2$, with the strict condition that $f(b) = b^2 < b$. In this case, we have

$$E(X(b)) = \frac{b^2(1-b^2)}{b-b^2} = b(1+b) = EX_2.$$

In this case, as b increases we see that $E(X(b))$ increases because the arrival probability increases faster than service completion increases, i.e. $f'(b) = 2b$.

REMARK 1.2.– *For the case where $a = f(b) = b^2$, with arrival probability increasing in a convex form of the service completion probability, the mean number in the system increases in a convex form as well.*

CASE 3: Let us consider another case where $f(b)$ is a linear decreasing function of b, say for example $a = f(b) = 1 - b$, with the strict condition that $1 - b < b$. In this case, we have

$$E(X(b)) = \frac{b(1-b)}{2b-1} = EX_3.$$

In this case, as b increases we see that $E(X(b))$ decreases because the arrival probability is decreasing in the same rate as the service completion rate increases, i.e. $f'(b) = -1$.

REMARK 1.3.– *However, for the case where $a = f(b) = 1 - b$, the arrival probability decreasing in a linear form of the service completion probability, the mean number in the system decreases in a strict convex form.*

We summarize the results of the mean number in the system for the three cases in Table 1.1, with **** representing infeasible situations because the traffic intensity is greater than 1.

1.2.2. *Service times dependent on interarrival times*

The case of the Geo/Geo/1 system in which the service times depend on the interarrival times can be studied using the same idea and techniques as in section 1.2.1. If we define the arrival probability as a and the service time completion probability as $b = g(a)$, it is straightforward to extend the ideas of that section directly. The queueing performance measures such as the mean number in the system, which we now define as $E(X(a))$ can be written as

$$E(X(a)) = \frac{a(1-a)}{g(a)-a},$$

and the stability condition given as $a < g(a)$. Since the procedures will be the same, they will not be discussed in this chapter.

b	EX_1	EX_2	EX_3
0.10	****	0.1100	****
0.20	****	0.2400	****
0.30	****	0.3900	****
0.40	****	0.5600	****
0.50	****	0.7500	****
0.60	****	0.9600	1.2000
0.70	1.3153	1.1900	0.5250
0.75	0.7875	1.3125	0.3750
0.80	0.5236	1.4400	0.2667
0.85	0.3502	1.5725	0.1821
0.90	0.2168	1.7100	0.1125
0.95	0.1032	1.8525	0.0528

Table 1.1. *Mean number in system for the three cases*

1.3. The PH/PH/1 case

In this section, the idea of Geo/Geo/1 system with interdependent interarrival and service times is generalized to the case of the PH/PH/1 system. However, first let us give a very brief review of the discrete PH distribution.

1.3.1. *A review of discrete PH distribution*

Consider a discrete time absorbing Markov chain with state space $X_n,\ n = 0, 1, 2, \cdots$, with $X_n = 0, 1, 2, \cdots, N$, where state 0 is an absorbing state. The transition matrix of this chain can be written as

$$P = \begin{bmatrix} 1 & \mathbf{0} \\ \mathbf{t} & T \end{bmatrix},$$

where $0 \leq T_{i,j} \leq 1, \ \forall (i,j), \ 0 \leq \sum_{j=1}^{N} T_{i,j} \leq 1, \ \forall i,$ with $\sum_{j=1}^{N} T_{i,j} < 1$ for at least one i. Also define $\mathbf{t} = \mathbf{1} - T\mathbf{1}$, where $\mathbf{1}$ is a column of ones.

There is a discrete random variable Y, which is said to have a PH distribution $(\boldsymbol{\alpha}, T)$ if one can write

$$y_k = Pr\{Y = k\} = \boldsymbol{\alpha} T^{k-1}\mathbf{t}, \ k \geq 1,$$

where $\boldsymbol{\alpha} = [\alpha_1, \ \alpha_2, \ \cdots, \ \alpha_N]$, with $0 \leq \alpha_n \leq 1, 1 \leq n \leq N$, $\boldsymbol{\alpha}\mathbf{1} \leq 1$ and $\alpha_0 = 1 - \boldsymbol{\alpha}\mathbf{1}$, $y_0 = \alpha_0$.

Several well-known discrete distributions can be represented as PH distributions. Examples include the geometric distribution, the negative binomial distribution, to name just a few. In addition, most discrete distributions can be reasonably approximated by discrete PH distribution (see Mészáros *et al.* 2014 and references therein).

It was shown in Alfa (2004) that any discrete distribution, with finite support, can be represented by PH distribution with elapsed time or remaining time format. For example, consider the interarrival $\mathbf{A}$ and let $a_i = Pr\{\mathbf{A} = i\}, \ i = 1, 2, \cdots, N$. In the elapsed time format, the matrix T has only its superdiagonal elements that are non-zero and every other element is zero, and the vector $\boldsymbol{\alpha} = [1, \ 0, \ 0, \ \cdots, \ ]$, and the matrix T is written as

$$T = \begin{bmatrix} 0 & \tilde{a}_1 & & & \\ & 0 & \tilde{a}_2 & & \\ & & \ddots & \ddots & \\ & & & 0 & \tilde{a}_{N-1} \\ & & & & 0 \end{bmatrix},$$

where $\tilde{a}_i = Pr\{\mathbf{A} \geq i+1 | \mathbf{A} \geq i\} = \frac{\sum_{j=i+1}^{N} a_i}{\sum_{j=i}^{N} a_j}$.

For the remaining time representation, the vector $\boldsymbol{\alpha} = [a_1, \ a_2, \ \cdots, \ a_N]$ is the vector of the discrete distribution and the matrix T is written as

$$T = \begin{bmatrix} 0 & 0 & \cdots & & \\ 1 & 0 & \cdots & & \\ 0 & 1 & 0 & \cdots & \\ \vdots & \ddots & \ddots & \ddots & \\ 0 & \cdots & \cdots & 1 & 0 \end{bmatrix}.$$

1.3.2. *The PH/PH/1 system*

Consider a PH/PH/1 system with interarrival times and service times that are represented by $(\boldsymbol{\alpha}, T)$ and $(\boldsymbol{\beta}, B)$, respectively, with means given as a^{-1} and b^{-1}. Note that $a^{-1} = \boldsymbol{\alpha}(I - T)^{-1}\mathbf{1}$ and $b^{-1} = \boldsymbol{\beta}(I - B)^{-1}\mathbf{1}$. At time $t \geq 0$, let X_t be the number of items in the system, Y_t the phase of interarrival times and Z_t the phase of the ongoing service. It is straightforward to see that $(X_t, Y_t, Z_t), t \geq 0$, is a DTMC with transition matrix P given as

$$P = \begin{bmatrix} F & C & & & \\ E & A_1 & A_0 & & \\ & A_2 & A_1 & A_0 & \\ & & \ddots & \ddots & \ddots \end{bmatrix},$$

where

$$F = T,\ C = (\mathbf{t}\boldsymbol{\alpha}) \otimes \boldsymbol{\beta},\ E = T \otimes \mathbf{b},\ A_0 = (\mathbf{t}\boldsymbol{\alpha}) \otimes B,$$

$$A_1 = T \otimes B + (\mathbf{t}\boldsymbol{\alpha}) \otimes (\mathbf{b}\boldsymbol{\beta}),\ A_2 = T \otimes (\mathbf{b}\boldsymbol{\beta}),\ \mathbf{t} = \mathbf{1} - T\mathbf{1},\ \mathbf{b} = \mathbf{1} - B\mathbf{1},$$

and where $\otimes$ is the Kronecker product. For two matrices U of order $n \times m$ and V of order $r \times k$, the product $U \otimes W$ is a matrix of order $nr \times mk$ (for more details, see Graham 1981).

It is well-known in the literature that for a stable PH/PH/1 system there is a matrix R, with spectral radius less than 1, which is the minimal non-negative solution to $R = A_0 + RA_1 + R^2A_2$ and the stationary distribution of the DTMC, which is given as $\boldsymbol{x} = [\boldsymbol{x}_0, \boldsymbol{x}_1, \boldsymbol{x}_2, \cdots]$, i.e. $\boldsymbol{x} = \boldsymbol{x}P,\ \boldsymbol{x}\mathbf{1} = 1$ can be obtained through the matrix-geometric theorem (Neuts 1981) as

$$\boldsymbol{x}_{i+1} = \boldsymbol{x}_i R,\ i \geq 1,$$

after appropriately obtaining the boundary solution $[\boldsymbol{x}_0, \boldsymbol{x}_1]$. For more details, see Alfa (2016).

The idea of the Geo/Geo/1 system discussed in section 1.2 can be extended to the PH/PH/1, but requires a more involved representation and analysis. Consider the case where $(\boldsymbol{\alpha}, T)$ depends on $(\boldsymbol{\beta}, B)$. The dependence can be captured by letting

$$\boldsymbol{\alpha} = g_\alpha(\boldsymbol{\beta}),\ \text{and } T = f_T(B),$$

with the condition that

$$0 < a < b < 1.$$

We see that the case of the Geo/Geo/1 is a special case where $(\boldsymbol{\beta}, B) = (1, b)$, hence $\boldsymbol{\alpha} = 1$ and $T = f_T(B) = f(b)$. Several functional forms of $\boldsymbol{\alpha} = g_\alpha(\boldsymbol{\beta})$, and $T = f_T(B)$ can be studied, but are beyond the scope of this chapter. However, a small example of a representation is presented as an illustration. Consider a simple case with T and B having the same size, i.e. $n \times n$. Define a doubly stochastic matrix V of order $n \times n$. We can select

$$\boldsymbol{\alpha} = g_\alpha(\boldsymbol{\beta}) = \boldsymbol{\beta}V, \text{ and } T = f_Y(B) = TV,$$

provided $b = (\boldsymbol{\beta}(I - B)^{-1}\mathbf{1})^{-1} > a = (\boldsymbol{\alpha}(I - T)^{-1}\mathbf{1})^{-1}$.

Even though this system is more complicated to analyze compared to the Geo/Geo/1 case, the idea is exactly the same and it can be studied numerically.

Next we consider a model in which there is a general service time distribution; however, there are several interarrival time distributions and the selection of the next interarrival time distribution depends on the service time distribution.

1.4. The model with multiple interarrival time distributions

We consider the case where the service times are independent and the interarrival times depend on the distribution of the service times, and not just one parameter such as the mean:

– let the service time, S, be of a general renewal type in discrete time with finite support. Let $s_j = Pr\{\mathrm{S} = j\}$, $j = 1, 2, \cdots, M$. Let us write $\mathrm{s} = [s_1, s_2, \cdots, s_M]$;

– let there be K interarrival times with the selection of the interarrival time depending on the service time. Let $A_{nk} \leq N_k < \infty$ be the interarrival time between the nth and the $(n-1)$st customer if the kth interarrival is selected. Let $a_j(k) = Pr\{A_{nk}\}_{n\to\infty} = j$, $j = 1, 2, \cdots, N_k$;

– by the results in Alfa (2004), we can represent each of the K interarrival times by discrete PH distribution and in this case chose the remaining time approach. Let $(\boldsymbol{\alpha}(k), T(k))$ be the PH representation for the kth interarrival time distribution of dimension N_k.

Next we define some parameters that will be used to relate the interarrival times to the service times.

1.4.1. *Preliminaries*

Define a $K \times M$ matrix E_{KM} such that its elements have values of 0 or 1 and in addition

– for $K \geq M$, we require that $E_{KM}\mathbf{1}_M = \mathbf{1}$, $\mathbf{1}_K^T E_{KM} \geq \mathbf{1}^T$ and $\mathbf{1}_K^T E_{KM}\mathbf{1}_M = M$;

– for $K \leq M$, we require that $E_{KM}\mathbf{1}_M \geq \mathbf{1}$, $\mathbf{1}_K^T E_{KM} = \mathbf{1}^T$ and $\mathbf{1}_K^T E_{KM}\mathbf{1}_M = M$.

Next we define a stochastic vector $\boldsymbol{\psi} = [\psi_1, \psi_2, \cdots, \psi_K]$, which is a function of s, i.e. $\boldsymbol{\psi} = \mathbf{f}(\mathbf{s})$, where $\mathbf{f} : \mathbf{R}^M \to \mathbf{R}^K$. We give some very simple examples of this type of mapping.

– Consider the case when $K = M$. A simple mapping of $\mathbf{f}$ could be $\psi_i = s_i, \forall i$, i.e.$\boldsymbol{\psi} = \mathbf{s}$. Another possible example could be $\boldsymbol{\psi} = \mathbf{s}E_{K,K}$. Finally, we could generate $\boldsymbol{\psi}$ from s using any argument possible, as long as $\boldsymbol{\psi}$ depends on s. For this simple example alone, there are $K!$ ways of generating the mapping by rearranging the elements of $\boldsymbol{\psi}$, if we keep the values of s_i unchanged, e.g.

$$\boldsymbol{\psi} = [s_2, s_M, s_1, s_3, \cdots, s_{M-1}].$$

– There are two possible cases when $K \neq M$.

- Consider the case where $K < M$, an example of the mapping here could be

$$\boldsymbol{\psi} = [s_1, s_2, \cdots, s_{K-1}, s_K + s_{K+1} + \cdots + s_M],$$

as an example only, and

- for the case of $K > M$, an example of the mapping here could be

$$\boldsymbol{\psi} = [s_1, s_2, \cdots, s_{M-1}, \frac{s_M}{K-M+1}, \frac{s_M}{K-M+1}, \cdots, \frac{s_M}{K-M+1}],$$

again just an example.

The K interarrival times can each be written as PH distribution $(\boldsymbol{\alpha}(k), T(k))$, where $k = 1, 2, \cdots, K$. Hence the resulting interarrival time, which is a function of the service time, is a PH distribution represented by $(\tilde{\boldsymbol{\alpha}}, T)$, where

$$T = \begin{bmatrix} T(1) & & & \\ & T(2) & & \\ & & \ddots & \\ & & & T(K) \end{bmatrix}, \text{ and } \tilde{\boldsymbol{\alpha}} = \boldsymbol{\psi} \star \boldsymbol{\alpha},$$

where $\boldsymbol{\alpha} = [\boldsymbol{\alpha}(1), \boldsymbol{\alpha}(2), \cdots, \boldsymbol{\alpha}(k)]$ and the operator $\star$ is the Khatri–Rao (K-R) product (Khatri and Rao 1968) , and here we have

$$\boldsymbol{\psi} \star \boldsymbol{\alpha} = [\psi_1 \boldsymbol{\alpha}(1), \psi_2 \boldsymbol{\alpha}(2), \cdots, \psi_K \boldsymbol{\alpha}(K)].$$

In applying K-R operator, here the first vector $\boldsymbol{\psi}$ is partitioned into K, 1×1 vectors, i.e. into scalars in order to properly apply the K-R operator.

This is better illustrated with an example.

Example

Consider a case of a service time S with $s_j = Pr\{S = j\}$, $j = 1, 2, 3, 4$, i.e. $M = 4$. Let $K = 3$, with the three interarrival PH distributions given as $(\boldsymbol{\alpha}(k), T(k))$, $k = 1, 2, 3$. Let $\boldsymbol{\psi} = [\psi_1, \psi_2, \psi_3] = [s_2, s_3 + s_4, s_1]$, as an example.

Then we have a system with service time distribution vector given as $\mathrm{s} = [s_1, s_2, s_3, s_4]$ and interarrival time distribution given as a PH with

$$T = \begin{bmatrix} T(1) & & \\ & T(2) & \\ & & T(3) \end{bmatrix}, \text{ and } \tilde{\boldsymbol{\alpha}} = \boldsymbol{\psi} \star \boldsymbol{\alpha}.$$

Numerical examples

Let $\mathrm{s} = [0.1,\ 0.3,\ 0.4,\ 0.2]$ and

$$(\boldsymbol{\alpha}(1), T(1)) = \left([0.1,\ 0.9],\ \begin{bmatrix} 0.3 & 0.6 \\ 0.5 & 0.1 \end{bmatrix} \right),$$

$$(\boldsymbol{\alpha}(2), T(2)) = \left([0.7,\ 0.3],\ \begin{bmatrix} 0.5 & 0.2 \\ 0.1 & 0.8 \end{bmatrix} \right),$$

$$(\boldsymbol{\alpha}(3), T(3)) = \left([0.1,\ 0.5,\ 0.4],\ \begin{bmatrix} 0.1 & 0.2 & 0.6 \\ 0.5 & 0 & 0.2 \\ 0.1 & 0.1 & 0.3 \end{bmatrix} \right).$$

Suppose we select three mappings of s with $\boldsymbol{\psi}_1 = [s_2, s_1, s_3 + s_4] = [0.3,\ 0.1,\ 0.6]$, $\boldsymbol{\psi}_2 = [s_1,\ s_3 + s_4,\ s_2] = [0.1,\ 0.6,\ 0.3]$, and $\boldsymbol{\psi}_3 = [s_4,\ s_2,\ s_1 + s_3] = [0.2,\ 0.3,\ 0.5]$. We can then generate two PH distributions that are both dependent on s, but with different mappings with both having the same T given as follows

$$T = \begin{bmatrix} T(1) & & \\ & T(2) & \\ & & T(3) \end{bmatrix},$$

but different $\tilde{\boldsymbol{\alpha}}_j = \boldsymbol{\psi}_j \star \boldsymbol{\alpha},\ j = 1, 2, 3.$

Let A_j be the interarrival times for the jth mapping and $P_j(k) = Pr\{A_j \geq k\}$. Table 1.2 summarizes the results.

k	1	2	3	4	5	6	7	8	9
$P_1(k)$	0.6490	0.4886	0.3311	0.2382	0.1704	0.1246	0.0919	0.688	0.0522
$P_2(k)$	0.7110	0.5526	0.4225	0.3361	0.2705	0.2210	0.1823	0.1516	0.1268
$P_3(k)$	0.6740	0.5128	0.3657	0.2751	0.2084	0.1613	0.1265	0.1006	0.0810

Table 1.2. *Complementary Cumulative distributions of interarrival times*

In future works, we study how the different arrangements of s affect the features and distribution and moments of the interarrival times. This can be used to show how one can control the arrival process and hence the queue performance by using different selections of the vector s.

1.4.2. *A queueing model with interarrival times dependent on service times*

Consider a single-server queue with service times distribution vector, given as s of M dimension and K independent interarrival times, from which a customer's next interarrival is selected based on the service time experienced by customers. It is clear that the true interarrival times is Markovian Arrival Process (MAP), which is constructed from the K interarrival times and the probability vector $\boldsymbol{\psi}$ of dimension K that was created by mapping service distribution as shown in section 1.4.1. The PH representation of interarrival times has $\kappa = k_1 + k_2 + \cdots + k_K$, where k_j is the number of phases of the jth interarrival time. We let the service times be represented by an elapsed time PH distribution with representation $(\boldsymbol{\beta}, B)$ of order M. Also let $\mathbf{b} = \mathbf{1} - B\mathbf{1}$.

At time $n = 0, 1, 2, 3, \cdots$, let X_n be the number of customers in the system, Y_n the phase of arrival and J_n the phase of the ongoing service. The stochastic process $\{(X_n, J_n, Y_n), n \geq 0\}$ is a DTMC with transition matrix P given as

$$P = \begin{bmatrix} F & C & & & \\ E & A_1 & A_0 & & \\ & A_2 & A_1 & A_0 & \\ & & \ddots & \ddots & \ddots \end{bmatrix},$$

where

$$F = T,\ C = (\mathbf{t}\tilde{\boldsymbol{\alpha}}) \otimes \boldsymbol{\beta},\ E = \mathbf{b} \otimes T$$

$$A_0 = B \otimes (\mathbf{t}\tilde{\boldsymbol{\alpha}}),\ A_1 = (\mathbf{b}\boldsymbol{\beta}) \otimes (\mathbf{t}\tilde{\boldsymbol{\alpha}}) + B \otimes T,\ A_2 = (\mathbf{b}\boldsymbol{\beta}) \otimes T.$$

This Markov chain is a simple Quasi-Birth-Death (QBD), which can be analyzed using the matrix-analytical methods for discrete time queues (Neuts 1981; Alfa 2016).

If this system is stable, which we will assume it is, then there is a unique probability vector $\boldsymbol{x} = [\boldsymbol{x}_0, \boldsymbol{x}_1, \boldsymbol{x}_2, \cdots]$ for which we have

$$\boldsymbol{x} = \boldsymbol{x}P,\ \boldsymbol{x}\mathbf{1} = 1.$$

Further, we have $\boldsymbol{x}_0 = [x_{0,1}, x_{0,2}, \cdots, x_{0,\kappa}]$, $\boldsymbol{x}_i = [\boldsymbol{x}_{i,1}, \boldsymbol{x}_{i,2}, \cdots, \boldsymbol{x}_{i,M}]$ and $\boldsymbol{x}_{i,j} = [x_{i,j,1}, x_{i,j,2}, \cdots, x_{i,j,\kappa}]$.

It is known that for a stable system, there exists a matrix R, with spectral radius less than 1, which is the minimal non-negative solution of the matrix quadratic equation

$$R = A_0 + RA_1 + R^2 A_2.$$

The R matrix has a counterpart stochastic matrix G for a stable system, a minimal non-negative solution to the matrix quadratic equation

$$G = A_2 + A_1 G + A_0 G^2,$$

and there is a simple relationship between the two as follows

$$R = A_0 (I - A_1 - A_0 G)^{-1}.$$

Given the matrix R, we can then use the matrix-geometric results of Neuts (1981) to obtain

$$\boldsymbol{x}_{i+1} = \boldsymbol{x}_i R, \; i \geq 1,$$

after solving for the boundary equations as

$$[\boldsymbol{x}_0, \, \boldsymbol{x}_1] = [\boldsymbol{x}_0, \, \boldsymbol{x}_1] \begin{bmatrix} B & C \\ E & A_1 + RA_2 \end{bmatrix}, \; [\boldsymbol{x}_0, \, \boldsymbol{x}_1]\mathbf{1} = 1.$$

This is then normalized by $\boldsymbol{x}_0\mathbf{1} + \boldsymbol{x}_1[I - R]^{-1} = 1$.

As one sees the matrix, R could be of huge dimension, so solving it using the well-known methods could still be time consuming. However, because of the structure presented by the matrix G, we could exploit it and then obtain R directly from G. Due to the structure of the matrix A_2, we see that the matrix G has only one column block of non-zero matrix, hence the computation of matrix G is more efficient. The matrix is of the form

$$G = \begin{bmatrix} G_1 & 0 \cdots 0 \\ G_2 & 0 \cdots 0 \\ \vdots & \vdots \cdots \vdots \\ G_M & 0 \cdots 0 \end{bmatrix}.$$

In what follows, we present a case of a single-server queueing model in discrete time, with interdependent interarrival and service times based on bivariate geometric distribution.

1.5. Interdependent interarrival and service times

Chao (1995) discussed the case of bivariate exponential distributions for joint interarrival and service times queues and showed the waiting time is monotonically decreasing in the dependency in increasing convex ordering sense. Sun and Basu (1995) introduced a class of bivariate geometric distribution originally presented by Hawkes (1972). Omey and Minkova (2013) presented a bivariate geometric distribution, which we extend to the general matrix structure. In what follows, we consider the case where the service and the interarrival times are interdependent and represented by a bivariate geometric distribution as in Omey and Minkova (2013).

Omey and Minkova (2013) consider a bivariate geometric distribution described by two random variables X (with two possible states S_1, S_2) and F. S_1 is success of type 1 and S_2 is success of type 2, and F is a failure. Let

$$Pr\{X = S_1\} = p, \; Pr\{X = S_2\} = q, \; Pr\{X = F\} = r,$$

where $0 \leq p, q, r \leq 1$ and $p + q + r = 1$. Let $T_j,\ j = 1, 2$ be defined as $T_j = min\{m : X = S_j\}$, then

$$P\{T_1 = m, T_2 = n\} = \begin{cases} pqr^{m-1}(1-q)^{n-m-1}, & n > m \geq 1, \\ 0, & m = n, \\ pqr^{n-1}(1-p)^{m-n-1}, & m > n \geq 1. \end{cases}$$

It is straightforward to see that if r is fixed, then as p increases q decreases and vice versa. Hence, there is negative correlation between type 1 and type 2 successes in that case. Equally, if we say p is fixed, then q can be allowed to increase or decrease at the expense of r. These types of effects of interarrival and service times have not received much attention in queueing theory; $Cov(S_1, S_2) = \frac{-1}{1-r}$. This makes sense as an increase in S_1 leads to a decrease in S_2 and vice versa.

1.5.1. *A discrete time queueing model with bivariate geometric distribution*

If we study a single-server queue in discrete time, with success of type 1 signifying an arrival and type 2 a service completion if there is a customer in the system, and letting a failure represent no event. Then keeping in mind that the event of having both type 1 and type 2 successes at the same time is zero, then our DTMC is

$$P = \begin{bmatrix} q+r & p & & & \\ q & r & p & & \\ & q & r & p & \\ & & \ddots & \ddots & \ddots \end{bmatrix}.$$

The system is stable if $p < q$. Letting the number in the system at steady state be X and defining $x_i = Pr\{X = i\}, i = 0, 1, 2, \cdots$ with $\boldsymbol{x} = [x_0, x_1, x_2, \cdots]$, then we have

$$\boldsymbol{x} = \boldsymbol{x}P, \ \ \boldsymbol{x}\mathbf{1} = 1.$$

Letting $\rho = \frac{p}{q}$ we end up with

$$x_i = (1-\rho)\rho^i, \ i = 0, 1, 2, \cdots.$$

Letting $X(z) = \sum_{i=0}^{\infty} x_i z^i, \ |z| \leq 1$, we have

$$X(z) = \frac{1-\rho}{1-\rho z},$$

and

$$E[X] = \frac{p}{q-p}.$$

REMARK 1.4.– *The bivariate geometric distribution is the discrete analogue of the bivariate exponential distribution. From the result of the mean number in the system, it is clear that by holding q constant and varying r, this mean is monotonically decreasing in the dependency in increasing convex ordering sense as expected. It is conjectured that this is also transferrable to the mean waiting time.*

Let W be the waiting time in the system, i.e. the time spent waiting in the queue, plus the service time. If we define $w_j = Pr\{W = j\}$, then the waiting time distribution can be written as

$$w_j = \sum_{i=0}^{j-1} x_i (r+p)^{j-i-1} q^{i+1} = \frac{q}{r+p} \sum_{i=0}^{j-1} x_i (r+p)^{j-i} q^i$$

$$= \frac{q-p}{r+p} \sum_{i=0}^{j-1} p^i (r+p)^{j-i} = \frac{q-p}{r}(r+p)[(1-\left(\frac{p}{r+p}\right)^j], \; j = 1, 2, 3, \cdots$$

1.5.2. *Matrix equivalent model*

Consider three square matrices D_0, D_1 and D_2, with $(D_k)_{i,j}$ capturing the probability of transition from state i to state j, with $k = 0, 1, 2$ event occurring. The event 1 is type 1 success, event 2 is type 2 success and event 0 is no success of either type. We assume that the matrix $D = D_0 + D_1 + D_2$ is a stochastic matrix, which is irreducible. Further, we define $\boldsymbol{\pi} = \boldsymbol{\pi} D$, $\boldsymbol{\pi}\mathbf{1} = 1$. Keeping in mind that the event of having both type 1 and type 2 successes at the same time is zero, then our DTMC is

$$P = \begin{bmatrix} D_0 + D_2 & D_1 & & & \\ D_2 & D_0 & D_1 & & \\ & D_2 & D_0 & D_1 & \\ & & \ddots & \ddots & \ddots \end{bmatrix}.$$

If the DTMC is positive recurrent, which we assume it is, then we have

$$\boldsymbol{x} = \boldsymbol{x} P, \; \boldsymbol{x}\mathbf{1} = 1,$$

where $\boldsymbol{x} = [\boldsymbol{x}_0, \boldsymbol{x}_1, \boldsymbol{x}_2, \cdots]$.

From the matrix-geometric theorem of Neuts (1981), we know that there is a matrix R, which is the minimal non-negative solution to the matrix quadratic equation

$$R = D_1 + RD_0 + R^2 D_2,$$

and because of the theorem and the special structure of the boundary of the DTMC we have

$$\boldsymbol{x}_i = \boldsymbol{\pi}(I - R)R^i, \; i = 0, 1, 2, 3, \cdots .$$

From this result, all the other performance measures such as the waiting time and busy period can be obtained.

1.6. Conclusion

In this chapter, we have introduced some queueing models that incorporate some form of interdependence between interarrival times and service times. We mainly considered the case where the interarrival times depend on the service times, before extending the idea to the interdependence of both. We show that the standard matrix-analytic methods can be used for analyzing such queues. Still there is a considerable amount of work to be done in the exact representations of the interdependencies when it comes to the phase type-based models. That will be the focus of future work on this subject.

1.7. Acknowledgements

The author thanks the reviewer for useful comments that have assisted in improving the presentation of the work. The author's research is supported in part by the NRF (South Africa) SARChI ASN Chair programme.

1.8. References

Alfa, A.S. (2004). Markov chain representations of discrete time distributions applied to queueing models. *Comp. Oper. Res.*, 31, 2365–2385.

Alfa, A.S. (2016). *Applied Discrete-Time Queues*, 2nd edition. Springer, New York.

Benes, V.E. (1957). On queues with Poisson arrivals. *Annal. Math. Statist.*, 28(3), 670–677.

Benes, V.E. (1960). Combinatory methods and stochastic Kolmogorov Equations in the theory of queues. *Trans. Amer. Math. Soc.*, 94(2), 282–294.

Chao, X. (1995). Monotone effect of dependency between interarrival and service times in a simple queueing system. *Oper. Res. Lett.*, 17, 47–51.

Cidon, I., Guerin, R., Khamisy, A., Sidi, M. (1993). Analysis of a correlated queue in a communication system. *IEEE Trans. Inform. Theory*, 39(2), 456–465.

Cidon, I., Guerin, R., Khamisy, A., Sidi, M. (1996). On queues with interarrival times proportional to service times. *Probab. Eng. Inform. Sci.*, 10, 87–107.

Graham, A. (1981). *Kronecker Products and Matrix Calculus: With Applications*. Dover Publications, New York.

Fendick, K.W., Saksena, V.R., Whitt, W. (1989). Dependence in packet queues. *IEEE Trans. Commun.*, 37(11), 1173–1183.

Hadidi, N. (1981a). Queues with partial correlation. *SIAM J. Appl. Math.*, 40, 467–475.

Hadidi, N. (1981b). Further result on queues with partial correlation. *Oper. Res.*, 33, 203–209.

Iyer, S.K., Manjunath, D. (2006). Queues with dependency between interarrival and service times using mixture of bivariates. *Stoch. Model.*, 22, 3–20.

John, F.L. (1963). Single server queues with dependent service and inter-arrival times. *J. Soc. Indust. Appl. Math.*, 11(3), 526–534.

Khatri, C.G., Rao, C.R. (1968). Solutions to some functional and their applications to characterization of probability distributions. *Sankhya*, 30, 167–180.

Mészáros, A., Papp, J., Telek, M. (2014). Fitting traffic traces with discrete canonical phase type distributions and Markov arrival processes. *Int. J. Appl. Math. Comput. Sci.*, 24(3), 453–470.

Muller, A. (2000). On the waiting times in queues with dependency between interarrival and service times. *Oper. Res. Lett.*, 26, 43–47.

Neuts, M.F. (1981). *Matrix-geometric Solutions in Stochastic Models – An Algorithmic Approach*. The Johns Hopkins University Press, Baltimore.

Omey, E., Minkova, L.D. (2013). Bivariate geometric distributions. *HUB Research Papers 2013/02, Economics and Business Science*. March.

Panda, G., Banik, A.D., Chaudhry, M.L. (2017). Stationary distributions of the $R^{[X]}/R/1$ cross-correlated queue. *Comm. Statistics – Theory and Methods*, 46(17), 8666–8689.

Sun, K., Basu, A.P. (1995). A characterization of a bivariate geometric distribution. *Stat. and Probabil. Lett.*, 23, 307–311.

2

Busy Period, Congestion Analysis and Loss Probability in Fluid Queues

Fabrice GUILLEMIN[1], **Marie-Ange REMICHE**[2] **and Bruno SERICOLA**[3]

[1] *Orange Labs, Lannion, France*

[2] *University of Namur, Belgium*

[3] *Inria Rennes – Bretagne Atlantique, Rennes, France*

Stochastic fluid flow models and in particular those driven by Markov chains have been intensively studied in the last two decades. Not only have they been proven to be efficient tools to mimic Internet traffic flow at a macroscopic level but they are useful tools in many areas of applications such as manufacturing systems or in actuarial sciences to cite but a few. This chapter proposes to focus on such a model in the context of performance analysis of a potentially congested system. The latter is modeled by means of a finite-capacity system whose content is described by a Markov-driven stable fluid flow. We step-by-step describe a methodology to compute exactly the loss probability of the system. Our approach is based on the computation of hitting probabilities jointly with the peak level reached during a busy period, both in the infinite and finite buffer case. Accordingly, we end up with differential Riccati equations that can be solved numerically. Moreover, we are able to characterize the complete distribution of both the duration of congestion and of the total information lost during such a busy period.

2.1. Introduction

In this chapter, we consider advanced queuing models arising in the performance evaluation of telecommunication networks, notably the Internet. The basic

Queueing Theory 1,
coordinated by Vladimir ANISIMOV, Nikolaos LIMNIOS. © ISTE Editions 2020.

assumption underlying these models is that information flows through the network as a fluid. The packet nature of traffic is thus ignored: a packet stream actually appears as a fluid flow. While this assumption may be too rough at the microflow level (for instance, a single Transmission Control Protocol (TCP) connection), it becomes very relevant when considering the superposition of a large number of microflows. From a mathematical point of view, this amounts to replacing discrete time arrival processes (arrivals of packets) with piecewise continuous arrival processes. Fluid flow models are very useful to study the performance of networks at a macroscopic level when aggregating a large number of flows. This is all the more relevant in current backbone networks. As matter of fact, those networks are equipped with very high speed transmission links, which have bit rates above 100 Gbit/s and are capable of aggregating millions of flows. This is why fluid flow models have gained ever growing attention in the past two decades in the study of telecommunication networks (see, for instance, Sericola (2001); Ahn and Ramaswami (2003, 2004); da Silva Soares and Latouche (2002, 2006); Guillemin and Sericola (2007); Ramaswami (1999, 2006) and Remiche (2005)).

The benefit of fluid flow models is in that it becomes easier to estimate key performance indicators characterizing the network. In particular, modern telecommunication networks heavily rely on statistical multiplexing, which allows a better usage of network capacities by exploiting the fluctuating nature of traffic flows. Typically, TCP or more recently QUIC transport protocols are based on end-to-end congestion protocols and generate sporadic flows (bursts of IP packets). Moreover, the large number of applications used today in networks (notably, video streaming) give rise to variable bit rate traffic. A direct consequence of statistical multiplexing is that a transmission link may be temporally congested. A key performance metric for characterizing statistical multiplexing in packet networks is then the volume of information lost during congestion periods of the network.

The fraction of lost information is a global performance measure reflecting congestion phenomena occurring on each transmission link of the network when performing statistical multiplexing. In practice, several packet flows are statistically multiplexed on the same transmission link, which is in general equipped with a buffer to store packets when the arrival rate is momentarily greater than the link transmission capacity. This buffer aims at preventing the loss of information as much as possible. However, such a buffer necessarily has finite capacity and the arrival rate can exceed the transmission capacity for a sufficiently long time period, causing transient overflow of the buffer. Such phenomena have become more and more frequent on the Internet since new versions of TCP or QUIC are ever more aggressive for the network; the slow start mechanism is sometimes disabled, leading to the transmission of bulks of data at very high rates.

In the following, we consider a single transmission link equipped with a buffer of finite or infinite capacity. We model the arrival of packets as a continuous process $\chi(t)$ denoting the input rate at time t and we denote by $r(t)$ the net input rate at time t (i.e. the difference between the instantaneous arrival and service rates). The link is congested at time t when $r(t) > 0$ and the buffer capacity (denoted by $x \leq \infty$) is reached. Denoting by $X(t)$ the volume of information in the buffer at time t, the fraction of lost information when $x < \infty$ is the long run ratio

$$\pi_{\text{loss}} = \lim_{t \to \infty} \frac{\displaystyle\int_0^t r(s) 1_{\{X(s)=x\}} ds}{\displaystyle\int_0^t \chi(s) ds},$$

where 1_A denotes the characteristic function of the set A (note that we can have $X(t) = x$ only if $r(t) \geq 0$). When the process $(r(t))$ is stationary and ergodic, the above equation can be rewritten as

$$\pi_{\text{loss}} = \frac{\mathrm{E}\left(r 1_{\{X=x\}}\right)}{\mathrm{E}(\chi)}, \qquad [2.1]$$

where χ, r and X denote the input rate, the net input rate and the buffer occupancy in the stationary regime, respectively.

The major difficulty for estimating the quantity π_{loss} is the computation of the joint probability distribution of the couple (r, X). One classical approach consists of approximating the quantity π_{loss} by the buffer overshoot probability $\mathrm{P}\{\overline{X} > x\}$, where $\overline{X}$ is the buffer occupancy when the buffer capacity is infinite and the system is stable (i.e. $\mathrm{E}(r) < 0$). For large x, large deviations techniques can be used to approximate this latter probability.

To study the quantity π_{loss} defined by equation [2.1], we develop a method of computing this quantity when the input process is modulated by a Markov process. More specifically, we assume that the input rate is of the form $\chi_{\varphi(t)}$, where the phase process $(\varphi(t))$ is a Markov process with a finite state space S and χ is some real-valued function from $S \to \mathbb{R}$. We specifically establish the law of the volume of fluid lost during an overflow period (i.e. a time period during which the occupancy of the buffer is x) as well as the duration of such a period.

The model considered is a stochastic fluid model and in particular a Markov modulated fluid queue. These models have widely been used in several domains of system performance evaluation such as manufacturing systems, communication networks, risk processes in insurance and environmental problems (see for instance

Bean *et al.* (2009) and the references therein). The transient and stationary behaviors of these models have been extensively studied and numerous performance metrics with their corresponding numerical algorithms have been proposed and derived in the literature (see Guillemin and Sericola (2012) for a long list of references on the subject).

Hitting probabilities for stochastic fluid models have been analyzed in previous studies (Bean *et al.* 2005a, 2008b, 2009); the relevant algorithms have been developed in Bean *et al.* (2005b, 2008). It is worth noting that O'Reilly and Palmowski (2013) consider similar performance metrics via the use of Laplace transforms. The key difference with that paper is that here we compute the distribution of the random variables under consideration instead of their Laplace transforms. In particular, we compute the joint distributions of the congestion durations and the lost volumes in successive congestion periods in the same busy period. We finally prove that the total congestion duration and lost volume in a typical stationary busy period have phase type distributions.

This chapter is divided into two main parts: in section 2.2, we analyze the infinite buffer case while the finite case is treated in section 2.3. Techniques developed in the infinite case will be of crucial help in the finite buffer case. Both cases are studied with the help of the hitting probabilities together with the maximum peak observed within a busy period. This permits us to obtain differential Riccati equations that can be solved numerically.

In section 2.2, we first introduce the model and define the quantities of interest, such as, among others, the maximum peak or the number of such peaks observed within a busy period. These probabilities are obtained step by step and allow us to finally establish the matrix differential Riccati equations for the probability that starting at any level u with an increasing buffer content, we observe at return to level u a given phase and a maximum peak level less than or equal to $x + u$. We next extend our approach to the analysis of time periods when the buffer occupancy is below a given threshold.

These latter results are first extended in section 2.3 to the finite buffer case. These permit us to obtain congestion metrics in terms of the complete distribution of the total lost information as well as the duration of congestion within a busy period. These finally allow us to characterize the π_{loss} probability.

2.2. Modeling a link under congestion and buffer fluctuations

We consider in this section a classical fluid queue with infinite buffering capacity. This allows us to describe the buffer fluctuations and introduce the notation and the variables necessary to study the fluid queue when the buffer is finite.

2.2.1. *Model description*

The input and service rates are controlled by a homogeneous Markov chain $\varphi = (\varphi(t),\ t \geq 0)$ taking values in the finite state space S, with infinitesimal generator T and initial probability distribution α. The process φ is also called the phase process and we denote by $X(t)$ the amount of fluid in the buffer at time t. It is well known that the process $((X(t), \varphi(t)),\ t \geq 0)$ forms a Markov process having a pair of continuous and discrete states.

Let χ_i be the input rate and η_i be the service rate of the queue when the Markov chain φ is in state i. We denote by r_i the net (or effective) input rate of state i, that is $r_i = \chi_i - \eta_i$ and we define the diagonal matrix of net input rates $R = \text{diag}(r_i,\ i \in S)$. We denote by S^0, S^- and S^+ the subsets of states $i \in S$ such that $r_i = 0$, $r_i < 0$ and $r_i > 0$, respectively. In the same way, we denote by R_0, R_- and R_+ the diagonal matrices $R_0 = \text{diag}(r_i,\ i \in S^0)$, $R_- = \text{diag}(-r_i,\ i \in S^-)$ and $R_+ = \text{diag}(r_i,\ i \in S^+)$. We clearly have $R_0 = 0$, the null matrix.

The numbers of states in S^0, S^+ and S^- are denoted by n_0, n_+ and n_-, respectively. By using this partition of states, we decompose the transition matrix T as

$$T = \begin{pmatrix} T_{--} & T_{-0} & T_{-+} \\ T_{0-} & T_{00} & T_{0+} \\ T_{+-} & T_{+0} & T_{++} \end{pmatrix}.$$

The Markov chain φ is supposed to be irreducible and we denote by $p = (p_i,\ i \in S)$ its stationary distribution. We thus have $pT = 0$ and $p\mathbf{1} = 1$, where $\mathbf{1}$ is the column vector with all entries equal to 1; its dimension is specified by the context.

We suppose that the stability condition for the fluid queue is satisfied, which means that

$$\lim_{t \longrightarrow \infty} \text{E}(r_{\varphi(t)}) = \sum_{i \in S} r_i p_i < 0.$$

This condition ensures that all the busy periods are of finite length a.s. and that the maximum level of the queue during every busy period is finite a.s. in the infinite buffer case.

Let us introduce the $(n_- + n_+) \times (n_- + n_+)$ matrix

$$Q = \begin{pmatrix} Q_{--} & Q_{-+} \\ Q_{+-} & Q_{++} \end{pmatrix},$$

where $Q_{--} = R_-^{-1}\left(T_{--} - T_{-0}T_{00}^{-1}T_{0-}\right)$, $Q_{-+} = R_-^{-1}\left(T_{-+} - T_{-0}T_{00}^{-1}T_{0+}\right)$,

$Q_{+-} = R_+^{-1}\left(T_{+-} - T_{+0}T_{00}^{-1}T_{0-}\right)$ and $Q_{++} = R_+^{-1}\left(T_{++} - T_{+0}T_{00}^{-1}T_{0+}\right)$.

These matrices allow us to restrict our problem to a Markov chain with state space $S^- \cup S^+$ with infinitesimal generator Q and effective input rates equal to -1 or 1. For instance, the matrix Q_{--} governs the transitions from $i \in S^-$ to $j \in S^-$ without any visit to S^+. More formally, as shown in Rubino and Sericola (1989) and Bean *et al.* (2005a), $e^{Q_{--}x}(i,j)$ is the probability, starting from state i, to reach state j with an accumulated reward equal to x and without leaving the set $S^0 \cup S^-$. Here, the accumulated reward corresponds to the amount of fluid generated from the effective input rates of the matrix R_-. A symmetric interpretation holds for matrix Q_{++}. Concerning the matrix Q_{-+}, the entry $Q_{-+}(i,j)$ is the rate, rescaled according to matrix R_-, at which state j is reached from state i either directly from state i or after some time spent in subset S^0. A symmetric interpretation holds for matrix Q_{+-}. Such a transformation consists of considering that the time spent in the zero rate states is immaterial.

Since we are concerned by quantities such as the minimum valley, the maximum level and hitting probabilities, the fluid queue with parameters (T, R) is equivalent to the fluid queue with parameters (Q, C), where C is the $(n_- + n_+) \times (n_- + n_+)$ matrix

$$C = \begin{pmatrix} -I & 0 \\ 0 & I \end{pmatrix},$$

where I is the identity matrix whose dimension is specified by the context of its use. This transformation has been also suggested in previous studies (Asmussen (1995); Rogers (1994)).

In view of the above observations, we will consider in the following a fluid queue driven by a Markov chain $\varphi = (\varphi(t),\ t \geq 0)$ with state space $S = S^- \cup S^+$, infinitesimal generator Q and effective input rates given by matrix C, i.e. equal to -1 or 1.

2.2.2. *Peaks and valleys*

Let us consider a busy period of a fluid queue. The buffer occupancy $X(t)$ is piecewise linear as illustrated in Figure 2.1(a) with peaks followed by valleys. Let M be the maximum peak level in the queue during a busy period. We, moreover, denote by H the fluid level at the minimum valley and by L the integer-valued random variable representing the index of the smallest valley, when valleys are numbered in the order of their occurrence, if they exist. Note that a busy period can have a single peak as illustrated in Figure 2.1(b).

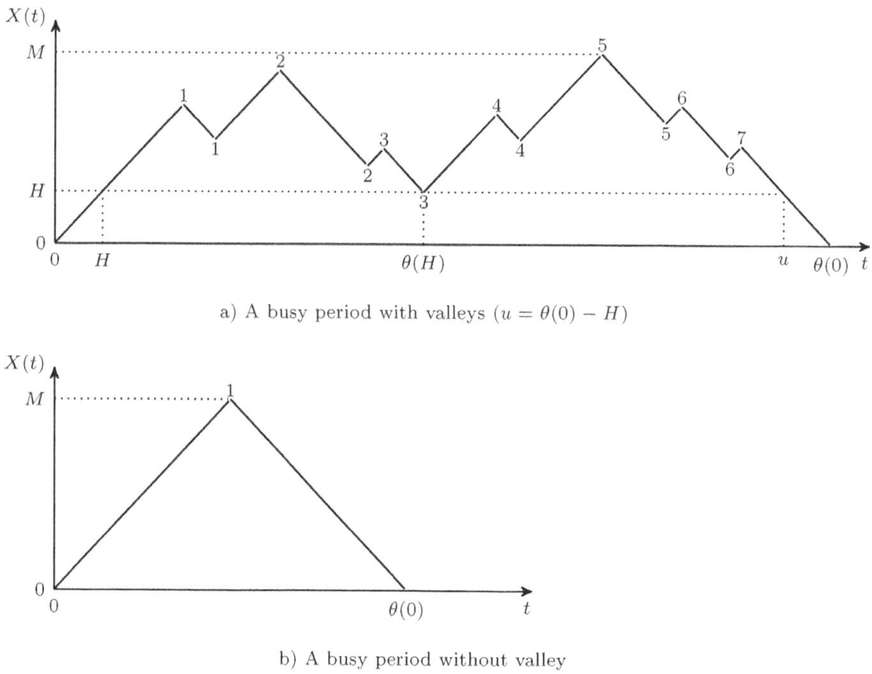

Figure 2.1. *A classical busy period*

For any $x \geq 0$, let $\theta(x)$ denote the first instant greater than x at which the fluid level is equal to x, that is,

$$\theta(x) = \inf\{t > x \mid X(t) = x\}. \quad [2.2]$$

When $X(0) = 0$ and $\varphi(0) \in S^-$, then, by definition, we have $\theta(0) = 0$.

When $X(0) = 0$ and $\varphi(0) \in S^+$, the instants 0 and $\theta(0)$ are, respectively, the initial and the final instants of the busy period. These variables are shown in Figure 2.1 in the case where the effective input rates are either equal to -1 or 1. We have numbered the successive peaks and valleys arising during a busy period.

For every $n \geq 2$, $\ell = 1, \ldots, n-1$ and $x \geq 0$, we consider the $n_+ \times n_-$ matrix $F(n, \ell, y)$ whose entries are defined, for $i \in S^+$ and $j \in S^-$, by

$$F_{i,j}(n, \ell, y) = \mathrm{P}\{\varphi(\theta(0)) = j, N = n, L = \ell, H \leq y \mid \varphi(0) = i, X(0) = 0\}, \quad [2.3]$$

where N is the number of peaks arising during a busy period. In the case $N = 1$, as shown in Figure 2.1(b), there is only one peak during the busy period and so, since there is no valley, we do not define the matrix for $n = 1$. In the path described in Figure 2.1(a), we have $N = 7$ and $L = 3$. $F_{i,j}(\ell, n, y)$ is the probability that a busy period, starting in phase i, contains n peaks, has its minimum valley index equal to ℓ, its minimum valley less than or equal to x and ends in phase j.

We also consider the $n_+ \times n_-$ matrix $\Psi(x, y)$ whose entries are defined, for $i \in S^+, j \in S^-, x \geq 0$ and $0 \leq y \leq x$, by

$$\Psi_{i,j}(x, y) = \mathrm{P}\{\varphi(\theta(0)) = j, M \leq x, H \leq y \mid \varphi(0) = i, X(0) = 0\},$$

and the $n_- \times n_+$ matrix $\Theta(x)$ whose entries are defined, for $i \in S^-$, $j \in S^+$ and $x > 0$, by

$$\Theta_{i,j}(x) = \mathrm{P}\{\varphi(\gamma(x)) = j \mid \varphi(0) = i, X(0) = x\},$$

where, for any $x \geq 0$, $\gamma(x)$ denotes the first positive instant at which the fluid level is equal to x, i.e.

$$\gamma(x) = \inf\{t > 0 \mid X(t) = x\}. \qquad [2.4]$$

The quantity $\Psi_{i,j}(x, y)$ is the probability that a busy period, starting in phase i, has its maximum peak less than or equal to x, its minimum valley less than or equal to y and ends in phase j. $\Theta_{i,j}(x)$ is the probability that a busy period, starting in phase i, ends in phase j.

Note that $\gamma(x)$ is the first time at which the fluid level is equal to x. The function $\theta(x)$ is used to detect the time at which the minimum valley H when $X_0 = 0$ (see Figure 2.1(a)) or the maximum peak M when $X_0 > 0$ (see Figure 2.2(a)) occurs.

By definition of H and M, when $X_0 = 0$, $\theta(H)$ is the second instant at which the fluid level is equal to H (see Figure 2.1(a)) and, when $X_0 > 0$, $\theta(M)$ is the second instant at which the fluid level is equal to M (see Figure 2.2(a)).

2.2.3. *Minimum valley height in a busy period*

For $i \in S^+, j \in S^-, n \geq 2, \ell = 1, \ldots, n-1$ and $y \geq 0$, we denote by $f_{i,j}(n, \ell, y)$ the joint density associated with the distribution $F_{i,j}(n, \ell, y)$ given by equation [2.3] and defined by $f_{i,j}(n, \ell, y) = dF_{i,j}(n, \ell, y)/dy$. We also introduce, for $n \geq 1$, the $n_+ \times n_-$ matrix $W(n)$ whose entries are defined, for $i \in S^+$ and $j \in S^-$, by

$$W_{i,j}(n) = \mathrm{P}\{\varphi(\theta(0)) = j, N = n \mid \varphi(0) = i, X(0) = 0\}.$$

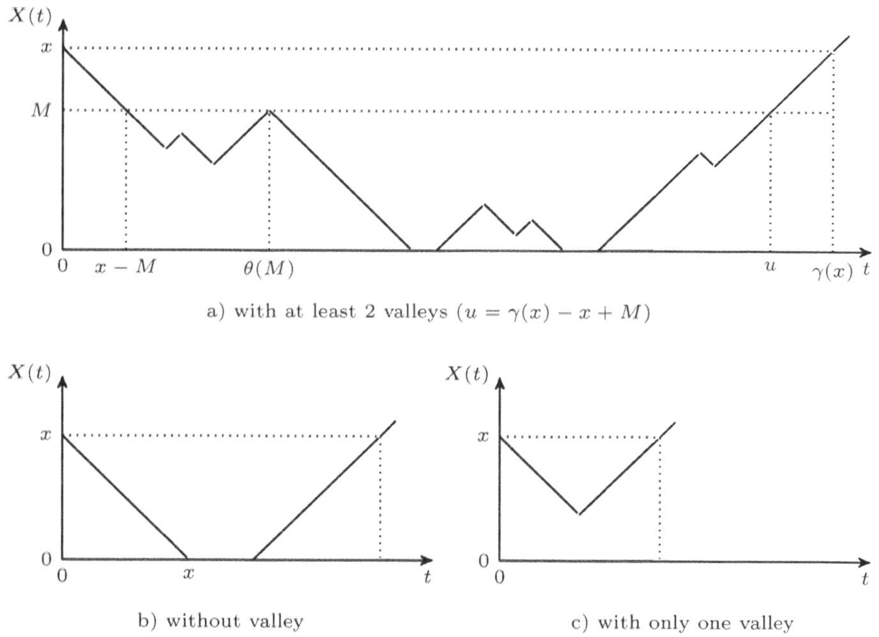

Figure 2.2. *A period under level x*

The quantity $W_{i,j}(n)$ is the probability that a busy period, starting in phase i, contains n peaks and ends in phase j. Summing and integrating the density $f_{i,j}(\ell, n, y)$ over the location and the value of the minimum valley, we obtain, for $n \geq 2$,

$$W_{i,j}(n) = \int_0^{\infty} \sum_{\ell=1}^{n-1} f_{i,j}(n, \ell, y)dy. \qquad [2.5]$$

The $n_+ \times n_-$ matrix containing the terms $f_{i,j}(n, \ell, y)$ is denoted by $f(n, \ell, y)$ and is given together with matrix $W(1)$ by the following theorem.

THEOREM 2.1.– We have

$$W(1) = \int_0^{\infty} e^{Q_{++}y} Q_{+-} e^{Q_{--}y} dy, \qquad [2.6]$$

and for $n \geq 2$

$$W(n) = \int_0^{\infty} e^{Q_{++}y} \sum_{\ell=1}^{n-1} W(\ell) Q_{-+} W(n-\ell) e^{Q_{--}y} dy. \qquad [2.7]$$

PROOF.– The proof follows the same arguments as those developed in a previous study (Bean *et al.* 2005a). To obtain the expression for $W(1)$, we consider a sample path starting from fluid level 0 in phase $i \in S^+$ and returning to level 0 in phase $j \in S^-$ with only one peak of height y. Such a sample path (see Figure 2.1(b)) can be decomposed into three stages as follows:

1) the phase process φ starts in phase $i \in S^+$ and reaches some state $k \in S^+$ at time y without leaving subset S^+. This means that the fluid level increases from level 0 to level y. As seen in the previous section, the corresponding probability is equal to $e^{Q_{++}y}(i,k)$;

2) since a peak (of height equal to y) occurs, this means that a transition of the phase process φ occurs from state k to some state $h \in S^-$. The corresponding transition rate is equal to $Q_{+-}(k,h)$;

3) starting from state $h \in S^-$, the process φ reaches state $j \in S^-$ at time y without leaving subset S^-. This means that the fluid level decreases from level y to level 0. As seen in the previous section, the corresponding probability is equal to $e^{Q_{--}y}(h,j)$.

We thus obtain

$$W_{i,j}(1) = \int_0^{\infty} \sum_{k \in S^+} \sum_{h \in S^-} e^{Q_{++}y}(i,k) Q_{+-}(k,h) e^{Q_{--}y}(h,j) dy,$$

that is

$$W(1) = \int_0^{\infty} e^{Q_{++}y} Q_{+-} e^{Q_{--}y} dy.$$

Let now $n \geq 2$ and $1 \leq \ell \leq n-1$. A typical example of such a sample path is shown in Figure 2.1(a). It is a path starting from fluid level 0 in phase $i \in S^+$ and returning to level 0 in phase $j \in S^-$ with $N = n$ peaks, with the minimum valley of height $H = y$ at position $L = \ell$. Such a sample path can be broken up into five stages as follows:

1) as we can observe, H denotes the minimum level as well as the amount of time needed to reach it. The phase process φ starts in phase $i \in S^+$ and reaches some state $k \in S^+$ at time $H = y$ without leaving subset S^+. This means that the fluid level increases from level 0 to level y. As seen in the previous section, the corresponding probability is equal to $e^{Q_{++}y}(i,k)$;

2) starting from phase $k \in S^+$ and level y, the fluid process returns for the first time to level y in some phase $h \in S^-$. By definition of H and L, the number of peaks between instants H and $\theta(H)$ is equal to ℓ. By the spatial homogeneity of the process, the corresponding probability is equal to $W_{k,h}(\ell)$;

3) since a valley (of height equal to y at time $\theta(H)$) occurs, this means that a transition of the phase process φ occurs from state $h \in S^-$ to some state $m \in S^+$. The corresponding transition rate is equal to $Q_{-+}(h, m)$;

4) once again, starting from phase $m \in S^+$ and level y, the fluid process returns for the first time to level y in some phase $v \in S^-$. By definition of H and L, the number of peaks between instants $\theta(H)$ and $\theta(0) - H$ is equal to $n - \ell$. By the spatial homogeneity of the process, the corresponding probability is equal to $W_{m,v}(n - \ell)$;

5) the phase process φ starts in phase $v \in S^-$ and level y. It reaches level 0 in state j without leaving subset S^- at time $\varphi(\theta(0))$. This means that the fluid level decreases from level y to level 0. As seen in the previous section, the corresponding probability is equal to $e^{Q_{--}y}(v, j)$.

We thus obtain

$$f_{i,j}(n, \ell, y) =$$

$$\sum_{k \in S^+} \sum_{h \in S^-} \sum_{m \in S^+} \sum_{v \in S^-} e^{Q_{++}y}(i, k) W_{k,h}(\ell) Q_{-+}(h, m) W_{m,v}(n - \ell) e^{Q_{--}y}(v, j),$$

that is $f(n, \ell, y) = e^{Q_{++}y} W(\ell) Q_{-+} W(n-\ell) e^{Q_{--}y}$. Equation [2.7] follows by using equation [2.5]. ■

Let Ψ denote $n_+ \times n_-$ the stochastic matrix whose entries are defined, for $i \in S^+$ and $j \in S^-$, by $\Psi_{i,j} = \text{P}\{\varphi(\theta(0)) = j \mid \varphi(0) = i, X(0) = 0\}$. $\Psi_{i,j}$ is the probability that the phase at the instant of the first return to the initial level is equal to j. By spatial homogeneity, this probability is independent of the value of the initial level. We have

$$\Psi = \sum_{n=1}^{\infty} W(n).$$

In the next theorem, we give a relation satisfied by matrix Ψ.

THEOREM 2.2.–

$$\Psi = \int_0^\infty e^{Q_{++}y} Q_{+-} e^{Q_{--}y} dy + \int_0^\infty e^{Q_{++}y} \Psi Q_{-+} \Psi e^{Q_{--}y} dy.$$

PROOF.– From equation [2.7], we have for $n \geq 2$

$$W(n) = \int_0^\infty e^{Q_{++}y} \sum_{\ell=1}^{n-1} W(\ell) Q_{-+} W(n-\ell) e^{Q_{--}y} dy.$$

By summation over n, we get, using the monotone convergence theorem,

$$\begin{aligned}
\Psi &= W(1) + \sum_{n=2}^{\infty} \int_0^\infty e^{Q_{++}y} \sum_{\ell=1}^{n-1} W(\ell) Q_{-+} W(n-\ell) e^{Q_{--}y} dy \\
&= W(1) + \int_0^\infty e^{Q_{++}y} \sum_{n=2}^{\infty} \sum_{\ell=1}^{n-1} W(\ell) Q_{-+} W(n-\ell) e^{Q_{--}y} dy \\
&= W(1) + \int_0^\infty e^{Q_{++}y} \sum_{\ell=1}^{\infty} \sum_{n=\ell+1}^{\infty} W(\ell) Q_{-+} W(n-\ell) e^{Q_{--}y} dy \\
&= W(1) + \int_0^\infty e^{Q_{++}y} \sum_{\ell=1}^{\infty} W(\ell) Q_{-+} \sum_{n=\ell+1}^{\infty} W(n-\ell) e^{Q_{--}y} dy \\
&= W(1) + \int_0^\infty e^{Q_{++}y} \Psi Q_{-+} \Psi e^{Q_{--}y} dy,
\end{aligned}$$

which is the desired result. ■

It is easily checked that matrix Ψ is the solution to the following matrix algebraic Riccati equation $Q_{++}\Psi + \Psi Q_{--} + \Psi Q_{-+} \Psi + Q_{+-} = 0$. This equation has been considered in several papers and several algorithms have been developed to compute Ψ, which is the minimal solution of this equation (see, for instance, Bean *et al.* (2005a), Bean *et al.* (2005b) and the references therein). An explicit expression for Ψ has been obtained in the previous study Samuelson *et al.* (2017) by means of an auxiliary matrix.

Let $F(x)$ denote the $n_+ \times n_-$ matrix whose entries are defined, for $i \in S^+$ and $j \in S^-$, by $F_{i,j}(x) = \mathrm{P}\{\varphi(\theta(0)) = j, H \leq x \mid \varphi(0) = i\}$. $F_{i,j}(x)$ is the probability that the phase at the instant of the first return to the initial level is equal to j with a minimum valley of height less than or equal to x. This probability is defined only when the minimum valley exists. The minimum valley exists if and only if the number of peaks N is greater than or equal to 2. Moreover, when it exists, we have $H > 0$. Thus, we define the event $H = 0$ to represent the non-existence of the minimum valley, i.e. the case where the number of peaks is equal to 1 (see Figure 2.1(b)). We then have the following result.

COROLLARY 2.1.– For every $x \geq 0$,

$$F(x) = \int_0^{\infty} e^{Q_{++}y} Q_{+-} e^{Q_{--}y} dy + \int_0^{x} e^{Q_{++}y} \Psi Q_{-+} \Psi e^{Q_{--}y} dy.$$

PROOF.– By definition, we have, for $i \in S^+$ and $j \in S^-$,

$$\begin{aligned} &F_{i,j}(x) \\ &= \mathrm{P}\{\varphi(\theta(0)) = j, H = 0 | \varphi(0) = i\} + \mathrm{P}\{\varphi(\theta(0)) = j, 0 < H \leq x | \varphi(0) = i\} \\ &= \mathrm{P}\{\varphi(\theta(0)) = j, N = 1 | \varphi(0) = i\} + \mathrm{P}\{\varphi(\theta(0)) = j, N \geq 2, H \leq x | \varphi(0) = i\} \\ &= W_{i,j}(1) + \int_0^x \sum_{n=2}^{\infty} \sum_{\ell=1}^{n-1} f_{i,j}(n, \ell, y) dy. \end{aligned}$$

Following the same lines used in the proof of theorem 2.2, we get the result. ■

2.2.4. *Maximum peak level in a busy period*

We consider in this section the maximum fluid level M reached during a busy period. A typical path is shown in Figure 2.1. For $i \in S^+$, $j \in S^-$, $x \geq 0$ and $0 \leq y \leq x$, we denote by $\psi_{i,j}(x, y)$ the marginal density associated with the distribution $\Psi_{i,j}(x, y)$, introduced in section 2.2.2, and defined by $\psi_{i,j}(x, y) = \partial \Psi_{i,j}(x, y)/\partial y$. For $x \geq 0$, we denote by $\Psi(x)$ the $n_+ \times n_-$ matrix containing the $\Psi_{i,j}(x)$ defined, for $i \in S^+$ and $j \in S^-$, by

$$\Psi_{i,j}(x) = \mathrm{P}\{\varphi(\theta(0)) = j, M \leq x \mid \varphi(0) = i, X(0) = 0\}. \quad [2.8]$$

$\Psi_{i,j}(x)$ is the probability that, starting from phase $i \in S^+$ and any level $u \geq 0$, the phase at the instant of the first return to the initial level u is equal to j and the maximum level is less than or equal to $x + u$. By the spatial homogeneity, this probability is independent of u, that is why we define $\Psi(x)$ only for $u = 0$. The $n_+ \times n_-$ matrix containing the terms $\psi_{i,j}(x, y)$ is denoted by $\psi(x, y)$ and is given by the following theorem.

THEOREM 2.3.– For $x \geq 0$ and $0 \leq z \leq x$, we have

$$\Psi(x, z) = \int_0^{x} e^{Q_{++}y} Q_{+-} e^{Q_{--}y} dy + \int_0^{z} e^{Q_{++}y} \Psi(x-y) Q_{-+} \Psi(x-y) e^{Q_{--}y} dy. \quad [2.9]$$

$$\Psi(x) = \int_0^{x} e^{Q_{++}y} Q_{+-} e^{Q_{--}y} dy + \int_0^{x} e^{Q_{++}y} \Psi(x-y) Q_{-+} \Psi(x-y) e^{Q_{--}y} dy. \quad [2.10]$$

PROOF.– We proceed as for the proof of theorem 2.1. The term $\Psi(x,0)$ corresponds to the case where there is only one peak (see Figure 2.1(b)), and thus no valley, which means, as defined in the previous section, that $H = 0$. To obtain the expression for $\Psi(x,0)$, we consider a sample path starting from fluid level 0 in phase $i \in S^+$ and returning to level 0 in phase $j \in S^-$ with only one peak of height y (with $y \leq x$). Such a sample path can be broken up into three stages as follows:

1) the phase process φ starts in phase $i \in S^+$ and reaches some state $k \in S^+$ at time y without leaving subset S^+. This means that the fluid level increases from level 0 to level y. As seen in the previous section, the corresponding probability is equal to $e^{Q_{++}y}(i,k)$;

2) since a peak (of height equal to y) occurs, this means that a transition of the phase process φ occurs from state k to some state $h \in S^-$. The corresponding transition rate is equal to $Q_{+-}(k,h)$;

3) starting from state $h \in S^-$, the process φ reaches state $j \in S^-$ at time y without leaving subset S^-. This means that the fluid level decreases from level y to level 0. As seen in the previous section, the corresponding probability is equal to $e^{Q_{--}y}(h,j)$.

We thus obtain

$$\Psi_{i,j}(x,0) = \int_0^x \sum_{k\in S^+} \sum_{h\in S^-} e^{Q_{++}y}(i,k)Q_{+-}(k,h)e^{Q_{--}y}(h,j)dy,$$

that is

$$\Psi(x,0) = \int_0^x e^{Q_{++}y}Q_{+-}e^{Q_{--}y}dy.$$

We consider now the case where a minimum valley exists, i.e. $H > 0$. A typical example of such a sample path is shown in Figure 2.1(a). It is a path starting from fluid level 0 in phase $i \in S^+$ and returning to level 0 in phase $j \in S^-$ with at least two peaks (i.e. $H > 0$), with the minimum valley and the maximum level less than or equal to x. Such a sample path can be broken up into five stages as follows:

1) the phase process φ starts in phase $i \in S^+$ and reaches some state $k \in S^+$ at time $H = y$ (with $y \leq x$) without leaving subset S^+. This means that the fluid level increases from level 0 to level y. As seen in the previous section, the corresponding probability is equal to $e^{Q_{++}y}(i,k)$;

2) starting from phase $k \in S^+$ and level y, the fluid process returns for the first time to level y in some phase $h \in S^-$, without exceeding level x. By the spatial homogeneity of the process, the corresponding probability is equal to $\Psi(x-y)$;

3) since a valley (of height equal to y at time $\theta(H)$) occurs, this means that a transition of the phase process φ occurs from state $h \in S^-$ to some state $m \in S^+$. The corresponding transition rate is equal to $Q_{-+}(h,m)$;

4) once again, starting from phase $m \in S^+$ and level y, the fluid process returns for the first time to level y in some phase $v \in S^-$ without exceeding level x. By the spatial homogeneity of the process, the corresponding probability is equal to $\Psi(x-y)$;

5) the phase process φ starts in phase $v \in S^-$ and level y. It reaches level 0 in state j without leaving subset S^-, at time $\varphi(\theta(0))$. This means that the fluid level decreases from level y to level 0. As seen in the previous section, the corresponding probability is equal to $e^{Q_{--}y}(v,j)$.

We thus obtain

$$\psi_{i,j}(x,y) =$$

$$\sum_{k\in S^+}\sum_{h\in S^-}\sum_{m\in S^+}\sum_{v\in S^-} e^{Q_{++}y}(i,k)\Psi_{k,h}(x-y)Q_{-+}(h,m)\Psi_{m,v}(x-y)e^{Q_{--}y}(v,j),$$

which can be rewritten as $\psi(x,y) = e^{Q_{++}y}\Psi(x-y)Q_{-+}\Psi(x-y)e^{Q_{--}y}$. Equation [2.9] follows by writing

$$\Psi(x,z) = \Psi(x,0) + \int_0^z \psi(x,y)dy.$$

Since $\Psi(x) = \Psi(x,x)$, equation [2.10] follows. ■

We have $\lim_{x\longrightarrow\infty}\Psi(x,z) = F(z)$ and $\lim_{x\longrightarrow\infty}\Psi(x) = \Psi$. We denote by $\Psi'(x)$ the derivative of $\Psi(x)$ with respect to x. We then have the following result.

THEOREM 2.4.– The function $\Psi(x)$ satisfies the following matrix differential Riccati equation

$$\Psi'(x) = Q_{++}\Psi(x) + \Psi(x)Q_{--} + \Psi(x)Q_{-+}\Psi(x) + Q_{+-}, \qquad [2.11]$$

with $\Psi(0) = 0$ as initial condition.

PROOF.– By definition of function $\Psi(x)$ in relation [2.8], the initial condition is trivially given by $\Psi(0) = 0$, since the maximum level M during a busy period is positive. Using a variable change, relation [2.10] can be written as

$$\Psi(x) = \int_0^x e^{Q_{++}y}Q_{+-}e^{Q_{--}y}dy + \int_0^x e^{Q_{++}(x-y)}\Psi(y)Q_{-+}\Psi(y)e^{Q_{--}(x-y)}dy.$$

In order to avoid excessively long expressions in the derivation of $\Psi'(x)$, we introduce the following notation:

$$\alpha(x) = \int_0^x e^{Q_{++}y} Q_{+-} e^{Q_{--}y} dy \text{ and } \beta(x) = \int_0^x e^{-Q_{++}y} \Psi(y) Q_{-+} \Psi(y) e^{-Q_{--}y} dy.$$

It is easy to check that $e^{Q_{++}x} Q_{+-} e^{Q_{--}x} - Q_{++}\alpha(x) - \alpha(x)Q_{--} = Q_{+-}$. Differentiating [2.10], we get

$$\begin{aligned}
\Psi'(x) &= e^{Q_{++}x} Q_{+-} e^{Q_{--}x} + Q_{++} e^{Q_{++}x} \beta(x) e^{Q_{--}x} \\
&\quad + e^{Q_{++}x} \left[\beta'(x) e^{Q_{--}x} + \beta(x) e^{Q_{--}x} Q_{--}\right] \\
&= e^{Q_{++}x} Q_{+-} e^{Q_{--}x} + Q_{++} \left[\Psi(x) - \alpha(x)\right] + e^{Q_{++}x} \beta'(x) e^{Q_{--}x} \\
&\quad + \left[\Psi(x) - \alpha(x)\right] Q_{--} \\
&= Q_{++}\Psi(x) + \Psi(x) Q_{--} + e^{Q_{++}x} \beta'(x) e^{Q_{--}x} + Q_{+-} \\
&= Q_{++}\Psi(x) + \Psi(x) Q_{--} + \Psi(x) Q_{-+} \Psi(x) + Q_{+-},
\end{aligned}$$

which is the desired result. ■

With the initial condition $\Psi(0) = 0$, the Cauchy–Lipschitz theorem ensures that the matrix differential Riccati equation [2.11] has a unique solution. This solution is thus given, for $x \geq 0$, by [2.8]. By definition of diagonal matrix C, we have

$$CQ = \begin{pmatrix} -Q_{--} & -Q_{-+} \\ Q_{+-} & Q_{++} \end{pmatrix}.$$

According to the decomposition $S = S^- \cup S^+$, we define the four matrices $A(x)$, $B(x)$, $C(x)$ and $D(x)$ occurring in the matrix e^{CQx} by writing

$$e^{CQx} = \begin{pmatrix} A(x) & B(x) \\ C(x) & D(x) \end{pmatrix}. \qquad [2.12]$$

The following theorem gives an expression of the solution $\Psi(x)$ to the matrix differential Riccati equation [2.11].

THEOREM 2.5.– For every $x \geq 0$, we have $\Psi(x) = C(x)A(x)^{-1}$.

PROOF.– Let us consider the following linear differential equation:

$$Y'(x) = (-Q_{--} - Q_{-+}\Psi(x))Y(x) \text{ with } Y(0) = I. \qquad [2.13]$$

The function Ψ being continuous, this linear system has a unique solution that is invertible. We now define the $n_+ \times n_-$ matrix $Z(x)$ by $Z(x) = \Psi(x)Y(x)$. Using this definition, equation [2.13] becomes

$$Y'(x) = -Q_{--}Y(x) - Q_{-+}Z(x). \qquad [2.14]$$

Differentiating $Z(x)$ with respect to x, we obtain from [2.11] and [2.14]

$$\begin{aligned} Z'(x) &= \Psi'(x)Y(x) + \Psi(x)Y'(x) \\ &= (Q_{++}\Psi(x) + \Psi(x)Q_{--} + \Psi(x)Q_{-+}\Psi(x) + Q_{+-})Y(x) \\ &\quad +\Psi(x)(-Q_{--}Y(x) - Q_{-+}Z(x)) \\ &= Q_{++}Z(x) + \Psi(x)Q_{-+}\Psi(x)Y(x) + Q_{+-}Y(x) - \Psi(x)Q_{-+}Z(x) \\ &= Q_{++}Z(x) + Q_{+-}Y(x). \end{aligned} \qquad [2.15]$$

Putting together equations [2.14] and [2.15], we obtain

$$\begin{pmatrix} Y'(x) \\ Z'(x) \end{pmatrix} = \begin{pmatrix} -Q_{--} & -Q_{-+} \\ Q_{+-} & Q_{++} \end{pmatrix} \begin{pmatrix} Y(x) \\ Z(x) \end{pmatrix} = CQ \begin{pmatrix} Y(x) \\ Z(x) \end{pmatrix},$$

with $Y(0) = I$ and $Z(0) = 0$. The solution to that equation is given by

$$\begin{pmatrix} Y(x) \\ Z(x) \end{pmatrix} = e^{CQx} \begin{pmatrix} I \\ 0 \end{pmatrix} = \begin{pmatrix} A(x) \\ C(x) \end{pmatrix},$$

which means that $Y(x) = A(x)$, $Z(x) = C(x)$ and thus, since $A(x)$ is invertible, we have $\Psi(x) = C(x)A(x)^{-1}$. ■

2.2.5. *Maximum peak under a fixed fluid level*

In this section, we still consider an infinite fluid queue and we are interested in time periods when the buffer occupancy is below a given threshold x.

We consider the first return to the initial level x when $x > 0$ and the initial phase is in S^-. More formally, we introduce the $n_- \times n_+$ matrix $G(x, y)$ whose entries are defined, for $i \in S^-$, $j \in S^+$, $x > 0$ and $0 \leq y \leq x$, by

$$G_{i,j}(x, y) = \mathrm{P}\{\varphi(\gamma(x)) = j, M \leq y \mid \varphi(0) = i, X(0) = x\},$$

where M denotes the maximum peak between instants 0 and $\gamma(x)$. When there are no peaks, we take as convention $M = 0$. The $n_- \times n_+$ matrix $\Theta(x)$ the entries $\Theta_{i,j}(x)$ of which have been defined, for $i \in S^-$, $j \in S^+$ and $x > 0$, in section 2.2.2, satisfies $\Theta(x) = G(x, x)$. By spatial homogeneity, we have, for $i \in S^+$ and $j \in S^-$, $\mathrm{P}\{\varphi(\gamma(x)) = j \mid \varphi(0) = i, X(0) = x\} = \Psi_{i,j}$. For $i \in S^-$ and $j \in S^+$, the problem is more complicated because of the influence of the boundary level zero, which makes the hitting probability $\Theta_{i,j}(x)$ dependent on x. A typical example of such a sample path is shown in Figure 2.2.

We denote by $g_{i,j}(x, y)$ the density associated with the distribution $G_{i,j}(x, y)$, i.e. $g_{i,j}(x, y) = \partial G_{i,j}(x, y)/\partial y$.

THEOREM 2.6.– For $x > 0$, we have

$$G(x, 0) = \int_0^x e^{Q_{--}y} Q_{-+} e^{Q_{++}y} dy + (-Q_{--})^{-1} e^{Q_{--}x} Q_{-+} e^{Q_{++}x} \quad [2.16]$$

and, for $0 \le y \le x$,

$$g(x, y) = e^{Q_{--}y} \Theta(x - y) Q_{+-} \Theta(x - y) e^{Q_{++}y}. \quad [2.17]$$

PROOF.– We proceed as for the proof of theorems 2.1 and 2.3. The term $G(x, 0)$ corresponds to the case where there are no peaks in the interval $[0, \gamma(x)]$, i.e. where $M = 0$. This situation corresponds to the paths shown in Figures 2.2(b) and (c). Figure 2.2(b) corresponds to the case where the height of the unique valley, denoted by H, is zero and Figure 2.2(c) corresponds to the case where H is positive. We thus have

$$\begin{aligned} G(x, 0) &= \mathrm{P}\{\varphi(\gamma(x)) = j, M = 0, H > 0 \mid \varphi(0) = i, X(0) = x\} \\ &\quad + \mathrm{P}\{\varphi(\gamma(x)) = j, M = 0, H = 0 \mid \varphi(0) = i, X(0) = x\}. \end{aligned}$$

The first term is the symmetric term of $\Psi(x, 0)$, so we easily get

$$\mathrm{P}\{\varphi(\gamma(x)) = j, M = 0, H > 0 \mid \varphi(0) = i, X(0) = x\} = \int_0^x e^{Q_{--}y} Q_{-+} e^{Q_{++}y} dy.$$

For the second term, which corresponds to the sample path in Figure 2.2(b), the phase process φ starts in phase $i \in S^-$ with a level x. It stays in subset S^- for a duration $y \ge x$, reaching some state $k \in S^-$ and thus with a level 0. Next a transition occurs from state $k \in S^-$ to state $h \in S^+$, with rate $Q_{+-}(k, h)$, and the process φ reaches state $j \in S^+$ at time x without leaving subset S^+. We thus have

$$\begin{aligned} &\mathrm{P}\{\varphi(\gamma(x)) = j, M = 0, H = 0 \mid \varphi(0) = i, X(0) = x\} \\ &\qquad = \int_x^\infty e^{Q_{--}y} dy Q_{-+} e^{Q_{++}x} = (-Q_{--})^{-1} e^{Q_{--}x} Q_{-+} e^{Q_{++}x}. \end{aligned}$$

We consider now the case where there is at least one peak, i.e. $M > 0$. A typical example of such a sample path is shown in Figure 2.2(a). Such a sample path can be broken up into five stages as follows:

1) the phase process φ starts in phase $i \in S^-$ and reaches some state $k \in S^-$ at time $x - M = y$ without leaving subset S^-. This means that the fluid level decreases from level x to level $x-y$. As seen in the previous section, the corresponding probability is equal to $e^{Q_{--}y}(i,k)$;

2) starting from phase $k \in S^-$ and level $x - y$, the fluid process returns for the first time to level $x - y$ in some phase $h \in S^+$, without exceeding level $x - y$. The corresponding probability is equal to $\Theta(x - y)$;

3) since a peak (of height equal to $x - y$ at time $\theta(M)$) occurs, this means that a transition of the phase process φ occurs from state $h \in S^+$ to some state $m \in S^-$. The corresponding transition rate is equal to $Q_{-+}(h,m)$;

4) once again, starting from phase $m \in S^-$ and level $x - y$, the fluid process returns for the first time to level $x - y$ in some phase $v \in S^+$ without exceeding level $x - y$. By the spatial homogeneity of the process, the corresponding probability is equal to $\Theta(x - y)$;

5) the phase process φ starts in phase $v \in S^+$ and level $x - y$. It reaches level x in state j without leaving subset S^+, at time $\varphi(\gamma(x))$. This means that the fluid level increases from level $x-y$ to level x. As seen in the previous section, the corresponding probability is equal to $e^{Q_{++}y}(v,j)$.

We thus obtain

$$g_{i,j}(x,y) =$$

$$\sum_{k\in S^-}\sum_{h\in S^+}\sum_{m\in S^-}\sum_{v\in S^+} e^{Q_{++}y}(i,k)\Theta_{k,h}(x-y)Q_{+-}(h,m)\Theta_{m,v}(x-y)e^{Q_{++}y}(v,j),$$

that is $g(x,y) = e^{Q_{--}y}\Theta(x-y)Q_{+-}\Theta(x-y)e^{Q_{++}y}$. ■

The matrices $\Theta(x)$ and $G(x,z)$ have been defined only for $x > 0$. Clearly, if we set $x = 0$, and thus $z = 0$, in these definitions, we obtain the zero matrix. In fact, we write $\Theta(0)$ for $\Theta(0^+)$ and $G(0,0)$ for $G(0^+,0)$ which means that $\Theta(0) = \lim_{x\longrightarrow 0}\Theta(x) = \lim_{x\longrightarrow 0} G(x,0) = G(0,0)$. This simply means that $x = 0$ is not a continuity point of these two functions.

The matrices $G(x,z)$ and $\Theta(x)$ are given by the following corollary.

COROLLARY 2.2.– For $x \geq 0$ and $0 \leq z \leq x$, we have

$$G(x,z) = \int_0^x e^{Q_{--}y} Q_{-+} e^{Q_{++}y} dy + (-Q_{--})^{-1} e^{Q_{--}x} Q_{-+} e^{Q_{++}x} + \int_0^z e^{Q_{--}y} \Theta(x-y) Q_{+-} \Theta(x-y) e^{Q_{++}y} dy. \quad [2.18]$$

$$\Theta(x) = \int_0^x e^{Q_{--}y} Q_{-+} e^{Q_{++}y} dy + (-Q_{--})^{-1} e^{Q_{--}x} Q_{-+} e^{Q_{++}x} + \int_0^x e^{Q_{--}y} \Theta(x-y) Q_{+-} \Theta(x-y) e^{Q_{++}y} dy. \quad [2.19]$$

PROOF.– It suffices to write

$$G(x,z) = G(x,0) + \int_0^z g(x,y) dy,$$

and $\Theta(x) = G(x,x)$. ■

The following result shows that $\Theta(x)$ satisfies a matrix differential Riccati equation.

THEOREM 2.7.– $\Theta(0) = (-Q_{--})^{-1} Q_{-+}$ and, for $x > 0$, we have

$$\Theta'(x) = Q_{--}\Theta(x) + \Theta(x)Q_{++} + \Theta(x)Q_{+-}\Theta(x) + Q_{-+}. \quad [2.20]$$

PROOF.– The proof is based on equation [2.19] and is thus quite similar to the proof of theorem 2.4. ■

Again, the Cauchy–Lipschitz theorem ensures that the matrix differential Riccati equation [2.20] with the initial condition $\Theta(0) = (-Q_{--})^{-1} Q_{-+}$ has a unique solution. The following theorem gives an expression of the solution $\Theta(x)$ to that equation. Note that $\Theta(0)$ is a stochastic matrix.

THEOREM 2.8.– For every $x \geq 0$, we have $\Theta(x) = Z(x)Y(x)^{-1}$, where the matrices $Y(x)$ and $Z(x)$ are given by

$$\begin{pmatrix} Z(x) \\ Y(x) \end{pmatrix} = e^{-CQx} \begin{pmatrix} \Theta(0) \\ I \end{pmatrix}. \quad [2.21]$$

PROOF.– Let us consider the following linear differential equation:

$$Y'(x) = (-Q_{++} - Q_{+-}\Theta(x))Y(x) \text{ and } Y(0) = I. \qquad [2.22]$$

The function $\Theta(x)$ being continuous, this linear system has a unique solution which is invertible. We now define the $n_- \times n_+$ matrix $Z(x)$ by $Z(x) = \Theta(x)Y(x)$. Using this definition, equation [2.22] becomes

$$Y'(x) = -Q_{++}Y(x) - Q_{+-}Z(x). \qquad [2.23]$$

Differentiating $Z(x)$ with respect to x, we obtain from [2.20] and [2.23]

$$\begin{aligned} Z'(x) &= \Theta'(x)Y(x) + \Theta(x)Y'(x) \\ &= (Q_{--}\Theta(x) + \Theta(x)Q_{++} + \Theta(x)Q_{+-}\Theta(x) + Q_{-+})Y(x) \\ &\quad +\Theta(x)(-Q_{++}Y(x) - Q_{+-}Z(x)) \\ &= Q_{--}Z(x) + \Theta(x)Q_{+-}\Theta(x)Y(x) + Q_{-+}Y(x) - \Theta(x)Q_{+-}Z(x) \\ &= Q_{--}Z(x) + Q_{-+}Y(x). \end{aligned} \qquad [2.24]$$

Putting together equations [2.23] and [2.24] we obtain

$$\begin{pmatrix} Z'(x) \\ Y'(x) \end{pmatrix} = \begin{pmatrix} Q_{--} & Q_{-+} \\ -Q_{+-} & -Q_{++} \end{pmatrix} \begin{pmatrix} Z(x) \\ Y(x) \end{pmatrix} = -CQ \begin{pmatrix} Z(x) \\ Y(x) \end{pmatrix},$$

with $Y(0) = I$ and $Z(0) = \Theta(0) = (-Q_{--})^{-1}Q_{-+}$. The solution to that equation is given by

$$\begin{pmatrix} Z(x) \\ Y(x) \end{pmatrix} = e^{-CQx} \begin{pmatrix} \Theta(0) \\ I \end{pmatrix},$$

which completes the proof. ■

Note that relation [2.21] can also be written as

$$\begin{pmatrix} Z(x) \\ Y(x) \end{pmatrix} = \begin{pmatrix} A(x) & B(x) \\ C(x) & D(x) \end{pmatrix}^{-1} \begin{pmatrix} \Theta(0) \\ I \end{pmatrix}.$$

2.3. Fluid queue with finite buffer

We suppose now that the fluid queue is of finite capacity and we denote that capacity by x.

2.3.1. *Congestion metrics*

Assume that the buffer is initially empty ($X(0) = 0$) and that $\varphi(0) \in S^+$, so that a busy period starts at time 0.

We characterize congestion by the distribution of the two random variables:

– the total time $\tau(x)$ during which the buffer is full in a busy period;

– the total quantity of fluid $V(x)$ lost in a busy period.

The random variables $\tau(x)$ and $V(x)$ are formally defined by

$$\tau(x) = \int_0^{\theta(0)} 1_{\{X(t)=x\}} dt \quad \text{and} \quad V(x) = \int_0^{\theta(0)} r_{\varphi(t)} 1_{\{X(t)=x\}} dt.$$

Note that $X(t) = x$ implies that $\varphi(t) \in S^+ \cup S^0$ and when $\varphi(t) \in S^0$, $r_{\varphi(t)} = 0$.

We consider the successive periods when the buffer capacity x is reached by process $(X(t))$. We denote by K the number of such periods occurring during a busy period of the buffer. Clearly, K is an integer valued random variable and if $K = 0$, which means that process never reaches level x during the busy period, then we have $V(x) = 0$. We thus consider the case when $K \geq 1$.

For every $k \geq 1$, we denote by γ_k the kth instant at which $X(\gamma_k) = x$ and $X(\gamma_k^-) < x$. We also denote by $\tau_k(x)$ the duration of the kth period during which the fluid level is equal to x. More precisely, if we set $\gamma_0 = 0$ and $\tau_0(x) = 0$, we have for every $k \geq 1$

$$\begin{aligned}
\gamma_k &= \inf\{t \in (\gamma_{k-1} + \tau_{k-1}(x), \theta(0)) \mid X(t) = x\}, \\
\tau_k(x) &= \inf\{t > \gamma_k \mid X(t) < x\} - \gamma_k.
\end{aligned}$$

We have $\gamma_1 = \gamma(x)$. In addition, γ_k also depends on x but we do not mention this dependence to simplify the notation. When $K \geq 1$, for $k \in \{1, \ldots, K\}$, we denote by $V_k(x)$ the volume of fluid lost during period $\tau_k(x)$. This volume is easily expressed by

$$V_k(x) = \int_{\gamma_k}^{\tau_k(x)+\gamma_k} r_{\varphi(t)} dt$$

and we have

$$\tau(x) = \sum_{k=1}^{K} \tau_k(x) \text{ and } V(x) = \sum_{k=1}^{K} V_k(x).$$

The above variables are shown in Figure 2.3 in the case when $K = 3$. To simplify the figure, we have chosen the non-zero effective input rates equal to -1 or 1. We have represented in this figure, above the axis $X(t) = x$, the successive volumes of fluid lost during the busy period.

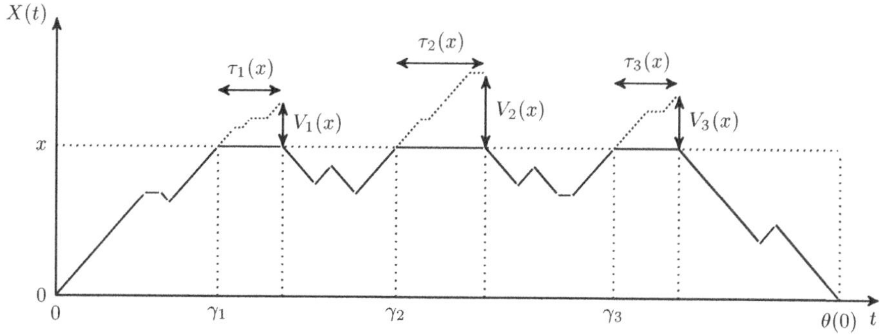

Figure 2.3. *A busy period with losses*

Note that, for every $k \geq 1$ and for every $j \in S^+$, we have

$$\varphi(\gamma_k) = j \Longrightarrow \varphi(\gamma_k) \in S^+ \Longleftrightarrow K \geq k. \qquad [2.25]$$

2.3.2. *Minimum valley height in a busy period*

We are first interested by the evaluation of the probability, starting from a phase $i \in S^+$ and a level z, to reach level z in phase j while staying above level z. This probability is denoted by $\Gamma_{i,j}(x, z)$, which is defined below.

We first introduce the $n_+ \times n_-$ matrix $K(x, y, z)$ whose entries are defined, for $i \in S^+, j \in S^-, x > 0$ and $0 \leq z \leq y \leq x$, by

$$K_{i,j}(x, y, z) = \text{P}\{\varphi(\gamma(z)) = j, H \leq y \mid \varphi(0) = i, X(0) = z\},$$

where H denotes the minimum valley between instants 0 and $\gamma(z)$. When there is no valley, we take as convention $H = 0$. This corresponds either to the situation in Figure 2.4(b), i.e. no peak, or to the situation in Figure 2.4(c), i.e. only one peak.

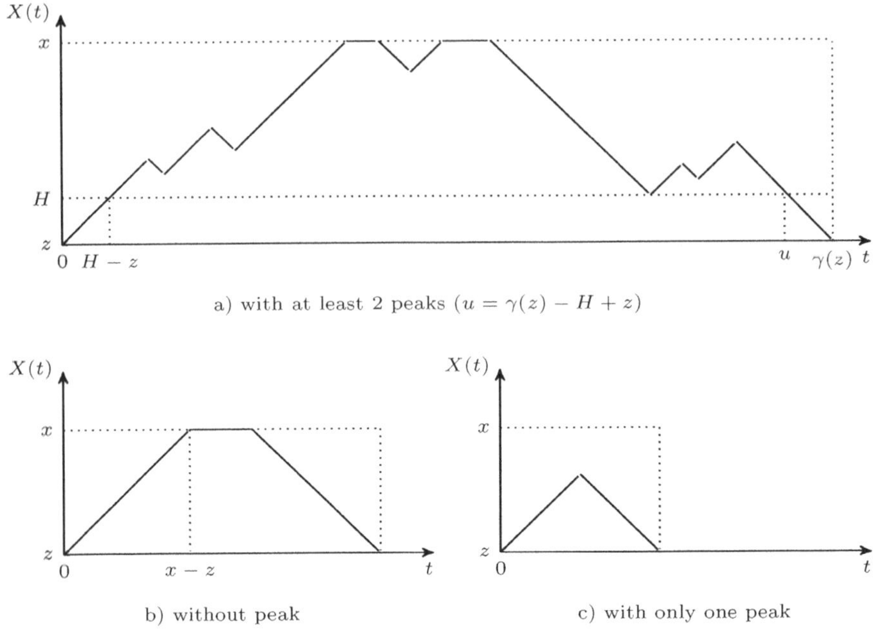

Figure 2.4. *A period of level greater than* z

By the spatial homogeneity, we have $K_{i,j}(x, y, z) = K_{i,j}(x - z, y - z, 0)$. The $n_+ \times n_-$ matrix $\Gamma(x, z)$ the entries of which are defined, for $i \in S^+$, $j \in S^-$, $x > 0$ and $0 \leq z \leq x$, by

$$\Gamma_{i,j}(x, z) = \mathrm{P}\{\varphi(\gamma(z)) = j \mid \varphi(0) = i, X(0) = z\},$$

satisfies $\Gamma(x, z) = K(x, x, z) = K(x - z, x - z, 0) = \Gamma(x - z, 0)$. The $n_+ \times n_-$ matrix $\Gamma(x)$ whose entries are defined, for $i \in S^+$, $j \in S^-$ and $x > 0$, by

$$\Gamma_{i,j}(x) = \mathrm{P}\{\varphi(\gamma(0)) = j \mid \varphi(0) = i, X(0) = 0\},$$

satisfies $\Gamma(x) = \Gamma(x, 0)$. A typical example of such paths is shown in Figure 2.4.

We denote by $k_{i,j}(x, y, z)$ the joint density associated with the joint distribution $K_{i,j}(x, y, z)$, i.e. $k_{i,j}(x, y, z) = \partial K_{i,j}(x, y, z)/\partial y$.

This situation is symmetric to the one studied in section 2.2.5 where we have considered the maximum peak between two successive visits to level x, starting with a negative net input rate and with a barrier at level 0. This barrier is natural but it is important to mention it for the comparison with the finite buffer case. Here, we consider the minimum valley between two successive visits to level z, starting with a positive net input rate and with a barrier at level x. When $z = 0$, if we take the symmetry in Figure 2.3 with respect to the line $X(t) = x/2$, we obtain a figure similar to Figure 2.2. Actually, if we exchange the indices $+$ and $-$ of the submatrices of Q, in the expression of $g(x, y)$ and $\Theta(x)$ we obtain, respectively, as we shall see, $k(x, y, 0)$ and $\Gamma(x)$.

THEOREM 2.9.– For $x > 0$ and $0 \leq z \leq y \leq x$, we have

$$K(x, z, z) = \int_z^x e^{Q_{++}(y-z)} Q_{+-} e^{Q_{--}(y-z)} dy + (-Q_{++})^{-1} e^{Q_{++}(x-z)} Q_{+-} e^{Q_{--}(x-z)}, \quad [2.26]$$

$$k(x, y, z) = e^{Q_{++}(y-z)} \Gamma(x-y) Q_{-+} \Gamma(x-y) e^{Q_{--}(y-z)}. \quad [2.27]$$

PROOF.– The proof is quasi-identical to the proof of theorem 2.6. ■

Using the spatial homogeneity, the matrices $K(x, y, z)$ and $\Gamma(x)$ are given by the following equations. For $x \geq 0$ and $0 \leq z \leq y \leq x$, we have

$$K(x, y, z) = K(x, z, z) + \int_z^y k(x, u, z) du. \quad [2.28]$$

For $x \geq 0$, we have $\Gamma(x) = \Gamma(x, 0) = K(x, x, 0)$, that is

$$\Gamma(x) = \int_0^x e^{Q_{++}y} Q_{+-} e^{Q_{--}y} dy + (-Q_{++})^{-1} e^{Q_{++}x} Q_{+-} e^{Q_{--}x} + \int_0^x e^{Q_{++}y} \Gamma(x-y) Q_{-+} \Gamma(x-y) e^{Q_{--}y} dy. \quad [2.29]$$

This equation is the symmetric version of equation [2.19]. We thus obtain, in the same way as for equation [2.19], the matrix differential Riccati equation

$$\Gamma'(x) = Q_{++}\Gamma(x) + \Gamma(x) Q_{--} + \Gamma(x) Q_{-+} \Gamma(x) + Q_{+-}. \quad [2.30]$$

This equation is identical to equation [2.11]. The only difference concerns the initial condition. Here, we have, from [2.29], $\Gamma(0) = (-Q_{++})^{-1} Q_{+-}$. Following the same lines used in the proof of theorem 2.5, we obtain $\Gamma(x) = Z(x) Y^{-1}(x)$, where

$$\begin{pmatrix} Y(x) \\ Z(x) \end{pmatrix} = e^{CQx} \begin{pmatrix} I \\ \Gamma(0) \end{pmatrix}.$$

Note that $\Gamma(0) = (-Q_{++})^{-1}Q_{+-}$ is a stochastic matrix. This can be written as

$$\begin{cases} Y(x) = A(x) + B(x)\Gamma(0) \\ Z(x) = C(x) + D(x)\Gamma(0) \end{cases}$$

which gives

$$\Gamma(x) = \left[C(x) + D(x)\Gamma(0)\right]\left[A(x) + B(x)\Gamma(0)\right]^{-1}. \qquad [2.31]$$

2.3.3. *Reduction of the state space*

As we did in section 2.2.2, we introduce the $(n_- + n_+) \times (n_- + n_+)$ matrix

$$Q = \begin{pmatrix} Q_{--} & Q_{-+} \\ Q_{+-} & Q_{++} \end{pmatrix},$$

where $Q_{--} = R_-^{-1}\left(T_{--} - T_{-0}T_{00}^{-1}T_{0-}\right)$, $Q_{-+} = R_-^{-1}\left(T_{-+} - T_{-0}T_{00}^{-1}T_{0+}\right)$, $Q_{+-} = R_+^{-1}\left(T_{+-} - T_{+0}T_{00}^{-1}T_{0-}\right)$ and $Q_{++} = R_+^{-1}\left(T_{++} - T_{+0}T_{00}^{-1}T_{0+}\right)$. As shown in previous studies (Bean *et al.* 2005a; Rubino and Sericola 1989), when considering quantities such as hitting probabilities and accumulated rewards during sojourn times in $S^+ \cup S^0$, the fluid queue with parameters (T, R) is equivalent to the fluid queue with parameters (Q, C), where C is the $(n_- + n_+) \times (n_- + n_+)$ matrix

$$C = \begin{pmatrix} -I & 0 \\ 0 & I \end{pmatrix}.$$

The distribution of $\tau_k(x)$, given that the phase at time γ_k is $j \in S^+$, is given by

$$\mathrm{P}\{\tau_k(x) > t \mid \varphi(\gamma_k) = j\} = \left(e^{At}\mathbf{1}\right)_j,$$

where A is the $(n_+ + n_0) \times (n_+ + n_0)$ matrix defined by

$$A = \begin{pmatrix} T_{00} & T_{0+} \\ T_{+0} & T_{++} \end{pmatrix}$$

and $\mathbf{1}$ is the column vector with all its entries equal to 1, its dimension being defined by the context. Since $j \in S^+$, we only need the entries of $e^{At}\mathbf{1}$ corresponding to the subset S^+, so we decompose vector $e^{At}\mathbf{1}$ with respect to subsets S^0 and S^+, by writing

$$e^{At}\mathbf{1} = \begin{pmatrix} \left(e^{At}\mathbf{1}\right)^0 \\ \left(e^{At}\mathbf{1}\right)^+ \end{pmatrix}.$$

We then have

$$\mathrm{P}\{\tau_k(x) > t \mid \varphi(\gamma_k) = j\} = \left(e^{At}\mathbf{1}\right)_j^+. \qquad [2.32]$$

Now, if we consider the volume of fluid $V_k(x)$ lost during period $\tau_k(x)$, given that the phase at time γ_k is j, $V_k(x)$ can also be seen as the accumulated reward during a sojourn of Markov chain $(\varphi(t))$ in subset $S^0 \cup S^+$, starting in state $j \in S^+$. Using the results of Rubino and Sericola (1989), we get, for $j \in S^+$,

$$\mathrm{P}\{V_k(x) > v \mid \varphi(\gamma_k) = j\} = \left(e^{Q_{++}v}\mathbf{1}\right)_j. \qquad [2.33]$$

Thus, in the following, we will consider a fluid queue driven by a Markov chain $(\varphi(t))$ with state space $S = S^- \cup S^+$, infinitesimal generator Q and effective input rates given by matrix C, i.e. equal to -1 or 1.

2.3.4. *Distributions of* $\tau_1(x)$ *and* $V_1(x)$

2.3.4.1. *Matrix differential equation*

In order to determine the distribution of $\tau_1(x)$ and $V_1(x)$, we need the distribution of the Markov chain $(\varphi(t))$ at time γ_1. We introduce, for $i \in S^+$ and $j \in S^+$,

$$M_{i,j}(x) = \mathrm{P}\{\varphi(\gamma_1) = j \mid \varphi(0) = i, X(0) = 0\}$$

and we denote by $M(x)$ the matrix with entries $M_{i,j}(x)$. By definition of matrix $\Psi(x)$, which has already been defined in the infinite buffer case by relation [2.8], for all $i \in S^+$, the quantity $\sum_{j \in S^-} \Psi_{i,j}(x)$ is the probability that the process does not hit level x in a busy period starting from state i and an empty buffer and $\sum_{j \in S^+} M_{i,j}(x)$ is the probability of the complementary event. Hence, for all $i \in S^+$, we have $\sum_{j \in S^-} \Psi_{i,j}(x) + \sum_{j \in S^+} M_{i,j}(x) = 1$. This can be written in matrix form as $\Psi(x)\mathbf{1} + M(x)\mathbf{1} = \mathbf{1}$, where we recall that the dimension of $\mathbf{1}$ is given by the context of its use.

Matrices $M(x)$ and $\Psi(x)$ are related to each other via a differential and an integral equation. As shown in theorems 2.3 and 2.4, matrix $\Psi(x)$ satisfies the integral equation [2.10] and the matrix differential Riccati equation [2.11]. Matrix $M(x)$ is then related to $\Psi(x)$ by the following result.

THEOREM 2.10.– For every $x \geq 0$, we have

$$M(x) = e^{Q_{++}x} + \int_0^x e^{Q_{++}y}\Psi(x-y)Q_{-+}M(x-y)dy, \qquad [2.34]$$

$$M'(x) = \left(Q_{++} + \Psi(x)Q_{-+}\right)M(x). \qquad [2.35]$$

PROOF.– The proof of [2.34] follows the arguments developed in the proof of theorem 2.3 and [2.35] is obtained by differentiating [2.34]. ■

2.3.4.2. *Distributions of the random variables*

The distributions of $\tau_1(x)$ and $V_1(x)$ are given by the following theorem.

THEOREM 2.11.– For every $i \in S^+$ and for all $v, t \geq 0$, we have

$$\mathrm{P}\{\tau_1(x) > t \mid \varphi(0) = i\} = \left(M(x)\left(e^{At}\mathbf{1}\right)^+\right)_i, \qquad [2.36]$$

$$\mathrm{P}\{V_1(x) > v \mid \varphi(0) = i\} = \left(M(x)e^{Q_{++}v}\mathbf{1}\right)_i, \qquad [2.37]$$

$$\mathrm{P}\{K \geq 1 \mid \varphi(0) = i\} = \left(M(x)\mathbf{1}\right)_i. \qquad [2.38]$$

PROOF.– Using the Markov property, we obtain, for every $i \in S^+$,

$$\begin{aligned}\mathrm{P}\{\tau_1(x) > t \mid \varphi(0) = i\} &= \sum_{j\in S^+} \mathrm{P}\{\tau_1(x) > t \mid \varphi(\gamma_1) \\ &= j\}\mathrm{P}\{\varphi(\gamma_1) = j \mid \varphi(0) = i\} \\ &= \sum_{j\in S^+} M_{i,j}(x)\left(e^{At}\mathbf{1}\right)_j^+ = \left(M(x)\left(e^{At}\mathbf{1}\right)^+\right)_i.\end{aligned}$$

The same argument leads to relation [2.37]. From equivalence [2.25], we get, by definition of $M_{i,j}(x)$, $\mathrm{P}\{K \geq 1 \mid \varphi(0) = i\} = \mathrm{P}\{\varphi(\gamma_1) \in S^+ \mid \varphi(0) = i\} = \left(M(x)\mathbf{1}\right)_i$. ■

In the next section, we study the distributions of the state initiating the successive idle and busy periods of the buffer.

2.3.5. *Sequences of idle and busy periods*

We decompose the initial probability distribution α of Markov chain $(\varphi(t))$ through the partition S^-, S^0, S^+ by writing $\alpha = (\alpha^-, \alpha^0, \alpha^+)$. Recall that we suppose that the buffer is initially empty, i.e. that $X(0) = 0$ with probability 1. For every $n \geq 1$, we denote by SI_n the state or the phase by which the nth idle period starts and we denote by SB_n the state or the phase by which the nth busy period starts. We then have $SI_1 \in S^- \cup S^0$ if $\alpha^0 \neq 0$, $SI_n \in S^-$, for $n \geq 2$ and $SB_n \in S^+$. We denote, respectively, by v_n and w_n the probability distributions of SI_n and SB_n, i.e.

$$w_n = \left(\mathrm{P}\{SB_n = j\},\ j \in S^+\right),$$

$$v_1 = \left(\mathrm{P}\{SI_1 = j\},\ j \in S^- \cup S^0\right) \text{ and, for } n \geq 2,\ v_n = \left(\mathrm{P}\{SI_n = j\},\ j \in S^-\right).$$

Recall that for $i \in S^+$ and $j \in S^-$, the probability $\Gamma_{i,j}(x)$ that a busy period of the buffer with capacity x ends in phase j given that it starts in phase i, that is

$$\Gamma_{i,j}(x) = \mathrm{P}\{\varphi(\theta(0)) = j \mid \varphi(0) = i\}$$

is given by relation [2.31]. Note that for every $x \geq 0$, we have $\Gamma(x)\mathbb{1} = \mathbb{1}$. Using the results of Rubino and Sericola (1989), we have

$$v_1 = \left(\alpha^- + \alpha^+\Gamma(x), \alpha^0\right) \text{ and } w_1 = \alpha^+ + (\alpha^-, \alpha^0)\left(-A'\right)^{-1} B', \qquad [2.39]$$

where matrices A' and B' are defined by

$$A' = \begin{pmatrix} T_{--} & T_{-0} \\ T_{0-} & T_{00} \end{pmatrix} \text{ and } B' = \begin{pmatrix} T_{-+} \\ T_{0+} \end{pmatrix}.$$

Defining $Q_{00} = T_{00} - T_{0-}T_{--}^{-1}T_{-0}$ and $Q_{0+} = T_{0+} - T_{0-}T_{--}^{-1}T_{-+}$, we have

$$\left(-A'\right)^{-1} = \begin{pmatrix} \left(-Q_{--}\right)^{-1} R_{-}^{-1} & \left(-Q_{--}\right)^{-1} R_{-}^{-1} T_{-0}\left(-T_{00}\right)^{-1} \\ \left(-Q_{00}\right)^{-1} T_{0-}\left(-T_{--}\right)^{-1} & \left(-Q_{00}\right)^{-1} \end{pmatrix} \qquad [2.40]$$

and

$$\left(-A'\right)^{-1} B' = \begin{pmatrix} \left(-Q_{--}\right)^{-1} Q_{-+} \\ \left(-Q_{00}\right)^{-1} Q_{0+} \end{pmatrix} \qquad [2.41]$$

which leads to

$$w_1 = \alpha^+ + \alpha^- \left(-Q_{--}\right)^{-1} Q_{-+} + \alpha^0 \left(-Q_{00}\right)^{-1} Q_{0+}. \qquad [2.42]$$

In the following theorem, we give the expressions of v_n and w_n for $n \geq 2$.

THEOREM 2.12.– For every $n \geq 2$, we have

$$v_n = \left(\alpha^- + \alpha^+\Gamma(x)\right) \mathcal{L}^{n-1}(x) + \alpha^0 \left(-Q_{00}\right)^{-1} Q_{0+}\Gamma(x)\mathcal{L}^{n-2}(x),$$

$$w_n = \left(\alpha^+ + \alpha^- \left(-Q_{--}\right)^{-1} Q_{-+} + \alpha^0 \left(-Q_{00}\right)^{-1} Q_{0+}\right) \mathcal{K}^{n-1}(x),$$

where matrices $\mathcal{L}(x)$ and $\mathcal{K}(x)$ are given by

$$\mathcal{L}(x) = \left(-Q_{--}\right)^{-1} Q_{-+}\Gamma(x) \text{ and } \mathcal{K}(x) = \Gamma(x) \left(-Q_{--}\right)^{-1} Q_{-+}.$$

PROOF.– For every $j \in S^-$, we have

$$\mathrm{P}\{SI_2 = j\} = \sum_{i \in S^- \cup S^0} \mathrm{P}\{SI_1 = i\} \left[\left(-A'\right)^{-1} B'\Gamma(x)\right]_{i,j},$$

which gives, in matrix notation, using [2.41] and [2.39],

$$\begin{aligned} v_2 &= v_1 \left(-A'\right)^{-1} B'\Gamma(x) \\ &= \left(\alpha^- + \alpha^+\Gamma(x)\right) \left(-Q_{--}\right)^{-1} Q_{-+}\Gamma(x) + \alpha^0 \left(-Q_{00}\right)^{-1} Q_{0+}\Gamma(x) \\ &= \left(\alpha^- + \alpha^+\Gamma(x)\right) \mathcal{L}(x) + \alpha^0 \left(-Q_{00}\right)^{-1} Q_{0+}\Gamma(x). \end{aligned}$$

For every $n \geq 3$ and $j \in S^-$, we have, by homogeneity,

$$\mathrm{P}\{SI_n = j\} = \sum_{i \in S^-} \mathrm{P}\{SI_{n-1} = i\} \left[\left(-A'\right)^{-1} B'\Gamma(x)\right]_{i,j}.$$

We thus get, in matrix notation, again using [2.41] and [2.39],

$$v_n = v_{n-1} \left(-Q_{--}\right)^{-1} Q_{-+}\Gamma(x) = v_{n-1}\mathcal{L}(x).$$

This leads, for $n \geq 2$, to $v_n = v_2\mathcal{L}^{n-2}(x)$, which is the desired result. In the same way, we have for every $n \geq 2$ and $j \in S^+$, using [2.41],

$$\begin{aligned}
\mathrm{P}\{SB_n = j\} &= \sum_{i\in S^+} \mathrm{P}\{SB_{n-1} = i\} \sum_{\ell\in S^-} \Gamma_{i,\ell}(x) \left[(-A')^{-1} B'\right]_{\ell,j} \\
&= \sum_{i\in S^+} \mathrm{P}\{SB_{n-1} = i\} \left[\Gamma(x) (-Q_{--})^{-1} Q_{-+}\right]_{i,j} \\
&= \sum_{i\in S^+} \mathrm{P}\{SB_{n-1} = i\} \mathcal{K}_{i,j}(x).
\end{aligned}$$

This gives in matrix notation, using [2.42], $w_n = w_{n-1}\mathcal{K}(x)$, that is $w_n = w_1\mathcal{K}^{n-1}(x)$. ■

The Markov chain $(\varphi(t))$ being finite and irreducible, matrices $\mathcal{L}(x)$ and $\mathcal{K}(x)$, which are stochastic matrices, are also irreducible and aperiodic. We thus have

$$\lim_{n\longrightarrow\infty} v_n = \pi^- \text{ and } \lim_{n\longrightarrow\infty} w_n = \pi^+, \qquad [2.43]$$

where π^- is a row vector of dimension n_-, which is the unique solution to the system $\pi^- = \pi^-\mathcal{L}(x)$, with $\pi^-\mathbf{1} = 1$ and π^+ is a row vector of dimension n_+, which is the unique solution to the system $\pi^+ = \pi^+\mathcal{K}(x)$, with $\pi^+\mathbf{1} = 1$.

In what follows, we suppose that the queue is in stationary regime, that is the distribution of $\varphi(0)$ is π^+. We thus have $\mathrm{P}\{\tau_1(x) > t\} = \pi^+M(x)\left(e^{At}\mathbf{1}\right)^+$, $\mathrm{P}\{V_1(x) > v\} = \pi^+M(x)e^{Q_{++}v}\mathbf{1}$ and $\mathrm{P}\{K \geq 1\} = \pi^+M(x)\mathbf{1}$. The above result gives the distribution of the first congestion event. In the next section, we compute the characteristics of several congestion events.

2.3.6. *Joint distributions of loss periods and loss volumes*

In this section, we study the joint distribution of the times $\tau_k(x)$ and the joint distribution of the volumes $V_k(x)$. To deal with these distributions, we need to evaluate the probability $H_{i,j}(x)$, starting from level x in phase $i \in S^-$, that the phase at the time of the first return to level x is equal to j, without hitting level 0. This probability can be interpreted as the symmetric probability of $\Psi(x)$ with respect to levels 0 and x. It is defined, for every $i \in S^-$ and $j \in S^+$, by

$$H_{i,j}(x) = \mathrm{P}\{\varphi(\gamma(x)) = j, X(s) > 0 \text{ for } 0 \leq s \leq \gamma(x) \mid \varphi(0) = i, X(0) = x\}.$$

THEOREM 2.13.– For every $x \geq 0$, we have

$$H(x) = \int_0^x e^{Q_{--}y} Q_{-+} e^{Q_{++}y} dy + \int_0^x e^{Q_{--}y} H(x-y) Q_{+-} H(x-y) e^{Q_{++}y} dy. \quad [2.44]$$

PROOF.– The proof is quite similar to the proof leading to equation [2.10]. It can also be seen as a particular case of the way in which equation [2.19] has been obtained since here we do not consider the paths hitting level 0. ■

In the same way we obtained equation [2.11], matrix $H(x)$ satisfies $H(0) = 0$ and

$$H'(x) = Q_{--}H(x) + H(x)Q_{++} + H(x)Q_{+-}H(x) + Q_{-+}. \quad [2.45]$$

Note that, by homogeneity of ϕ, we also have, for every $i \in S^-$, $j \in S^+$ and $k \geq 1$,

$$H_{i,j}(x) = \mathrm{P}\{\phi(\gamma_{k+1}) = j \mid \phi(\gamma_k + \tau_k) = i\}.$$

This relation is used to prove the next lemma. We introduce the matrix B defined by

$$B = \begin{pmatrix} T_{+-} \\ T_{0-} \end{pmatrix}.$$

LEMMA 2.1.– For every $k \geq 1$, $v \geq 0$ and $i, j \in S^+$, we have

$$\mathrm{P}\{V_k(x) > v, \phi(\gamma_{k+1}) = j \mid \phi(\gamma_k) = i\} = \left(e^{Q_{++}v}(-Q_{++})^{-1}Q_{+-}H(x)\right)_{i,j}$$

and

$$\mathrm{P}\{\tau_k(x) > t, \phi(\gamma_{k+1}) = j \mid \phi(\gamma_k) = i\} = \left(\left[e^{At}(-A)^{-1}B\right]_{+-} H(x)\right)_{i,j},$$

where matrix $\left[e^{At}(-A)^{-1}B\right]_{+-}$ is the block $n_+ \times n_-$ of matrix $e^{At}(-A)^{-1}B$.

PROOF.– The homogeneity of Markov chain ϕ implies that, for every $k \geq 1$,

$$\mathrm{P}\{V_k(x) > v, \phi(\gamma_{k+1}) = j \mid \phi(\gamma_k) = i\} = \mathrm{P}\{V_1(x) > v, \phi(\gamma_2) = j \mid \phi(\gamma_1) = i\},$$

so we set $k = 1$. Note that the entry (i, ℓ) of matrix $e^{Q_{++}v}(-Q_{++})^{-1}Q_{+-}$ is (see Rubino and Sericola (1989)) the probability, starting from $i \in S^+$, that the accumulated reward during a sojourn of ϕ in $S^+ \cup S^0$ is greater than v and that the first state of S^- visited after this sojourn is ℓ. Conditioning on the phase at time $\gamma_1 + \tau_1$ and using the Markov property, we get

$$\begin{aligned}
&\mathrm{P}\{\phi(\gamma_2) = j, V_1(x) > v \mid \phi(\gamma_1) = i\}\\
&= \sum_{\ell \in S^-} \mathrm{P}\{\phi(\gamma_2) = j \mid \phi(\gamma_1 + \tau_1) = \ell\}\mathrm{P}\{\phi(\gamma_1 + \tau_1) = \ell, V_1(x) > v \mid \phi(\gamma_1) = i\}\\
&= \sum_{\ell \in S^-} H_{\ell,j} \left(e^{Q_{++}v}(-Q_{++})^{-1}Q_{+-}\right)_{i,\ell} = \left(e^{Q_{++}v}(-Q_{++})^{-1}Q_{+-}H(x)\right)_{i,j}.
\end{aligned}$$

To get the second relation, we proceed in the same way noting that the entry (i, ℓ) of matrix $\left[e^{At}(-A)^{-1}B\right]_{+-}$ is (see Rubino and Sericola (1989)) the probability, starting from $i \in S^+$, that a sojourn of ϕ in $S^+ \cup S^0$ is greater than t and that the first state of S^- visited after this sojourn is ℓ. ■

Note that we have the following relations:

$$\left[(-A)^{-1}B\right]_{+-} = (-Q_{++})^{-1}Q_{+-} \text{ and } (-Q_{++})^{-1}Q_{+-}\mathbb{1} = \mathbb{1}. \quad [2.46]$$

THEOREM 2.14.– For every $k \geq 1$ and $v_1, \ldots, v_k \geq 0$, we have

$$\begin{aligned}
&\mathrm{P}\{V_1(x) > v_1, V_2(x) > v_2, \ldots, V_k(x) > v_k\}\\
&\qquad = \pi^+ M(x) \left(\prod_{\ell=1}^{k-1} e^{Q_{++}v_\ell}(-Q_{++})^{-1}Q_{+-}H(x)\right) e^{Q_{++}v_k}\mathbb{1}. \quad [2.47]
\end{aligned}$$

PROOF.– Note that by convention the product is equal to 1 for $k = 1$. Let us define $F_{i,j}(k)$, for every $k \geq 1$ and $i, j \in S^+$, by

$$F_{i,j}(k) = \mathrm{P}\{V_1(x) > v_1, \ldots, V_k(x) > v_k, \phi(\gamma_{k+1}) = j \mid \phi(0) = i\}.$$

Conditioning on $\phi(\gamma_k)$ and using the Markov property, we have

$$\begin{aligned}
F_{i,j}(k) &= \sum_{\ell \in S^+} \mathrm{P}\{V_1(x) > v_1, \ldots, V_k(x) > v_k, \phi(\gamma_{k+1}) = j, \phi(\gamma_k) = \ell \mid \phi(0) = i\}\\
&= \sum_{\ell \in S^+} F_{i,\ell}(k-1)\mathrm{P}\{V_k(x) > v_k, \phi(\gamma_{k+1}) = j \mid \phi(\gamma_k) = \ell\}.
\end{aligned}$$

Denoting by $F(k)$ the matrix $[F_{i,j}(k)]_{i,j\in S^+}$, we obtain, using Lemma 2.1,

$$F(k) = F(k-1)e^{Q_{++}v_k}(-Q_{++})^{-1}Q_{+-}H(x),$$

and thus,

$$F(k) = F(1)\prod_{\ell=2}^{k} e^{Q_{++}v_\ell}(-Q_{++})^{-1}Q_{+-}H(x).$$

Matrix $F(1)$ is obtained in the same way, using the Markov property and lemma 2.1, as

$$\begin{aligned}
F_{i,j}(1) &= \sum_{\ell\in S^+} \mathrm{P}\{V_1(x) > v_1, \phi(\gamma_2) = j, \phi(\gamma_1) = \ell \mid \phi(0) = i\} \\
&= \sum_{\ell\in S^+} \mathrm{P}\{\phi(\gamma_1) = \ell \mid \phi(0) = i\}\mathrm{P}\{V_1(x) > v_1, \phi(\gamma_2) = j \mid \phi(\gamma_1) = \ell\} \\
&= \sum_{\ell\in S^+} M_{i,\ell}(x)\left[e^{Q_{++}v_1}(-Q_{++})^{-1}Q_{+-}H(x)\right]_{\ell,j}.
\end{aligned}$$

Finally, we have

$$F(k) = M(x)\prod_{\ell=1}^{k} e^{Q_{++}v_\ell}(-Q_{++})^{-1}Q_{+-}H(x).$$

To obtain the joint distribution of $V_1(x),\dots,V_k(x)$, we use relation [2.33], the previous relation and the Markov property, by writing

$$\begin{aligned}
&\mathrm{P}\{V_1(x) > v_1, V_2(x) > v_2, \dots, V_k(x) > v_k \mid \phi(0) = i\} \\
&\quad= \sum_{j\in S^+} \mathrm{P}\{V_1(x) > v_1, \dots, V_{k-1}(x) > v_{k-1}, \phi(\gamma_k) = j, V_k(x) > v_k \mid \phi(0) = i\} \\
&\quad= \sum_{j\in S^+} F_{i,j}(k-1)\mathrm{P}\{V_k(x) > v_k \mid \phi(\gamma_k) = j\} \\
&\quad= \sum_{j\in S^+} F_{i,j}(k-1)\left(e^{Q_{++}v_k}\mathbf{1}\right)_j = \left[F(k-1)e^{Q_{++}v_k}\mathbf{1}\right]_i.
\end{aligned}$$

Since the distribution of $\phi(0)$ is π^+, we get by unconditioning

$$\mathrm{P}\{V_1(x) > v_1, V_2(x) > v_2, \dots, V_k(x) > v_k\} = \pi^+F(k-1)e^{Q_{++}v_k}\mathbf{1},$$

which completes the proof. ■

Similar arguments are used to get the joint distribution of the $\tau_k(x)$.

THEOREM 2.15.– For every $k \geq 1$ and $t_1, \ldots, t_k \geq 0$, we have

$$\mathrm{P}\{\tau_1(x) > t_1, \tau_2(x) > t_2, \ldots, \tau_k(x) > t_k\}$$
$$= \pi^+ M(x) \left(\prod_{\ell=1}^{k-1} \left(e^{At_\ell}(-A)^{-1}B\right)_{+-} H(x) \right) \left(e^{At_k}\mathbf{1}\right)^+. \quad [2.48]$$

Note that the joint distribution of the τ_k, given by [2.48], can be rewritten as

$$\mathrm{P}\{\tau_1(x) > t_1, \ldots, \tau_k(x) > t_k\} = a(x) \left(\prod_{\ell=1}^{k-1} e^{At_\ell}(-A)^{-1}BR(x) \right) e^{At_k}\mathbf{1}, \quad [2.49]$$

where $a(x)$ is the row vector defined by $a(x) = (\pi^+ M(x)\ 0)$, 0 being here the null row vector of dimension n_0, and $R(x)$ is the matrix defined by $R(x) = (H(x)\ 0)$, 0 being here the $n_- \times n_0$ null matrix.

COROLLARY 2.3.– For every $k \geq 1$ and $v, t \geq 0$, we have

$$\mathrm{P}\{V_k(x) > v\} = \pi_+ M(x) \left((-Q_{++})^{-1}Q_{+-}H(x)\right)^{k-1} e^{Q_{++}v}\mathbf{1}, \quad [2.50]$$

$$\mathrm{P}\{\tau_k(x) > t\} = \pi_+ M(x) \left((-Q_{++})^{-1}Q_{+-}H(x)\right)^{k-1} \left(e^{At}\mathbf{1}\right)^+. \quad [2.51]$$

PROOF.– To get relation [2.50], put $v_1 = \cdots = v_{k-1} = 0$ and $v_k = v$ in [2.47]. To get relation [2.51], put $t_1 = \cdots = t_{k-1} = 0$ and $t_k = t$ in [2.48] and observe that $\left((-A)^{-1}B\right)_{+-} = (-Q_{++})^{-1}Q_{+-}$. ■

COROLLARY 2.4.– For every $k \geq 0$, we have

$$\mathrm{P}\{K > k\} = \pi^+ M(x) \left((-Q_{++})^{-1}Q_{+-}H(x)\right)^k \mathbf{1}. \quad [2.52]$$

PROOF.– For every $k \geq 0$, we have $V_{k+1}(x) > 0 \Longleftrightarrow \tau_{k+1}(x) > 0 \Longleftrightarrow K > k$. We thus get the result using either [2.50] or [2.51]. ■

As we did for the joint distribution of the τ_k, obtained in [2.49], the distribution of K, given by [2.52], can be rewritten as

$$\mathrm{P}\{K > k\} = a(x) \left((-A)^{-1} BR(x)\right) \mathbf{1}. \quad [2.53]$$

2.3.7. *Total duration of losses and volume of information lost*

We determine in this section the distribution of the total loss duration $\tau(x)$ in a busy period and the distribution of the total volume of information lost $V(x)$ in a busy period. We have seen in section 2.3.1 that we have

$$\tau(x) = \sum_{k=1}^{K} \tau_k(x) \text{ and } V(x) = \sum_{k=1}^{K} V_k(x).$$

To get the distribution of $\tau(x)$ and $V(x)$, we need some results about absorbing Markovian arrival processes.

2.3.7.1. *Absorbing Markovian arrival process*

The Markovian arrival process is well known in the queuing literature and was introduced in a previous study by Neuts (1979). We consider here a variant of this process in which we add an absorbing state. This process is defined as a continuous-time Markov chain $\{(N(t), J(t)),\ t \geq 0\}$, with $N(0) = 0$, on the state space $S = \{(a, 0)\} \cup (\mathbb{N} \times \{1, \ldots, m\})$, with $m \geq 1$ and where state $(a, 0)$ is absorbing. Its infinitesimal generator G is given by

$$G = \begin{pmatrix} 0 & 0 & 0 & 0 & 0 & 0 & 0 & \cdots \\ V & D_0 & D_1 & 0 & 0 & 0 & 0 & \cdots \\ V & 0 & D_0 & D_1 & 0 & 0 & 0 & \cdots \\ V & 0 & 0 & D_0 & D_1 & 0 & 0 & \cdots \\ \vdots & \vdots & \vdots & \vdots & \vdots & \ddots & \ddots & \ddots \end{pmatrix},$$

where the first row and the first column correspond to the absorbing state a. Matrices D_0 and D_1 are non-zero (m, m) matrices and column vector V is a non-zero vector with dimension m. The matrices G being an infinitesimal generator, we have $D_0\mathbb{1} + D_1\mathbb{1} + V = 0$. We moreover assume that all the states, except state $(a, 0)$, are transient. This implies in particular that matrix D_0 is invertible. For $\ell \geq 0$, we define the subset of states $S_\ell = \{(\ell, 1), \ldots, (\ell, m)\}$. As shown in a previous study by Sericola (2013), we set $\xi_0 = 0$ and we denote, for all $n \geq 1$, by ξ_n the instant at which either the nth arrival occurs or the absorbing state is reached. Variable ξ_n is then defined, for all $n \geq 0$, by $\xi_n = \inf\{t \geq 0 \mid N(t) \in \{n, a\}\}$. The nth interarrival time Γ_n is then defined, for $n \geq 1$, by $\Gamma_n = \xi_n - \xi_{n-1}$. We also define Y as the number of subsets S_ℓ visited by the process until absorption, i.e.

$$Y = \sum_{\ell=1}^{\infty} \mathbb{1}_{\{\Gamma_\ell > 0\}}.$$

We then have $\xi_n = \xi_Y$ for $n \geq Y$ and thus $\Gamma_Y > 0$ and $\Gamma_n = 0$ for $n \geq Y+1$. Since $N(0) = 0$, the initial probability distribution is concentrated on the states $(0,1), \ldots, (0,m)$. We denote by β the row vector of dimension m defined by $\beta_i = \mathrm{P}\{J(0) = i\}$, for $i = 1, \ldots, m$. We thus have $\beta_1 + \cdots + \beta_m = 1$. The joint distribution of $\Gamma_1, \ldots, \Gamma_k$ is given, following the results of a previous study by Sericola (2013), by

$$\mathrm{P}\{\Gamma_1 > t_1, \ldots, \Gamma_k > t_k\} = \beta \left[\prod_{\ell=1}^{k-1} e^{D_0 t_\ell} \left(-D_0\right)^{-1} D_1 \right] e^{D_0 t_k} \mathbf{1}. \qquad [2.54]$$

We then have, for every $k \geq 0$, $\Gamma_{k+1} > 0 \Longleftrightarrow Y > k$. By taking $t_1 = \cdots = t_{k-1} = 0$ and $t_k = t$ in [2.54], we get, for every $k \geq 1$ and $t \geq 0$,

$$\mathrm{P}\{\Gamma_k > t\} = \beta \left(\left(-D_0\right)^{-1} D_1 \right)^{k-1} e^{D_0 t} \mathbf{1},$$

that is

$$\mathrm{P}\{Y > k\} = \beta \left(\left(-D_0\right)^{-1} D_1 \right)^{k} \mathbf{1}. \qquad [2.55]$$

We denote by Γ the total time spent by the process in the transient states, that is $\Gamma = \inf\{t \geq 0 \mid N(t) = a\}$. By definition of Y, we have

$$\Gamma = \sum_{k=1}^{Y} \Gamma_k = \sum_{k=1}^{\infty} \Gamma_k.$$

The distribution of Γ is given, for every $t \geq 0$, by $\mathrm{P}\{\Gamma > t\} = (\beta, 0, 0, \ldots) e^{Lt} \mathbf{1}$, where

$$L = \begin{pmatrix} D_0 & D_1 & 0 & 0 & 0 & 0 & \cdots \\ 0 & D_0 & D_1 & 0 & 0 & 0 & \cdots \\ 0 & 0 & D_0 & D_1 & 0 & 0 & \cdots \\ \vdots & \vdots & \vdots & \vdots & \ddots & \ddots & \ddots \end{pmatrix}.$$

It is easily checked that for every $k \geq 0$, we have

$$L^k \mathbf{1} = \begin{pmatrix} (D_0 + D_1)^k \mathbf{1} \\ (D_0 + D_1)^k \mathbf{1} \\ \vdots \end{pmatrix}.$$

We then get

$$P\{\Gamma > t\} = \beta e^{(D_0+D_1)t}\mathbb{1}. \qquad [2.56]$$

2.3.7.2. *Distribution of $\tau(x)$ and $V(x)$*

Using the notation $a(x)$ and $R(x)$ introduced in [2.49], we obtain the following result.

THEOREM 2.16.– For every $v, t \geq 0$, we have

$$P\{V(x) > v\} = \pi^+ M(x) e^{(Q_{++}+Q_{+-}H(x))v}\mathbb{1}, \qquad [2.57]$$

$$P\{\tau(x) > t\} = a(x) e^{(A+BR(x))t}\mathbb{1}. \qquad [2.58]$$

PROOF.– The joint distribution of the V_k, which is given by [2.47], is equal to the joint distribution of the Γ_k given by [2.54] by taking $\beta = \pi^+ M(x)$, $D_0 = Q_{++}$ and $D_1 = Q_{+-}H(x)$. With these values, the distribution of K, given by [2.52], and the distribution of Y, given by [2.55], are identical. We thus have, according to [2.56],

$$P\{V(x) > v\} = \pi^+ M(x) e^{(Q_{++}+Q_{+-}H(x))v}\mathbb{1}.$$

The joint distribution of the τ_k given by [2.49] is equal to the joint distribution of the Γ_k given by [2.54] by taking $\beta = a(x)$, $D_0 = A$ and $D_1 = BR(x)$. With these values, the distribution of K, given by [2.53], and the distribution of Y, given by [2.55], are identical. We thus have, according to [2.56], $P\{\tau(x) > t\} = a(x)e^{(A+BR(x))t}\mathbb{1}$. ■

2.3.7.3. *Loss probability*

Let $E(IP)$ and $E(BP)$ be the mean durations of an idle and a busy period, respectively, in stationary regime and recall that p is the stationary distribution of Markov chain $(\varphi(t))$. The mean input rate in stationary regime is then $E(\chi) = \sum_{i \in S} p_i \chi_i$, where we recall that χ_i is the input rate in the queue when the modulating Markov chain $(\varphi(t))$ is in state i. The quantity π_{loss} defined by equation [2.1] is then given by

$$\pi_{\text{loss}} = \frac{1}{E(IP) + E(BP)} \frac{E(V(x))}{E(\chi)}. \qquad [2.59]$$

From relation [2.57], we have

$$E(V(x)) = \pi^+ M(x) \left(-Q_{++} - Q_{+-}H(x)\right)^{-1} \mathbb{1}. \qquad [2.60]$$

The mean duration of an idle period is given, using [2.40] and [2.43], by

$$\mathrm{E}(IP) = \left(\pi^{-}, 0\right)\left(-A'\right)^{-1}\mathbf{1} = \pi^{-}\left(-Q_{--}\right)^{-1}R_{-}^{-1}\left(I + T_{-0}\left(-T_{00}\right)^{-1}\right)\mathbf{1}. \quad [2.61]$$

The mean durations $\mathrm{E}(BP)$ and $\mathrm{E}(IP)$ are related by the relation

$$\mathrm{P}\{X = 0\} = \frac{\mathrm{E}(IP)}{\mathrm{E}(IP) + \mathrm{E}(BP)},$$

where X is the fluid level in stationary regime. We then have from [2.59]

$$\pi_{\text{loss}} = \frac{\mathrm{P}\{X = 0\}\mathrm{E}(V(x))}{\mathrm{E}(IP)\mathrm{E}(\chi)}. \quad [2.62]$$

The distribution of X, and more precisely the quantity $\mathrm{P}\{X = 0\}$, has been obtained (da Silva Soares and Latouche 2006) using matrix analytic methods.

To conclude this section, we note that the matrices $\Psi(x)$, $M(x)$, $H(x)$ and $\Gamma(x)$ are equivalent to the matrices $G(0, x)$, $H(0, x)$, $H(x, x)$ (Bean *et al.* 2005a) and Ψ_0 (Bean *et al.* 2009), respectively, and can efficiently be evaluated using the explicit expressions in previous studies (Bean *et al.* 2008b, corollary 2 and Bean *et al.* 2009, theorem 3) (substitute $s = 0$) and the quadratic algorithms in Bean *et al.* (2005b, 2008), and without the use of numerical inversion of the Laplace–Stieltjes transforms.

Matrices $\Psi(x)$, $H(x)$ and $\Gamma(x)$ are expressed here in terms of matrix differential Riccati equations that can be solved using a classical method, which consists of computing the exponential of matrix CQx and then in solving a linear system whose parameters are blocks of matrix $\exp(CQx)$. Matrix $M(x)$, see equation [2.35], is the solution to a system of ODE with non-constant coefficient matrix that can easily be solved numerically using a classical Runge–Kutta fourth-order method available in Matlab, for instance.

2.4. Conclusion

In this chapter, we have first deeply analyzed congestion when the buffer content is described by means of a Markov modulated fluid flow model in the stationary regime. In order to obtain such performance metrics, we have analyzed hitting probabilities together with the maximum peak observed within a busy period in both the infinite and finite buffer case.

The distributions of congestion metrics, defined as the total amount of lost information or the duration of congestion within a busy period, have then been obtained. We established that these distributions are of phase type. Finally, we proposed an exact solution to compute the metric π_{loss} as defined by equation [2.1].

2.5. References

Ahn, S., Ramaswami, V. (2003). Fluid flow models and queues – a connection by stochastic coupling. *Stoch. Models*, 19(3), 325–348.

Ahn, S., Ramaswami, V. (2004). Transient analysis of fluid flow models via stochastic coupling to a queue. *Stoch. Models*, 20(1), 71–101.

Asmussen, S. (1995). Stationary distributions for fluid flow models with or without Brownian noise. *Stoch. Models*, 11(1), 21–49.

Bean, N.G., O'Reilly, M.M., Taylor, P.G. (2005a). Hitting probabilities and hitting times for stochastic fluid flows. *Stoch. Process. Their Appl.*, 115, 1530–1556.

Bean, N.G., O'Reilly, M.M., Taylor, P.G. (2005b). Algorithms for the first return probabilities for stochastic fluid flows. *Stoch. Models*, 21(1), 149–184.

Bean, N.G., O'Reilly, M.M., Taylor, P.G. (2008a). Algorithms for the Laplace-Stieltjes transforms of the first return probabilities for stochastic fluid flows. *Methodol. Comput. Appl. Probab.*, 10(3), 381–408.

Bean, N.G., O'Reilly, M.M. (2008b). Performance measures of a multi-layer Markovian fluid model. *Ann. Oper. Res.*, 160(1), 99–120.

Bean, N.G., O'Reilly, M.M., Taylor, P.G. (2009). Hitting probabilities and hitting times for stochastic fluid flows: The bounded model. *Probab. Eng. Inform. Sc.*, 23, 121–147.

da Silva Soares, A., Latouche, G. (2002). Further results on the similarity between fluid queues and QBDs. In *Proceedings of the 4th International Conference on Matrix-Analytic Methods (MAM'4)*, Latouche, G., Taylor, P. (eds), 89–106. World Scientific, Singapore.

da Silva Soares, A., Latouche, G. (2006). Matrix-analytic methods for fluid queues with finite buffers. *Perform. Eval.*, 63(4), 295–314.

Guillemin, F., Sericola, B. (2007). Stationary analysis of a fluid queue driven by some countable state space Markov chain. *Methodol. Comput. Appl. Probab.*, 9, 521–540.

Guillemin, F., Sericola, B. (2012). On the Fluid Queue Driven by an Ergodic Birth and Death Process. In *Telecommunications Networks – Current Status and Future Trends*, Ortiz, J. (ed.), InTech, Available at: http://www.intechopen.com/books/telecommunications-networks-current-status-and-future-trends/on-the-fluid-queue-driven-by-an-ergodic-birth-and-death-process.

Neuts, M.F. (1979). A versatile Markovian point process. *J. Appl. Probab.*, (16)3, 764–779.

O'Reilly, M.M., Palmowski, Z. (2013). Loss rates for stochastic fluid models. *Perform. Eval.*, 70, 593–606.

Ramaswami, V. (1999). Matrix analytic methods for stochastic fluid flows. In *Proceedings of the 16th International Teletraffic Congress: Teletraffic Engineering in a Competitive World (ITC'16)*, Smith, D., Hey, P. (eds), 1019–1030. Elsevier, The Netherlands.

Ramaswami, V. (2006). Passage times in fluid models with application to risk processes. *Methodol. Comput. Appl. Probab.*, 8, 497–515.

Remiche, M.-A. (2005). Compliance of the token-bucket model with Markovian traffic. *Stoch. Models*, 21, 615–630.

Rogers, L.C.G. (1994). Fluid models in queueing theory and Wiener-Hopf factorization of Markov chains. *Ann. Appl. Probab.*, 4(2), 390–413.

Rubino, G., Sericola, B. (1989). Accumulated reward over the n first operational periods in fault-tolerant computing systems. Technical Report 1028, INRIA.

Samuelson, A., O'Reilly, M., Bean, N. (2017). On the generalized reward generator for stochastic fluid models: A new equation for Ψ. *Stoch. Models*, 33(4), 495–523.

Sericola, B. (2001). A finite buffer fluid queue driven by a Markovian queue. *Queueing Syst.*, 38(2), 213–220.

Sericola, B. (2013). *Markov Chains: Theory, Algorithms and Applications*. ISTE Ltd, London and John Wiley & Sons, New York.

3

Diffusion Approximation of Queueing Systems and Networks

Dimitri KOROLIOUK[1] **and Vladimir S. KOROLIUK**[2]
[1] *Institute of Telecommunications and Global Information Space, National Academy of Sciences of Ukraine, Kiev, Ukraine*
[2] *Institute of Mathematics, National Academy of Sciences of Ukraine, Kiev, Ukraine*

Markov processes with locally independent increments in Euclidean phase space R^N are considered as mathematical models of Markov queueing systems and networks. Average and diffusion approximation schemes with a small series parameter $\varepsilon > 0$ are developed and applied for some concrete queueing systems and networks.

3.1. Introduction

The main problem in the analysis of stochastic queueing systems and networks is to select a class of Markov processes which is adequate for the corresponding queueing processes.

An auxiliary problem is to introduce a series parameter in such a way as to ensure adequate normalization of service processes.

The analytical problem is to choose an adequate method for asymptotic analysis of queueing processes in the diffusion approximation scheme.

Queueing Theory 1,
coordinated by Vladimir ANISIMOV, Nikolaos LIMNIOS.

For this purpose, the Markov processes with locally independent increments with normalizing series parameter $\varepsilon \to 0$ ($\varepsilon > 0$) are considered as adequate processes.

In asymptotical analysis of diffusion approximation schemes, the generator characterization for Markov queueing process is used, as well as the relative limit theorems in series schemes (Korolyuk and Limnios 2005; Korolyuk and Korolyuk 1999).

Limit theorems for switching processes (average and diffusion approximation), applied to the models of queueing systems and networks (branching processes as well), were studied in work by V.V. Anisimov (1995, 2002). A special subclass of switching – recurrent processes of semi-Markov type – was studied and described in a monograph by Anisimov (2008, see also the references therein).

3.2. Markov queueing processes

Markov processes $\zeta(t),\ t \geq 0$, with locally independent increments (Korolyuk and Korolyuk 1999) in Euclidean phase space R^N can be determined by a generator L on Banach space $\mathcal{B}_{R^N}$ of real-valued bounded measurable functions $\varphi(u),\ u \in R^N$, with sup-norm $||\varphi(u)|| := \sup_{u \in R^N} |\varphi(u)|$, as follows:

$$L\varphi(u) = \int_{R^N} Q(u,\, dz)[\varphi(u+z) - \varphi(u)] + C(u)\varphi'(u) + \frac{1}{2}Tr[B(u)\varphi''(u)]. \quad [3.1]$$

The kernel $Q(u, dz)$ defining intensities of values of jumps is supposed to be a positive bounded measure:

$$Q(u,\, R^N) =: q(u) \in \mathfrak{B}_{R^N}.$$

The vector function $C(u) = (c_k(u),\ 1 \leq k \leq N)$ sets the velocity of the deterministic drift $\rho(t)$, defined by a solution of the evolutional equation

$$d\rho(t)/dt = C(\rho(t)). \quad [3.2]$$

The second term in [3.1] means the scalar product of two vector functions:

$$C(u)\varphi'(u) := \sum_{k=1}^{N} c_k(u) d\varphi(u)/du_k.$$

The diffusion component of Markov process is defined by generator

$$L_0\varphi(u) := \frac{1}{2}Tr[B(u)\varphi''(u)] = \frac{1}{2}\sum_{k,r=1}^{N} b_{kr}(u)d^2\varphi(u)/du_k du_r. \qquad [3.3]$$

The covariance matrix of diffusion is

$$B(u) := [b_{kr}(u)\ ;\ \ 1 \leq k,\ \ r \leq N].$$

The deterministic drift $\rho(t)$ can be considered as a part of a diffusion process with generator

$$L^0\varphi(u) := C(u)\varphi'(u) + \frac{1}{2}Tr[B(u)\varphi''(u)]. \qquad [3.4]$$

The increments of Markov process $\zeta(t)$ on a small time interval Δt can be represented as follows:

$$\Delta\zeta(t) := \zeta(t + \Delta t) - \zeta(t) \simeq \Delta\nu(t) + \Delta\rho(t) + \Delta w(t),$$

where $\nu(t)$ is a Markov jump process with generator

$$Q\varphi(u) := \int_{R^N} Q(u,\, dz)[\varphi(u + z) - \varphi(u)]. \qquad [3.5]$$

The diffusion process $w(t)$ is defined by generator [3.3].

The homogeneous (in time) Markov jump process $\nu(t)$, defined by generator [3.3], is considered as a starting point for mathematical modeling of a queueing system.

3.3. Average and diffusion approximation

3.3.1. *Average scheme*

A queueing process in the average normalized scheme is considered in the following normalization:

$$\nu^{\varepsilon}(t) := \varepsilon\nu^{\varepsilon}(t/\varepsilon). \qquad [3.6]$$

That is, the generator of Markov process $\nu^\varepsilon(t)$ has the following representation:

$$Q^\varepsilon \varphi(u) = \varepsilon^{-1} \int_{R^N} Q^\varepsilon(u, dz)[\varphi(u + \varepsilon z) - \varphi(u)]. \qquad [3.7]$$

The main condition in the average approximation scheme is imposed on the asymptotical behavior of the first moment of jumps.

THEOREM 3.1.– (Averaging). Let the following approximation condition hold:

$$C^\varepsilon(u) := \int_{R^N} z Q^\varepsilon(u, dz) = C(u) + \theta^\varepsilon(u), \qquad [3.8]$$

with a negligible residual term

$$||\theta^\varepsilon(u)|| \to 0\,, \;\; as \;\; \varepsilon \to 0, \qquad [3.9]$$

and suppose the existence of a positive solution of equation

$$d\rho/dt = C(\rho(t))\,, \;\; \rho(0) = \rho_0 \geq 0. \qquad [3.10]$$

Then the normalized queueing process [3.6] converges in distribution, as $\varepsilon \to 0$, to a solution $\rho(t)$ of equation [3.10] with initial value $\rho_0 = \mathcal{P}\lim_{\varepsilon \to 0} \varepsilon \nu^\varepsilon(0)$:

$$\varepsilon \nu^\varepsilon(t/\varepsilon) \Rightarrow \rho(t) \;\; as \;\; \varepsilon \to 0. \qquad [3.11]$$

COROLLARY 3.1.– Let the initial value $\rho_0 \geq 0$ be an equilibrium point for the velocity $C(u)$:

$$C(\rho_0) = 0.$$

Then the queueing process in the average normalized scheme has an equilibrium point as well:

$$\varepsilon \nu^\varepsilon(t/\varepsilon) \Rightarrow \rho_0 \;\; as \;\; \varepsilon \to 0$$

uniformly on every finite time interval $t \in [0, T]$.

PROOF.– (Proof of theorem 3.1). Under the approximation condition [3.8], the generator [3.7], acting on the continuously differentiable function $\varphi(u)$, has the following asymptotic representation:

$$Q^{\varepsilon}\varphi(u) = \mathbb{C}\varphi(u) + \theta^{\varepsilon}(u), \qquad [3.12]$$

where the generator

$$\mathbb{C}\varphi(u) = C(u)\varphi'(u) \qquad [3.13]$$

determines the evolutional equation [3.10]. The residual term $\theta^{\varepsilon}(u)$ in [3.12] (different from the term in [3.8]) satisfies the negligible condition [3.9]. Now using martingale characterization of Markov process

$$\mu_t^{\varepsilon} = \varphi(\nu^{\varepsilon}(t)) - \int_0^t Q^{\varepsilon}\varphi(\nu^{\varepsilon}(s))ds \qquad [3.14]$$

and asymptotical representation [3.12], we get the following asymptotical form for martingale [3.14]:

$$\mu_t^{\varepsilon} = \varphi(\nu^{\varepsilon}(t)) - \int_0^t \mathbb{C}\varphi(\nu^{\varepsilon}(s))ds + \psi_t^{\varepsilon},$$

where the residual term ψ_t^{ε} satisfies the condition

$$E[\sup_{0\le t\le T} |\psi_t^{\varepsilon}|] \to 0 \ \ as \ \ \varepsilon \to 0. \qquad [3.15]$$

It is easy to verify that the square characteristic of martingale [3.14] admits the representation

$$\langle \mu^{\varepsilon} \rangle_t = \int_0^t \zeta^{\varepsilon}(s)ds \qquad [3.16]$$

with a random function $\zeta^{\varepsilon}(s)$, $s \geq 0$, satisfying the condition

$$E[\sup_{0\le s\le T} |\zeta^{\varepsilon}(s)|] \to 0 \ \ as \ \ \varepsilon \to 0. \qquad [3.17]$$

Hence, in the previous study (Korolyuk and Korolyuk 1999, theorem 3.4), the weak convergence [3.11] takes place. The limit process $\rho(t)$ is defined by a solution of equation

$$d\varphi(\rho(t))/dt = C\varphi(\rho(t)) \tag{3.18}$$

which is equivalent to evolutional equation [3.10] for the limit process $\rho(t)$. □

3.3.2. *Diffusion approximation scheme*

A Markov queueing process in a diffusion approximation scheme is considered in the following centered and normalized form:

$$\zeta^\varepsilon(t) := \varepsilon\nu^\varepsilon(t/\varepsilon^2) - \varepsilon^{-1}\rho(t), \tag{3.19}$$

where the Markov jump process $\nu^\varepsilon(t)$ is determined by generator

$$Q^\varepsilon\varphi(u) := \int_{R^N} Q^\varepsilon(u, dz)[\varphi(u + \varepsilon z) - \varphi(u)]. \tag{3.20}$$

The centered function $\rho(t)$ is defined by a solution of evolutional equation

$$d\rho(t)/dt = C(\rho(t)). \tag{3.21}$$

The more general centering scheme with the velocity $C^\varepsilon(u)$ that depends on a series parameter ε is motivated by a real queueing system with bounded input. In spite of the fact that the Markov jump process $\nu^\varepsilon(t)$ is homogeneous in time, the normalized centered process [3.19] is heterogeneous, but in a special way, due to the deterministic centered function $\rho(t)$. Meanwhile, the two components, homogeneous in time Markov process $\zeta^\varepsilon(t), \rho(t), t \geq 0$, can be used in the diffusion approximation scheme.

LEMMA 3.1.– The generator of the coupled Markov process $\zeta^\varepsilon(t), \rho(t), t \geq 0$, has the following representation:

$$L^\varepsilon\varphi(u, v) = L_0^\varepsilon\varphi(u, v) + C(v)\varphi'_v(u, v), \tag{3.22}$$

where

$$L_0^\varepsilon\varphi(u, v) := \varepsilon^{-2}Q^\varepsilon(v)\varphi(u, v) + \varepsilon^{-1}C(v)\varphi'_v(u, v), \tag{3.23}$$

and the generator $Q^\varepsilon(v)$ acts in the following way:

$$Q^\varepsilon(v)\varphi(u) := \int_{R^N} Q^\varepsilon(v + \varepsilon u,\, dz)[\varphi(u + \varepsilon z) - \varphi(u)]. \qquad [3.24]$$

PROOF OF LEMMA 3.1.– The increments of the first component in [3.19] have the following representation:

$$\Delta\zeta^\varepsilon(t) := \zeta^\varepsilon(t + \Delta) - \zeta^\varepsilon(t) = \varepsilon\Delta\nu^\varepsilon(t/\varepsilon^2) - \varepsilon^{-1}\Delta\rho(t).$$

According to equation [3.21], one has

$$\Delta\rho(t) := \rho(t + \Delta) - \rho(t) = C(\rho(t)\Delta + o(\Delta).$$

Hence

$$\Delta\zeta^\varepsilon(t) = \varepsilon\Delta\nu^\varepsilon(t/\varepsilon^2) - \varepsilon^{-1}C(\rho(t))\Delta + o(\Delta).$$

Now the conditional expectation

$$\begin{aligned} &E_{u,v}\varphi(\ \zeta^\varepsilon(t + \Delta),\, \rho(t + \Delta)) := \\ &\qquad := E[\varphi(\zeta^\varepsilon(t + \Delta),\, \rho(t + \Delta)) \,|\, \zeta^\varepsilon(t) = u,\, \rho(t) = v] \end{aligned}$$

can be estimated as follows:

$$\begin{aligned} &E_{u,v}\varphi(\zeta^\varepsilon(t + \Delta),\, \rho(t + \Delta)) = \\ &= E_u\varphi(u + \varepsilon\Delta\nu^\varepsilon(t/\varepsilon^2), v) + C(v)[\varphi'_v(u, v) - \varphi'_u(u, v)]\Delta + o(\Delta). \end{aligned} \qquad [3.25]$$

Note that, under the condition $\zeta^\varepsilon(t) = u$, $\rho(t) = v$, according to the definition of normalized centered process [3.19], we have:

$$\varepsilon^2\nu^\varepsilon(t/\varepsilon^2) = v + \varepsilon u.$$

Therefore, the first term in [3.25] can be estimated as follows:

$$\begin{aligned} &E_{u,v}\ \varphi(\zeta^\varepsilon(t + \Delta),\, \rho(t + \Delta)) = \\ &\qquad = E_u[\varphi(u + \varepsilon\Delta\nu^\varepsilon(t/\varepsilon^2), v) \,|\, \varepsilon^2\nu^\varepsilon(t/\varepsilon^2) = v + \varepsilon u] = \\ &\qquad = \varphi(u, v) + \varepsilon^{-2}Q^\varepsilon(v)\varphi(u, v)\Delta + o(\Delta), \end{aligned} \qquad [3.26]$$

with generator $Q^{\varepsilon}(v)$ defined in [3.24]. The relations [3.25] and [3.26] imply [3.22]–[3.24]. Note that the intensity kernel $Q^{\varepsilon}(u, dz)$ defines the intensity of jump values for the process $\nu^{\varepsilon}(t)$ under the condition

$$\varepsilon^2 \nu^{\varepsilon}(t) = u$$

for a fixed parameter u. The diffusion approximation for a normalized centered queueing process [3.19] is realized under asymptotical representation for infinitesimal characteristics, defined by generator [3.22]–[3.24].

THEOREM 3.2.– (Diffusion Approximation). Let the intensity of jump values of Markov queueing process $\nu^{\varepsilon}(t)$ admit asymptotical representation of the first and second moments, as $\varepsilon \to 0$:

$$b_{\varepsilon}(v, u) := \int_{R^N} z Q^{\varepsilon}(v + \varepsilon u,\, dz) = b(v) + \varepsilon b_1(v, u) + \varepsilon \theta^{\varepsilon}(v, u), \qquad [3.27]$$

$$B_{\varepsilon}(v, u) := \int_{R^N} z z^* Q^{\varepsilon}(v + \varepsilon u,\, dz) = B(v) + \theta^{\varepsilon}(v, u), \qquad [3.28]$$

where the residual terms $\theta^{\varepsilon}(v, u)$ in both formulas satisfy the negligibility condition

$$\|\theta^{\varepsilon}(v, u)\| \to 0 \ \ as \ \ \varepsilon \to 0. \qquad [3.29]$$

Let the velocity of the centered function $C^{\varepsilon}(v)$ admit asymptotical representation

$$C^{\varepsilon}(v) = C(v) + \varepsilon C_1(v) + \varepsilon \theta^{\varepsilon}(v) \qquad [3.30]$$

with a negligible term $\theta^{\varepsilon}(v)$:

$$\|\theta^{\varepsilon}(v)\| \to 0 \ \ as \ \ \varepsilon \to 0.$$

Then, under the additional balance condition

$$b(v) = C(v), \qquad [3.31]$$

the normalized centered queueing process $\zeta^{\varepsilon}(t)$, defined in [3.19], converges, in distribution, to a diffusion:

$$\zeta^{\varepsilon}(t) \Rightarrow \zeta^0(t) \ \ as \ \ \varepsilon \to 0.$$

The generator of the limit process $\zeta^0(t)$ is then defined as:

$$L_t^0\varphi(u) = [b_1(\rho(t),u) - C_1(\rho(t))]\varphi'(u) + \frac{1}{2}Tr[B(\rho(t))\varphi''(u)], \qquad [3.32]$$

where the centered function $\rho(t)$ is defined by a solution of evolutional equation

$$d\rho(t)/dt = C(\rho(t)) \qquad [3.33]$$

with initial value

$$\rho_0 = \mathcal{P}\lim_{\varepsilon\to 0} \varepsilon^2 \nu^\varepsilon(0). \qquad [3.34]$$

The initial value of the limit diffusion is defined by the condition

$$\zeta^0(0) = \mathcal{P}\lim_{\varepsilon\to 0} \zeta^\varepsilon(0).$$

COROLLARY 3.2.– Let the intensity kernel have the following asymptotical representation:

$$Q^\varepsilon(v,dz) = Q^0(v,dz) + \varepsilon Q_1(v,dz), \qquad [3.35]$$

and the kernel $Q^0(v,dz)$ be differentiable by v. Then the asymptotical representation [3.27] has the following form:

$$\begin{array}{l} b(v) = \int_{R^N} zQ^0(v,dz), \\ b_1(v,u) = b_1(v) + ub'(v)\,, \quad b_1(v) := \int_{R^N} zQ_1(v,dz). \end{array} \qquad [3.36]$$

COROLLARY 3.3.– Let the velocity $C(v)$ of the centered function have an equilibrium point $\rho_0 \geq 0$: $C(\rho_0) = 0$. Then under the initial conditions [3.34] and [3.35], the limit diffusion of the Ornstein–Uhlenbeck type process is defined by generator

$$L_0^0\varphi(u) = a(u)\varphi'(u) + \frac{1}{2}Tr[B\varphi''(u)], \qquad [3.37]$$

where the drift function is linear with respect to u:

$$a(u) = a_0 + a_1 u\,, \quad a_0 = b^1(\rho_0)\,, \quad a_1 = b'(\rho_0), \qquad [3.38]$$

and the covariance matrix is constant:

$$B = B(\rho_0). \qquad [3.39]$$

PROOF.– (Proof of theorem 3.2). The asymptotic representation of generators [3.22]–[3.24] is considered on the set $\mathcal{D}$ of real-valued, twice continuously differentiable functions $\varphi(u, v)$ with compact support.

Taking into account asymptotical representations [3.27] and [3.28] of the first two moments of the jump component and [3.30] for the velocity of centered function, we obtain

$$\begin{aligned} L_0^\varepsilon\varphi(u,v) :&= \varepsilon^{-2}Q^\varepsilon(v)\varphi(u,v) - \varepsilon^{-1}C^\varepsilon(v)\varphi'_u(u,v) \\ &= L_0^0\varphi(u,v) + \varepsilon^{-1}[b(v) - C(v)]\varphi'_u(u,v) + \theta^\varepsilon(u,v), \end{aligned}$$

$$L_0^0\varphi(u) = [b_1(v,u) - C_1(v)]\varphi'_u(u,v) + \frac{1}{2}Tr[B(v)\varphi''_u(u,v)]. \qquad [3.40]$$

As before, the residual term $\theta^\varepsilon(u, v)$ satisfies the negligiblility condition [3.29]. The balance condition [3.31] yields the following asymptotical representation for generator [3.22]:

$$L_0^\varepsilon\varphi(u,v) = L_0^0\varphi(u,v) + C(v)\varphi'_v(u,v) + \theta^\varepsilon(u,v).$$

Now the martingale characterization of the coupled Markov process $\zeta^\varepsilon(t)$, $\rho(t)$, $t \geq 0$, is the following:

$$\mu_t^\varepsilon = \varphi(\zeta^\varepsilon(t), \rho(t)) - \int_0^t L^\varepsilon\varphi(\zeta^\varepsilon(s), \rho(s))ds, \qquad [3.41]$$

which can be transformed into

$$\begin{aligned} \mu_t^\varepsilon = \varphi(\ &\zeta^\varepsilon(t), \rho(t)) - \\ &- \textstyle\int_0^t [L_0^0\varphi(\zeta^\varepsilon(s), \rho(s)) + C(\rho(s))\varphi'_v(\zeta^\varepsilon(s), \rho(s))]ds + \psi_t^\varepsilon \end{aligned}$$

with a residual negligible term ψ_t^ε, $\varepsilon > 0$, satisfying the condition

$$\sup_{\varepsilon\leq\varepsilon_0} E \sup_{0\leq t\leq T} |\psi_t^\varepsilon| \to 0 \ as \ \varepsilon_0 \to 0. \qquad [3.42]$$

One has to place the following representation of the square characteristic of martingale [3.41]:

$$\langle\mu^\varepsilon\rangle_t = \int_0^t \zeta^\varepsilon(s)ds \qquad [3.43]$$

with a random function $\zeta^\varepsilon(s)$, $s \geq 0$, which satisfies the compactness condition: for every fixed T, there exists $\delta > 0$ such that

$$E \sup_{0 \leq t \leq T} |\zeta^\varepsilon(s)|^{1+\delta} \leq C < \infty. \qquad [3.44]$$

Hence, by the Pattern theorem (Korolyuk and Korolyuk 1999, Theorem 3.4), the following weak convergence (in distribution) takes place:

$$(\zeta^\varepsilon(t),\, \rho(t)) \Rightarrow (\zeta^0(t),\, \rho(t)) \quad as \ \varepsilon \to 0.$$

The limit two-component Markov process $\zeta^0(t)$, $\rho(t)$, $t \geq 0$, is determined by the generator

$$L^0\varphi(u,v) = L_0^0\varphi(u,v) + C(v)\varphi'_v(u,v). \qquad [3.45]$$

To complete the proof of theorem 3.2, note that the coupled Markov process $\zeta^0(t)$, $\rho(t)$, $t \geq 0$, with the first component determined by generator [3.32], is defined by generators [3.37] and [3.45]. □

NOTE 3.1.– Using the centered drift function $\rho^\varepsilon(t)$, that depends on the series parameter ε, the drift term in [3.32], which is not dependent on u, can be removed. But it is not possible to remove the drift part of the limit diffusion completely.

NOTE 3.2.– The considered class of Markov processes contains "density-dependent population processes" for which the diffusion approximation without a drift component was investigated in the book by Ethier and Kurtz (1986). The limit diffusion is not ergodic despite the fact that the initial process is ergodic.

3.3.3. *Stationary distribution*

The diffusion approximation of stationary Markov jump processes provides the approximation of their stationary distributions. It means that the stationary distribution density $\pi^\varepsilon(du)$, $\varepsilon > 0$, of a normalized and centered queueing process weakly converges, as $\varepsilon \to 0$, to stationary distribution density $\pi^0(du)$ of the limit diffusion process.

Essential aspects of this problem are considered in the book by Ethier and Kurtz (1986, Section 4.9). The main idea is to verify the relative compactness of stationary densities $\pi^\varepsilon(du)$, $\varepsilon > 0$. The most useful approach involves Lyapunov functions $V(u)$, which satisfy the following conditions:

– $V(u) > 0$ in some vicinity of zero point $u = 0$;

– $V(0) = 0$;

– $V(u) \to \infty$ as $\|u\| \to \infty$.

In addition, $V(u)$ is an *excessive* function for generator L_0^0 of the limit diffusion process:

$$L_0^0 V(u) \leq 0.$$

For a given diffusion process, the Itô formula provides a more direct approach (Ethier and Kurtz 1986, Section 4.9, Problem 35).

A weak convergence of stationary distributions is considered for the normalized queueing process with equilibrium point $\rho \geq 0$:

$$\zeta^\varepsilon(t) = \varepsilon \nu^\varepsilon(t/\varepsilon^2) - \varepsilon^{-1}\rho. \qquad [3.46]$$

The generator of Markov process $\zeta^\varepsilon(t)$ has the following form:

$$L_0^\varepsilon \varphi(u) = \varepsilon^{-2} Q^\varepsilon(\rho)\varphi(u), \qquad [3.47]$$

where

$$Q^\varepsilon(\rho)\varphi(u) = \int_{R^N} Q^\varepsilon(\rho + \varepsilon u, dz)[\varphi(u + \varepsilon z) - \varphi(u)]. \qquad [3.48]$$

The limit diffusion process in this case, under the conditions of theorem 3.2, is determined by the generator

$$L_0^0 \varphi(u) = a(u)\varphi'(u) + \frac{1}{2} Tr[B\varphi''(u)], \qquad [3.49]$$

where

$$a(u) = b^1(\rho, u) \text{ and } B = B(\rho). \qquad [3.50]$$

THEOREM 3.3.– (Convergence of stationary distributions). Consider a Lyapunov function $V(u)$ for generator [3.49], which satisfies the condition:

$$L_0^0 V(u) \leq C < 0. \qquad [3.51]$$

Then the stationary densities $\pi^\varepsilon(du)$, $\varepsilon > 0$, of normalized queueing process [3.46], defined by generators [3.47], converge weakly to stationary density $\pi^0(du)$ of the limit diffusion process, defined by generator [3.49].

PROOF.– (Proof of theorem 3.3). At first, we establish the weak compactness of stationary densities $\pi^\varepsilon(du)$, $\varepsilon > 0$. According to the definition of the generator, we have the following representation:

$$E_u V(\zeta^\varepsilon(t+h)) = V(\zeta^\varepsilon(t)) + hL_0^\varepsilon V(\zeta^\varepsilon(t)) + o(h). \qquad [3.52]$$

Under the conditions of theorem 5.2, we obtain the following asymptotical representation:

$$L_0^\varepsilon V(u) = L_0^0 V(u) + \theta^\varepsilon(u) \qquad [3.53]$$

with negligible term $\theta^\varepsilon(u)$ that satisfies the condition [3.9].

Putting [3.51]–[3.53] together, we get the following inequality:

$$E_u V(\zeta^\varepsilon(t+h)) \leq V(\zeta^\varepsilon(t)) + h[C + o_h(\varepsilon)]$$

where $o_h(\varepsilon) \to 0$ as $\varepsilon \to 0$. Hence, for enough small h and ε, the following inequality takes place:

$$E_u V(\zeta^\varepsilon(t+h)) \leq V(\zeta^\varepsilon(t)). \qquad [3.54]$$

It follows that $V(\zeta^\varepsilon(t))$, $t \geq 0$, is a supermartingale:

$$E_u V(\zeta^\varepsilon(t)) \leq E_u V(\zeta^\varepsilon(0)) = V(u)$$

for all $t > 0$.

Let $R_k^N := \{u : \|u\| > k\}$ and $V_k := \inf_{u \in R_k^N} V(u)$. Due to the property of the Lyapunov function, one obtains:

$$V_k \to \infty \text{ as } k \to \infty.$$

Now the weak compactness of stationary distribution π^ε, $\varepsilon > 0$, is a consequence of the estimate, obtained by using Chebyshev's inequality for transition probabilities:

$$P_t^\varepsilon(u, R_k^N) := \mathcal{P}\{\zeta^\varepsilon(t) \in R_k^N \mid \zeta^\varepsilon(0)\} \leq \\ \leq V(u)/V_k \to 0 \text{ as } k \to \infty$$

for all enough small $\varepsilon > 0$.

But by the property of stationary distribution one has:

$$\pi^\varepsilon(R_k^N) = \lim_{t\to\infty} P_t^\varepsilon(u, R_k^N) \le V(u)/V_k \to 0 \text{ as } k \to \infty.$$

Therefore, there exists a convergent sequence

$$\lim_{\varepsilon_n\to 0} \int_{R^N} \pi^{\varepsilon_n}(du)\varphi(u) = \int_{R^N} \pi^0(du)\varphi(u). \qquad [3.55]$$

We have to establish now that the measure $\pi^0(du)$ is a stationary distribution of the limit diffusion process, defined by generator [3.49].

Let $P_t^0(u, A) := \mathcal{P}\{\zeta^0(t) \in A \,|\, \zeta^0(0) = u\}$ be the transition probabilities of the limit diffusion process defined by generator [3.49].

According to theorem 3.2, the convergence

$$\lim_{\varepsilon\to 0} \int_{R^N} \varphi(z)P_t^\varepsilon(u, dz) = \int_{R^N} \varphi(z)P_t^0(u, dz) \qquad [3.56]$$

takes place uniformly for $t \in [0, T]$ for any finite function $\varphi(u)$ with compact support. Consider a measure defined as

$$\mu_T^\varepsilon(u, A) := \int_0^T P_t^\varepsilon(u, A)dt$$

and the functional

$$\mu_T^\varepsilon(u, \varphi) := \int_{R^N} \varphi(z)\mu_T^\varepsilon(u, dz).$$

The convergence [3.56] implies the following limit representation:

$$\lim_{\varepsilon\to 0} \mu_T^\varepsilon(u, \varphi) = \mu_T^0(u, \varphi) = \int_{R^N} \varphi(z) \int_0^T P_t^0(u, dz)dt. \qquad [3.57]$$

Now we calculate

$$\int_{R^N} \pi^\varepsilon(du)\mu_T^\varepsilon(u, \varphi) = T \int_{R^N} \pi^\varepsilon(du)\varphi(u). \qquad [3.58]$$

Using the relative compactness of the densities $\pi^\varepsilon(du)$, $\varepsilon > 0$, established above, we get by [3.55] and [3.57] the following relations:

$$\lim_{\varepsilon\to 0}\int_{R^N}\pi^{\varepsilon_n}(du)\mu_T^{\varepsilon_n}(u,\varphi) = \int_{R^N}\pi^0(du)\mu_T^0(u,\varphi)$$

and by [3.58]

$$\lim_{\varepsilon\to 0}\int_{R^N}\pi^{\varepsilon_n}(du)\mu_T^{\varepsilon_n}(u,\varphi) = T\int_{R^N}\pi^0(du)\varphi(u).$$

Putting the last two limits together, we obtain

$$\int_{R^N}\pi^0(du)\int_{R^N}\varphi(z)\int_0^T P_t^0(u,dz)dt = T\int_{R^N}\pi^0(dz)\varphi(z).$$

It follows that

$$\pi^0(dz) = \frac{1}{T}\int_0^T dt\int_{R^N}\pi^0(du)P_t^0(u,dz).$$

Hence, $\pi^0(dz)$ is the stationary measure of the limit diffusion process, defined by the limit generator [3.49]. The proof of theorem 3.3 is completed. □

The construction of Lyapunov function $V(u)$ can also be realized in another way.

Let us consider the limit diffusion process, defined by generators [3.37]–[3.39] with a constant covariance matrix and linear drift function with respect to u. Such a limit diffusion process is used in numerous applications.

LEMMA 3.2.– A diffusion process on real line R has the generator

$$L_0^0\varphi(u) = a(u)\varphi'(u) + \frac{1}{2}B\varphi''(u) \qquad [3.59]$$

where the drift function has the following form:

$$a(u) = a_0 - a_1 u\,,\quad a_1 > 0. \qquad [3.60]$$

Then the Lyapunov function for the generator [3.59] can be chosen as follows:

$$V(u) = V_0\int_0^u e^{b(z)}dz\left[V_1 - \int_0^z e^{-b(y)}dy\right], \qquad [3.61]$$

where the constants $V_0 > 0$ and V_1 satisfy the inequality

$$V_1 > \int_0^\infty e^{-b(y)} dy. \qquad [3.62]$$

Here

$$b(z) := -2 \int_0^z a(u)du/B^2. \qquad [3.63]$$

PROOF.– (Proof of Lemma 3.2). Note that the integral in right-hand side of [3.62] is finite due to assumptions [3.60] and [3.63]. Using the first and second derivatives of the function [3.61], it is easy to calculate

$$L_0^0 V(u) = -V_0 B^2/2 < 0.$$

Hence, the function $V(u)$, given by [3.61], is excessive for the generator [3.59]. □

3.4. Markov queueing systems

There are two possibilities to construct diffusion approximations for real stochastic systems. Either to apply directly the pattern limit theorem 3.2 to a real queueing system and, then, compute local characteristics of the limit diffusion process by formulas [3.36], or to formulate some intermediate algorithm for a certain class of queueing systems. The second way is chosen.

3.4.1. *Collective limit theorem in* R^1

Consider a normalized queueing process

$$\zeta^\varepsilon(t) = \varepsilon \nu^\varepsilon(t/\varepsilon^2) - \varepsilon^{-1}\rho(t) \qquad [3.64]$$

where the queueing process $\nu^\varepsilon(t)$ determines the number of customers in the system including that which have to be served at time t.

The positive valued centering function $\rho(t), t \geq 0$, is supposed to be continuously differentiable. For simplicity, assume that the customers arrive and leave the system one by one.

The basic assumption is that the intensities of input and service times, under the condition $\nu^\varepsilon(t) = k$, depend on the parameter k in the following way:

$$\lambda_\varepsilon(k) = \lambda(\varepsilon^2 k) + \varepsilon\lambda_1(\varepsilon^2 k),$$
$$\mu_\varepsilon(k) = \mu(\varepsilon^2 k) + \varepsilon\mu_1(\varepsilon^2 k) \quad [3.65]$$

with some given functions $\lambda(v)$, $\mu(v)$, $\lambda_1(v)$ and $\mu_1(v)$, $v \in R^1$.

Note that the connection between the variables $k = \nu^\varepsilon(t)$, $v = \rho(t)$ and $u = \zeta^\varepsilon(t)$ is represented as follows:

$$\varepsilon^2 k = v + \varepsilon u.$$

Therefore, the intensities of jumps of Markov process [3.64], under the condition $\zeta^\varepsilon(t) = u$, have the following representation:

$$Q^\varepsilon(v + \varepsilon u,\ +1) = \lambda(v + \varepsilon u) + \varepsilon\lambda_1(v + \varepsilon u),$$
$$Q^\varepsilon(v + \varepsilon u,\ -1) = \mu(v + \varepsilon u) + \varepsilon\mu_1(v + \varepsilon u). \quad [3.66]$$

THEOREM 3.4.– (Collective limit theorem). Let the intensities of input and service times of the queueing process $\nu^\varepsilon(t)$ be set by relations [3.65]–[3.66] with continuously differentiable functions $\lambda(u)$ and $\mu(u)$, having a bounded first derivative, and continuous functions $\lambda_1(u)$ and $\mu_1(u)$. Let there exist a positive solution of evolutional equation

$$d\rho(t)/dt = C(\rho(t)), \quad [3.67]$$

$$C(v) := \lambda(v) - \mu(v) \quad [3.68]$$

with initial condition $\rho(0) = \rho_0 \geq 0$. And let the initial value of the queueing process converge, in probability, as follows:

$$\zeta^\varepsilon(0) = \varepsilon\nu^\varepsilon(0) - \varepsilon^{-1}\rho_0 \Rightarrow \zeta^0 \quad as\ \varepsilon \to 0. \quad [3.69]$$

Then the normalized queueing process [3.64] converges weakly as $\varepsilon \to 0$ to a diffusion process $\zeta^0(t)$:

$$\varepsilon\nu^\varepsilon(t/\varepsilon^2) - \varepsilon^{-1}\rho(t) \Rightarrow \zeta^0(t) \quad as\ \varepsilon \to 0.$$

The generator of the limit diffusion process $\zeta^0(t)$ is determined by the relation:

$$L_t^0\varphi(u) = a(t,u)\varphi'(u) + \frac{1}{2}B(\rho(t))\varphi''(u), \tag{3.70}$$

where

$$a(t,u) := uC'(\rho(t)) + b_1(\rho(t)), \tag{3.71}$$

$$b_1(v) := \lambda_1(v) - \mu_1(v)\ , \ \ B(v) := \lambda(v) + \mu(v). \tag{3.72}$$

The initial condition is $\zeta^0(0) = \zeta^0$.

PROOF.– (Proof of theorem 3.4). First, we calculate the local characteristics of the coupled Markov process $\zeta^\varepsilon(t)$, $\rho(t)$, $t \geq 0$, using the representation [3.66] of jump intensities and the formulas [3.27] and [3.28]. The jump intensities [3.66] are transformed as follows:

$$Q^\varepsilon(v + \varepsilon u,\ +1) = \lambda(v) + \varepsilon[u\lambda'(v) + \lambda_1(v)] + \varepsilon\theta^\varepsilon(v,u),$$
$$Q^\varepsilon(v + \varepsilon u,\ -1) = \mu(v) + \varepsilon[u\mu'(v) + \mu_1(v] + \varepsilon\theta^\varepsilon(v,u)$$

with residual negligible terms $\theta^\varepsilon(v,u)$ satisfying the condition [3.29]. By formulas [3.27], [3.35] and [3.36], we obtain

$$b(v) = \lambda(v) - \mu(v)\ , \ \ b_1(v) = \lambda_1(v) - \mu_1(v).$$

By formula [3.28], we get

$$B(v) = \lambda(v) + \mu(v).$$

At last, by balance condition [3.31]

$$C(v) = b(v) = \lambda(v) - \mu(v).$$

We finish the proof of theorem 5.4 by applying theorem 3.2. □

NOTE 3.3.– In the stationary regime with equilibrium point ρ:

$$C(\rho) = \lambda(\rho) - \mu(\rho) = 0, \tag{3.73}$$

the limit diffusion process is of the Ornstein–Uhlenbeck type with generator

$$L^0\varphi(u) = a(u)\varphi'(u) + \frac{1}{2}B\varphi''(u),$$

where

$$\begin{aligned} a(u,\rho) &= a_1(\rho) - ua(\rho), \\ a(\rho) &= \lambda'(\rho) - \mu'(\rho)\ , \ \ a_1(\rho) = \lambda_1(\rho) - \mu_1(\rho). \end{aligned} \qquad [3.74]$$

The *diffusion approximation algorithm of Markov queueing systems* is formulated in the collective limit theorem 3.4. Under given intensities of input and service times represented in the form [3.65], the centered function $\rho(t)$ is determined by a solution of evolutional equation [3.67] with velocity function, defined by [3.68]. The generator of the limit diffusion is defined by [3.70] with drift function given by [3.71] and covariance function given by [3.72]. The initial value of the limit diffusion is defined by [3.69].

3.4.2. *Systems of M/M type*

A single queueing system with unbounded queue is defined by input intensity λ and service time μ having exponential distributions.

COROLLARY 3.4.– Under the condition of heavy traffic $\lambda > \mu$, the normalized queueing process

$$\zeta^\varepsilon(t) = \varepsilon\nu^\varepsilon(t/\varepsilon^2) - \varepsilon^{-1}\rho(t) \qquad [3.75]$$

with centering function

$$\rho(t) = \rho_0 + (\lambda - \mu)t,$$

where ρ_0 is determined under the condition

$$\varepsilon\nu^\varepsilon(0) - \varepsilon^{-1}\rho_0 \Rightarrow \zeta^0 \quad as\ \varepsilon \to 0, \qquad [3.76]$$

converges weakly to a diffusion process ζ^0 with zero mean and variance

$$B = \lambda + \mu.$$

COROLLARY 3.5.– An unbounded queueing system, with the rate of Poisson arrivals λ and the intensity of exponential distributed service time μ, has $M/M/\infty$ type. The normalized queueing process [3.75] with centering function

$$\rho(t) = \lambda/\mu + (\rho_0 - \lambda/\mu)e^{-\mu t}, \tag{3.77}$$

where ρ_0 is determined by initial condition [3.76] and converges weakly, as $\varepsilon \to 0$, to a diffusion process $\zeta^0(t)$, is defined by generator [3.70] with drift function

$$a(t, u) = -u\mu \tag{3.78}$$

and variance

$$B(\rho(t)) = 2\lambda + (\rho_0\mu - \lambda)e^{-\mu t}. \tag{3.79}$$

3.4.3. *Repairman problem*

An extensive literature is devoted to this problem, for example, Iglehart (1965). There are n identical working devices that operate independently, m spare devices and r repairing facilities. If one of n operating devices fails, it transfers to a repair facility, being replaced by one of the spare devices. There are at most n devices operating at any time.

The working time of every device has exponential distribution with intensity λ. If r, or more, devices are being repaired, the failed device has to join a queue at the repair facility and wait for service. Let $\nu_n(t)$ be a number of devices undergoing or waiting for repair at a time instant t.

The diffusion approximation of queueing process $\nu_n(t)$ in the repairman problem depends on the correlations between parameters n, m and r.

There are considered to be three different variants.

COROLLARY 3.6.– (Iglehart (1965)). Let the following condition be fulfilled:

$$m = nm_0\ , \quad r = nr_0\ , \quad m_0 > r_0, \tag{3.80}$$

where m_0 and r_0 are fixed numbers and, in addition,

$$r_0\mu < \lambda. \tag{3.81}$$

Then the normalized queueing process

$$\zeta_n(t) = [\nu_n(t) - n\rho]/\sqrt{n} \qquad [3.82]$$

with centering constant

$$\rho = 1 + m_0 - r_0\mu/\lambda,$$

converges weakly, as $n \to \infty$, to a diffusion process $\zeta^0(t)$ with drift function $a(u) = -u\lambda$ and variance $B = 2r_0\mu$.

PROOF.– A normalized queueing process can be represented as follows:

$$\zeta_n(t) = \sqrt{n}[\kappa_n(t) - \rho],$$

where $\kappa_n(t) := \nu_n(t)/n$. This scaled process satisfies the inequality

$$0 \le \kappa_n(t) \le 1 + m_0.$$

The intensities of jumps of queueing process $\nu_n(t)$ are different on the intervals

$$[0, r)\,,\ \ [r, m)\,,\ \ [m, n+m],$$

namely

$$\lambda_n(k) = \begin{cases} n\lambda, & \text{if } k \le m, \\ (n + m - v)\lambda, & \text{if } k > m \end{cases} \qquad [3.83]$$

and

$$\mu_n(k) = \begin{cases} k\mu, & \text{if } k \le r, \\ r\mu, & \text{if } k > r. \end{cases} \qquad [3.84]$$

According to basic assumption [3.65] of theorem 3.4, the intensity functions are transformed to the next expression (considering the substitution $\varepsilon^2 k = v$, $\varepsilon = 1/\sqrt{n}$):

$$\lambda_\varepsilon(k) = \lambda(v) = \begin{cases} \lambda, & \text{if } v \le m_0, \\ (1 + m_0 - v)\lambda, & \text{if } v > m_0. \end{cases}$$

and

$$\mu_\varepsilon(k) = \begin{cases} v\mu, & \text{if } v \leq r_0, \\ r_0\mu, & \text{if } v > r_0. \end{cases}$$

By assumptions [3.80] and [3.81] of corollary 3.6 and taking into account the condition

$$\nu_n(t) = k\ ,\ \ \zeta_n(t) = u,$$

the normalized correlation has the form

$$\varepsilon^2 k = v = \rho + \varepsilon u.$$

Therefore, the intensity functions are considered on the intervals

$$\lambda(v) = \begin{cases} \lambda, & \text{if } \varepsilon u \leq -(1 - r_0\mu/\lambda), \\ (1 + m_0 - v)\lambda, & \text{if } \varepsilon u > -(1 - r_0\mu/\lambda), \end{cases}$$

and

$$\mu(v) = \begin{cases} v\mu, & \text{if } \varepsilon u \leq -(1 - r_0\mu/\lambda) - (m_0 - r_0), \\ r_0\mu, & \text{if } \varepsilon u > -(1 - r_0\mu/\lambda) - (m_0 - r_0). \end{cases}$$

The condition [3.80] of corollary 3.6 specifies the intensity functions on the interval

$$\varepsilon u \geq -(1 - r_0\mu/\lambda),$$

that is

$$\lambda(v) = (1 + m_0 - v)\lambda\ ,\ \ \mu(v) = r_0\mu.$$

Now, by theorem 3.4, we calculate the parameters of the limit diffusion:

$$C(v) = \lambda(v) - \mu(v) = (1 + m_0 - r_0\mu/\lambda - v)\lambda = (\rho - v)\lambda,$$

with drift function

$$a(u) = uC'(\rho) = -u\lambda,$$

and variance

$$B = B(\rho) = \lambda(\rho) + \mu(\rho) = 2r_0\mu. \qquad \square$$

3.5. Markov queueing networks

3.5.1. *Collective limit theorems in* R^N

A Markov network is determined by a finite number N of nodes. In every node there is a Markov queueing system. The customers are transferred from one node to another according to the marching matrix:

$$P = [p_{kr}\ ;\ \ 1 \leq k\ ,\ \ r \leq N]$$

where p_{kr} is probability of a custom transference from kth node to rth node.

The queueing process

$$\nu(t) = (\nu_k(t)\ ;\ \ 1 \leq k \leq N)$$

is a vector function with components $\nu_k(t)$ of customers in kth node at time instant t. So the queueing process $\nu(t)$ takes values in the Euclidean space R^N of the dimension N.

Consider an open network that means there exists at least one k for which

$$p_{k0} := 1 - \sum_{r=1}^{N} p_{kr} > 0.$$

In other words, a customer leaves the kth node with probability p_{k0}. The network temporal evolution is determined by intensity vector functions $\lambda(u) = \lambda_k(u)$; $1 \leq k \leq N$ and $\mu(u) = \mu_k(u)$; $1 \leq k \leq N$ where $u = (u_k)$: $1 \leq k \leq N) \in N$. The customers arrive from outside at the kth node with exponentially distributed time intensity $\lambda_k(u)$, and the exponentially distributed time of service at the kth node with intensity $\mu_k(u)$, $1 \leq k \leq N$. Without loss of generality, assume that the virtual passages are not possible: $p_{kk} = 0$ for all k. The diffusion approximation of queueing process begins by introduction of a small series parameter $\varepsilon > 0$. The main assumption is that the intensity functions depend on ε in the following form:

$$\lambda_\varepsilon(u) = \lambda(\varepsilon^2 u)\ ,\ \ \mu_\varepsilon(u) = \mu(\varepsilon^2 u),$$

where u=(u_n; $1 \leq n \leq N$) is a vector of values of the queueing process $\nu(t)$. The normalized queueing process is considered in a similar form as for the queueing system

$$\zeta^\varepsilon(t) = \varepsilon\nu^\varepsilon(t/\varepsilon^2) - \varepsilon^{-1}\rho(t) \qquad [3.85]$$

with some deterministic positive-valued vector function

$$\rho(t) = (\rho_n(t)\ ;\ 1 \leq n \leq N).$$

We use the following notation. A vector with the superscript "*d*" is a diagonal matrix, for example:

$$\mu^d := [\mu_k \delta_{kr}\ ;\ 1 \leq k,\ r \leq N]$$

where $\delta_{kr} = 0$, if $k \neq r$, and $\delta_{kk} = 1$ are Kronecker symbols.

THEOREM 3.5.– (Collective). Assume the following conditions:

a) the vector functions $\lambda(u)$ and $\mu(u)$ are continuously differentiable with bounded first derivatives;

b) there exists a unique positive solution of differential equation

$$d\rho(t)/dt = c(\rho(t))\ ,\ \ \rho(0) = \rho_0, \qquad [3.86]$$

where

$$c(u) := \lambda(u) + \mu^*(u)[P - I]; \qquad [3.87]$$

c) the marching matrix $P = [p_{kr}\ ;\ 1 \leq k,\ r \leq N]$ is irreducible and invertible;

d) the initial values of the normalized queueing process converge in probability: $\zeta^\varepsilon(0) \Rightarrow \zeta^0$ as $\varepsilon \to 0$.

Then the normalized queueing process converges in probability

$$\zeta^\varepsilon(t) = \varepsilon\nu^\varepsilon(t/\varepsilon^2) - \varepsilon^{-1}\rho(t) \Rightarrow \zeta^0(t) \quad as\ \varepsilon \to 0,$$

The limit Ornstein–Uhlenbeck diffusion process $\zeta^0(t),\ t \geq 0$ is defined by generator

$$L_t^0\varphi(u) = u^* c'(\rho(t))\varphi'(u) + \frac{1}{2}TrB(\rho(t))\varphi''(u). \qquad [3.88]$$

The covariance matrix $B(u)$ is determined as follows:

$$B(u) = c^d(u) - [\mu^d(u)Q + Q^*\mu^d(u)], \qquad [3.89]$$

where

$$Q := P - I.$$

PROOF.– (Proof of theorem 3.5). First of all, it is necessary to calculate the intensities of transfers of the normalized queueing process $\zeta^\varepsilon(t)$. Let us introduce the notation

$$\delta_k := (\delta_{kr}\ ;\ \ 1 \leq r \leq N),$$

where δ_{kr} is the Kronecker symbol. Note that the normalized form of queueing process [3.85] determines the following relation between the variables $\kappa = \nu^\varepsilon(t/\varepsilon^2)$, $v = \rho(t)$ and $u = \zeta^\varepsilon(t)$:

$$\varepsilon^2 \kappa = v + \varepsilon u.$$

Now it is possible to determine the jump intensities of the normalized Markov queueing process $\zeta^\varepsilon(t)$:

$$\begin{aligned} Q^\varepsilon(u,\, \delta_k) &= \lambda_k(v + \varepsilon u), \\ Q^\varepsilon(u,\, -\delta_k) &= p_{k0}\mu_k(v + \varepsilon u), \\ Q^\varepsilon(u,\, \delta_r - \delta_k) &= p_{kr}\mu_k(v + \varepsilon u). \end{aligned} \qquad [3.90]$$

The first formula gives the intensity of increase by one of the kth components $\nu_k(t)$: a customer arrives at the kth node of the network.

The second formula gives the intensity of decrease by one of the kth components $\nu_k(t)$: a customer leaves the network.

The third formula gives the intensity of increase by one of rth the components $\nu_r(t)$ and decrease by one of the kth components: a customer transfers from the kth node to the rth node of the network.

So the first two moments of jumps of the normalized queueing process [3.85] can be calculated. Using formula [3.27], the first moment in the following form is obtained:

$$\begin{aligned} b^\varepsilon(v, u) = &\sum_{k=1}^{N} \delta_k[\lambda_k(v + \varepsilon u) - p_{k0}\mu_k(v + \varepsilon u)] + \\ &+ \sum_{k,r=1}^{N} (\delta_r - \delta_k) p_{kr}\mu_k(v + \varepsilon u). \end{aligned}$$

Transforming the second term, one has

$$\begin{aligned} &\sum_{k,r=1}^{N} (\delta_r - \delta_k) p_{kr}\mu_k(v + \varepsilon u) = \\ &= \sum_{k=1}^{N} \delta_k \left[\sum_{r=1}^{N} p_{rk}\mu_r(v + \varepsilon u) - (1 - p_{k0})\mu_k(v + \varepsilon u) \right]. \end{aligned}$$

So the first moment has the following representation:

$$b^{\varepsilon}(v,u) = \sum_{k=1}^{N} \delta_k \left[\lambda_k(v+\varepsilon u) - \mu_k(\rho(t)+\varepsilon u) + \sum_{r=1}^{N} p_{rk}\mu_r(v+\varepsilon u)\right]$$

or, using notation [3.87], in the vector form:

$$b^{\varepsilon}(v,u) = c(v+\varepsilon u).$$

Using Taylor's formula, the asymptotic representation of the first moment is:

$$b^{\varepsilon}(v,u) = c(v) + \varepsilon u^* c'(v) + \varepsilon\theta_{\varepsilon}(v,u),$$

where $\|\theta_{\varepsilon}(v,u)\| \to 0$ as $\varepsilon \to 0$.

Hence, according to the notation in [3.27],

$$b(v) = c(v), b^1(v,u) = u^* c'(v).$$

By theorem 3.2, it can be concluded that the centering function $\rho(t)$ is determined by the solution of equation [3.86] and the drift function of the limit diffusion process is $b(v,u) = u * c'(v)$. The second moment of jumps of the normalized queueing process [3.85] with intensities of jumps [3.90] is calculated in a more complicated way. Using formula [3.28]

$$\begin{aligned} B^{\varepsilon}(v,u) = &\sum_{k,r=1}^{N} \delta_k\delta_r^*[\lambda_k(v+\varepsilon u) + p_{k0}\mu_k(v+\varepsilon u)] + \\ &+ \sum_{k,r=1}^{N} (\delta_r - \delta_k)(\delta_r^* - \delta_k^*)p_{kr}\mu_k(v+\varepsilon u). \end{aligned}$$

The second term is transformed as follows:

$$\begin{aligned} &\sum_{k,r=1}^{N} (\ \delta_r - \delta_k)(\delta_r^* - \delta_k^*)p_{kr}\mu_k(v+\varepsilon u) = \\ &\quad = \sum_{k,r=1}^{N} \delta_k\delta_k^* \left[(1-p_{k0})\mu_k(v+\varepsilon u) + \sum_{r=1}^{N} p_{rk}\mu_r(v+\varepsilon u)\right] - \\ &\qquad - \sum_{k,r=1}^{N} \delta_r\delta_k^*[p_{kr}\mu_k(v+\varepsilon u) + p_{rk}\mu_r(v+\varepsilon u)]. \end{aligned}$$

So the second moment has the following form:

$$B^{\varepsilon}(v,u) = \sum_{k,r=1}^{N} \delta_k \delta_k^* \left[\mu_k(v+\varepsilon u) + \lambda_k(v+\varepsilon u) + \sum_{r=1}^{N} p_{rk}\mu_r(v+\varepsilon u) \right] -$$

$$- \sum_{k,r=1}^{N} \delta_r \delta_k^* [p_{kr}\mu_k(v+\varepsilon u) + p_{rk}\mu_r(v+\varepsilon u)],$$

or, using notation [3.89], the second moment is:

$$B^{\varepsilon}(v,u) = B(v+\varepsilon u).$$

Applying the Taylor formula, it is seen that asymptotic representation of the second moment has the following expression:

$$B^{\varepsilon}(v,u) = B(v) + \theta_{\varepsilon}(v,u).$$

Using theorem 3.2, the covariance matrix of the limit diffusion process is defined by relation [3.89]. The proof of theorem 3.5 is completed. □

3.5.2. *Markov queueing networks*

Two types of Markov queueing networks and a superposition of independent Markov processes on a finite-dimensional phase state space are considered.

Network of $[M|M|\infty]^N$ type: the Markov queueing systems of type $M|M|\infty$ are considered in each of N nodes of the network. Let the intensities of arrival times at the nodes of the network be determined by a vector $\lambda(u) = (\lambda_k;\, 1 \le k \le N)$ and let the intensities of service times in nodes of the network be determined by a vector $\mu(u) = (\mu_k;\, 1 \le k \le N)$.

COROLLARY 3.7.– The normalized queueing process of a network of $[M|M|\infty]^N$ type

$$\zeta^{\varepsilon}(t) = \varepsilon\nu^{\varepsilon}(t/\varepsilon^2) - \varepsilon^{-1}\rho(t)$$

with the centering function is determined by the solution of the equation

$$d\rho(t)/dt = c(\rho(t))\,, \quad \rho(0) = \rho_0, \qquad [3.91]$$

where

$$c(u) := \lambda + u^*Q\,,\ \ Q := \mu^d(P - I), \tag{3.92}$$

under the conditions (c) and (d) of theorem 3.5 weakly converges as $\varepsilon \to 0$ to an Ornstein–Uhlenbeck diffusion process with drift function

$$a(u) = u^*Q \tag{3.93}$$

and covariance matrix

$$B(t) = c^d(\rho(t)) - [\rho^d(t)Q + Q^*\rho^d(t)] \tag{3.94}$$

PROOF.– For the network of $[M|M|\infty]^N$ type, the intensity functions $\lambda(u)$ and $\mu(u)$ in theorem 3.5 are determined as follows:

$$\lambda(u) = (\lambda_k\,;\ 1 \le k \le N)\,;\ \mu(u) = (u_k\mu_k\,;\ 1 \le k \le N).$$

So, by formulas [3.86], [3.87], [3.91] and [3.92] one obtains [3.93] and [3.94], taking into account [3.88] and [3.89]. □

Network of $[M|M|1|\infty]^N$ type: the Markov queueing systems of $M|M|1|\infty$ type with intensities of arrival times λ_k and service times μ_k, $1 \le k \le M$, are located in the nodes of the network.

COROLLARY 3.8.– The normalized queueing process of $[M|M|\infty]^N$ type network

$$\zeta^\varepsilon(t) = \varepsilon\nu^\varepsilon(t/\varepsilon^2) - \varepsilon^{-1}\rho(t)$$

with centering function $\rho(t) = \rho_0 + Ct$, $C = \lambda + \mu^*Q$, where $Q := [P - I]$, weakly converges, as $\varepsilon \to 0$, to a diffusion process with zero mean and covariance matrix

$$B = C^d - \mu^d[Q + Q^*]. \tag{3.95}$$

PROOF.– By theorem 3.5, using the intensity vector function

$$\lambda(u) = (\lambda_k\,;\ 1 \le k \le N)\,;\ \mu(u) = (\mu_k\,;\ 1 \le k \le N),$$

according to formulas [3.86] and [3.87], the equation for the centered function is

$$d\rho/dt = C\,,\ \ C = \lambda + Q.$$

Hence, the centering function is determined as $\rho(t) = \rho_0 + Ct$. By formulas [3.88] and [3.89], we can conclude that the drift function of the limit diffusion process is identically equal to zero, and the covariance matrix is given by [3.95]. □

3.5.3. *Superposition of Markov processes*

There are n independent ergodic Markov processes $\kappa_i(t)$, $1 \leq i \leq n$ on a finite state space $\{0, 1, ..., N\}$ with generator matrix

$$Q = [q_{kr} \,,\ \ 0 \leq k \,,\ r \leq N].$$

Introduce the counting process:

$$\nu^{(n)}(t) = (\nu_k^{(n)}(t) \,;\ \ 1 \leq k \leq N)$$

with components $\nu_k^{(n)}(t)$, which describe the number of Markov processes $\kappa_i(t)$, $1 \leq i \leq n$ in the state $(k \,,\ 1 \leq k \leq N)$. Let $\rho = (\rho_k,\ 1 \leq k \leq N)$ be the stationary distribution of Markov processes:

$$\rho Q = 0 \,,\ \sum_{k=1}^{N} \rho_k = 1.$$

Introduce the notations:

$$\rho^0 := (\rho_k \,,\ 1 \leq k \leq N) \,,\ \ q^0 := (q_{0k} \,,\ 1 \leq k \leq N),$$

$$Q^0 := [q_{kr} \,,\ \ 0 \leq k \,,\ r \leq N] \,,\ \ u^0 := \sum_{k=1}^{N} u_k.$$

COROLLARY 3.9.– The normalized counting process

$$\zeta^{(n)}(t) = [\nu^{(n)}(t) - n\rho^0]/\sqrt{n} \qquad [3.96]$$

weakly converges, as $n \to \infty$, to an Ornstein–Uhlenbeck diffusion process with drift function

$$a(u) = u^* Q^0 - u^0 q^0 \qquad [3.97]$$

and covariance matrix

$$D = -[Q^0 + Q^{0*}] \qquad [3.98]$$

PROOF.– The counting process [3.96] can be considered as a normalized queueing process for a Markov network with marching matrix

$$P = [p_{kr},\ 0 \le k,\ r \le N],\ \ p_{kr} := q_{kr}/q_k,\ \ q_k := \sum_{k=1}^{N} q_{kr}.$$

Consider the normalized counting process [3.96] in the following form:

$$\zeta^{(n)}(t) = \sqrt{n}[\nu^{(n)}(t)/n - \rho^0].$$

Note that

$$\sum_{k=1}^{N} \nu_k^{(n)}(t) = n.$$

Hence, the following relation exists between the two variables $u = \zeta^{(n)}(t)$ and $\kappa = \kappa^{(n)}(t)$:

$$u = \sqrt{n}(\kappa - \rho^0),$$

or

$$\kappa = \rho^0 + \varepsilon u\ ,\ \ \varepsilon := 1/\sqrt{n}.$$

Besides, $\sum_{k=0}^{N} u_k = 0$, i.e. $u_0 = -\sum_{k=1}^{N} u_k$. Now determine the intensities of jumps of Markov process [3.96] under the condition

$$\zeta^{(n)}(t) = u\ ,\ \ \text{or}\ \ \kappa^{(n)}(t) = \kappa.$$

We leave it to the readers to complete the proof of corollary 3.9 similarly as in the proof of theorem 3.5.

3.6. Semi–Markov queueing systems

As indicated in the Introduction, the main problem is to choose an adequate method for asymptotic analysis of queueing processes in diffusion approximation schemes.

We will assume that the initial queueing process, scaled by a small series parameter $\varepsilon \to 0$ ($\varepsilon > 0$), is set in the same way as in the Markov case:

$$\zeta^\varepsilon(t) = \varepsilon\nu(t/\varepsilon^2) - \varepsilon^{-1}\rho\,,\ t \geq 0. \tag{3.99}$$

However, now for simplicity, the centering constant $\rho > 0$ sets the equilibrium of the system evolution rate

$$C(u) := \lambda - u\mu\,,\quad u \in R\,,\quad C(\rho) = 0.$$

The parameter

$$\lambda = 1/g\,,\quad g := E\theta_{n+1} = \int_0^\infty \overline{G(t)}dt$$

sets the average intensity of input flow of claims with a distribution function of intervals between the adjacent arrivals

$$G(t) := \mathcal{P}\{\theta_{n+1} \leq t\},\ n \geq 0. \tag{3.100}$$

The parameter $\mu > 0$ determines the intensity of the claims service in the system, which have an exponential distribution of service intervals.

For simplicity, we consider a semi-Markov queueing system of $SM \mid M \mid \infty$ type. It means that the claims in the system are characterized by distribution function [3.100] with average intensity $\lambda = 1/g$, $g := E\theta_{n+1}$, $n \geq 0$.

The intensity $\mu > 0$ defines the exponential distribution of the claims processing time. The symbol "∞" means that the queueing system has an unbounded stock of claims.

The general scheme of asymptotic analysis of queueing systems is based on a *compensating operator* determined by the extended Markov process

$$\begin{gathered}\zeta_n^\varepsilon := \zeta^\varepsilon(\tau_n^\varepsilon)\,,\ \nu_n^\varepsilon := \varepsilon\nu(\tau_n^\varepsilon)\,,\ \tau_n^\varepsilon := \varepsilon^2\tau_n\,,\ n \geq 0,\\ \theta_{n+1} := \tau_{n+1} - \tau_n,\ n \geq 0,\ \tau_0 = 0,\end{gathered} \tag{3.101}$$

Note the following property:

$$\nu(\tau_n^\varepsilon) = \eta(\tau_n^\varepsilon) + 1\,,\quad n \geq 0. \tag{3.102}$$

Hence the extended Markov process [3.101] is defined by the conditional distributions of increments

$$\Delta\zeta^\varepsilon_{n+1} := \zeta^\varepsilon_{n+1} - \zeta^\varepsilon_n\,,\ \ \Delta\nu^\varepsilon_{n+1} := \nu^\varepsilon_{n+1} - \nu^\varepsilon_n\,,\ \ n \geq 0, \tag{3.103}$$

under the condition

$$\zeta^\varepsilon_n = \varepsilon u\,,\ \ \varepsilon\nu^\varepsilon_n = \rho + \varepsilon u. \tag{3.104}$$

The compensating operator of the extended Markov process [3.101] is determined as follows:

$$L^\varepsilon\varphi(u) = \varepsilon^{-2}g^{-1}[G^\varepsilon\Lambda^\varepsilon - I]\varphi(u)\,,\ \ \Lambda^\varepsilon\varphi(u) := \lambda\varphi(u+\varepsilon), \tag{3.105}$$

$$\begin{aligned} G^\varepsilon\varphi(u) :&= -\int_0^\infty d\overline{G(t)}\Gamma^\varepsilon_{\varepsilon^2 t}(\rho+\varepsilon u)\varphi(u) \\ &= \varepsilon^2\Gamma^\varepsilon\int_0^\infty \overline{G(t)}\Gamma^\varepsilon_{\varepsilon^2 t}(\rho+\varepsilon u)\varphi(u), \end{aligned} \tag{3.106}$$

where the semigroup $\Gamma^\varepsilon_{\varepsilon^2 t}(\rho+\varepsilon u)$ is defined by generator

$$\Gamma^\varepsilon(\rho+\varepsilon u)\varphi(u) = \varepsilon^{-1}[\rho+\varepsilon u]\mu\varphi'(u).$$

The asymptotic analysis of the compensating operator of the extended Markov process [3.105], as $\varepsilon \to 0$ ($\varepsilon > 0$), is given in the next theorem.

THEOREM 3.6.– Assume the following conditions:

1) the intensity of input claims $\lambda = 1/g$ is bounded and $\lambda > 0$;

2) the accelerated intensity of service time $\mu_\varepsilon := \mu/\varepsilon > 0$;

3) there exists the average limit

$$\varepsilon\rho(t/\varepsilon) \Rightarrow \rho\,,\ \ \varepsilon \to 0,$$

where the equilibrium $\rho > 0$ is determined by a balance condition for the velocity function

$$C(\rho) = \lambda - \rho\mu = 0\,,\ \ C(u) := \lambda - u\mu\,,\ \ u \in R.$$

Then there exists a limit generator L^0:

$$L^\varepsilon\varphi(u) \to L^0\varphi(u)\,,\ \ \varepsilon \to 0\,,\ \ \varphi(u) \in C^3(R), \tag{3.107}$$

$$L^0\varphi(u) := -u\mu\varphi'(u) + \frac{1}{2}B\varphi''(u), \tag{3.108}$$

$$B = \lambda/2(\lambda^2 g_2 - 1)\ ,\ \ g_2 := E\theta_{n+1}^2.$$

COROLLARY 3.10.– The normalized compensating operator [3.101] weakly converges to a diffusion Ornstein–Uhlenbeck process $\zeta^0(t)$:

$$\zeta^\varepsilon(t) \Rightarrow \zeta^0(t)\ ,\ \ 0 \le t \le T\ ,\ \ T > 0, \tag{3.109}$$

with generator [3.107], that is

$$d\zeta^0(t) = -\mu\zeta^0(t)dt + \sigma dW(t), \tag{3.110}$$

$$\sigma := B^{1/2}.$$

PROOF.– (Proof of theorem 3.6). The following identity is used:

$$G^\varepsilon\Lambda^\varepsilon - I = G^\varepsilon - I + \Lambda^\varepsilon - I + (G^\varepsilon - I)(\Lambda^\varepsilon - I). \tag{3.111}$$

We use the following asymptotic representations of the terms in [3.111]:

$$\varepsilon^{-2}g^{-1}[G^\varepsilon - I]\varphi(u) = -\varepsilon^{-1}\rho\mu\varphi'(u) - u\mu\varphi'(u) + \frac{1}{2}B\varphi''(u) + o(\varepsilon),$$

$$\varepsilon^{-2}g^{-1}[\Lambda^\varepsilon - I]\varphi(u) = \varepsilon^{-1}\lambda\varphi'(u) + \frac{\lambda}{2}\varphi''(u) + o(\varepsilon),$$

and also

$$\varepsilon^{-2}g^{-1}[G^\varepsilon - I][\Lambda^\varepsilon - I]\varphi(u) = -\lambda\varphi''(u) + o(\varepsilon).$$

Therefore, the following representation takes place:

$$L^\varepsilon\varphi(u) = L^0\varphi(u) + R_\varepsilon\varphi(u), \tag{3.112}$$

with a negligible term

$$R_\varepsilon\varphi(u) \to 0\ ,\ \ \varepsilon \to 0\ ,\ \ \varphi(u) \in C^3(R).\ \square$$

More general models of semi-Markov queueing systems are considered in the previous study by Korolyk *et al.* (2008).

3.7. Acknowledgements

The authors warmly thank Prof. V.V. Anisimov for his kind invitation to participate in the book.

3.8. References

Anisimov, V.V. (1995). Switching processes: Averaging principle, diffusion approximation and applications. *Acta Appl. Math.* 40(2), 95–141.

Anisimov, V.V. (2002). Diffusion approximation in overloaded switching queueing models. *Queueing Systems*, 40, 141–180.

Anisimov, V.V. (2008). *Switching Processes in Queueing Models*. ISTE Ltd, London and John Wiley & Sons, New York.

Ethier, S.N., Kurtz, T.G. (1986). *Markov Processes: Characterization and Convergence*. Wiley, Hoboken.

Iglehart, D.L. (1965). Limit diffusion approximation for many-server queue and the repairman problem. *J. Appl. Prob.* 2, 429–441.

Korolyk, V.S., Mishkoy, G.H., Mamonova, A., Griza, I. (2008). Queueing system evolution in phase merging scheme. *Buletinul Academiei De Stiinte a Republicii Moldova. Matematica* 3(58), 83–88.

Korolyuk, V.S., Korolyuk, V.V. (1999). *Stochastic Models of Systems*. Kluwer, Dodrecht.

Korolyuk, V.S., Limnios, N. (2005). *Stochastic Systems in Merging Phase Space*. World Scientific, London.

4

First-come First-served Retrial Queueing System by Laszlo Lakatos and its Modifications

Igor Nikolaevich KOVALENKO[†]
V.M. Glushkov Institute of Cybernetics, National Academy of Sciences of Ukraine, Kiev, Ukraine

The author suggests some modifications to a first-come first-served retrial queueing (FCFS RQ) system introduced by Hungarian mathematician Laszlo Lakatos in 1994.

In particular, two models of RQ systems are introduced. In both of them, the arrival of a one server queue is a renewal process (t_n) and service times of customers are independent identical distributed random variables (i.i.d.r.v.) Y_n. On the arrival of a customer, a time interval Γ_n is reserved for its service. If the interval $(t_n, t_n + Y_n)$ has no common point with $\bigcup_{i<n} \Gamma_i$, then this interval is defined as Γ_n; otherwise, Γ_n is chosen as an interval situated from the right-hand side of all the Γ_i, $i < n$. The length of the interval Γ_n equals Y_n (model 1) or a newly generated variable Y'_n (model 2), whereas the left-hand side of Γ_n is controlled by a renewal process of orbit times.

These models seem to increase the output of the queueing system in comparison with the Lakatos-type queueing system.

4.1. Introduction

Beginning with the 1950s, a new branch of queueing theory – RQ system – was introduced to the theory and engineering practice. The main peculiarity of such

Queueing Theory 1,
coordinated by Vladimir ANISIMOV, Nikolaos LIMNIOS. © ISTE Editions 2020.

systems consists of the existence of a number of so-called orbits to which a rejected customer is directed so that it can make periodical attempts (retrials) until a time of vacation of the service facility. Well known monographs and surveys on retrial queues include Artalejo and Gomez-Corral (2008); Bocharov *et al.* (2004); Falin and Templeton (1997); Artalejo (2010) and others. As a disciple and co-author with a distinguished scientist Boris V. Gnedenko, I would like to refer to some relevant works from B.V. Gnedenko's scientific school: Falin (2008, 2010); Afanasyeva (1984); Afanasyeva and Gnedenko (1985); Anisimov (1999); Anisimov and Artalejo (2001); Anisimov and Kurtulush (2001). Most of these works are connected to applied problems. For instance, L.G. Afanasyeva studied RQ models of aviation flows.

4.2. A contribution by Laszlo Lakatos and his disciples

In 1994, the Hungarian mathematician Laszlo Lakatos, on an order from aviation traffic researcher Vaclav Ceric, introduced a new type of retrial queue (Lakatos 1994), with Markovian input and service time, constant orbit time and moreover FCFS queueing discipline. Applying a suitable embedded Markov chain, Lakatos derived an elegant ergodicity condition of this chain as follows:

$$\frac{\lambda}{\mu} < \frac{e^{-\lambda T}\left(1 - e^{-\mu T}\right)}{1 - e^{-\lambda T}}$$

where λ is the arrival rate, μ is the service rate and T is a constant orbit time of a customer.

Moreover, Lakatos derived the ergodic distribution of the embedded Markov chain (k_n) denoting a queue size just before the arrival of the nth customer.

For further works by the school of Lakatos, see Lakatos (2002, 2006); Langaris (1999); Kárász and Farcas (2005); Lakatos (2018); Lakatos *et al.* (2013); Rogiest *et al.* (2006, 2007).

4.3. A contribution by E.V. Koba

My disciple Elena V. Koba has investigated a number of generalizations of Lakatos-type queueing systems. She studied $GI/G/1$ retrial FCFS systems with generally distributed orbit times for customers. The stability conditions of such systems were derived. The main point of Koba's argument consists of the following. Let W_n denote the waiting time of the nth customer. Then, under usual convergence conditions,

$$\lim_{x\to\infty} E\{W_{n+1} - W_n | W_n = x\} = E\left\{Y_n + U_n - (t_{n+1} - t_n)\right\},$$

where t_n are arrival epochs and Y_n are service times of customers. U_n is an excess of a renewal process controlling orbit intervals. Thus, one can prove that this limit is negative.

In particular, if an orbit time is a non-latticed random variable V_n, then the sequence is stable as soon as

$$EY_n - E\left(t_{n+1} - t_n\right) + \frac{1}{2}EV_n^2/EV_n < 0.$$

A more general case of Markovian dependence of variables is considered as well (see Koba (2002, 2004)).

4.4. An Erlangian and hyper-Erlangian approximation for a Laszlo Lakatos-type queueing system

In the preceding section, Koba's results on stability/ergodicity conditions for $GI/G/1$ RQ systems were outlined. It is interesting to note that in some cases, it is possible to derive elementary analytical equations for such conditions. Indeed, one such example is studied in a paper by Kovalenko (2018).

Consider a $GI/G/1$ FCFS RQ system with a constant orbit time T. We denote by X an interarrival time and by Y a service time. Assume that they possess complementary distribution functions (d.f.)

$$\bar{A}\left(x\right) = \sum_{i=0}^{l-1} \frac{\lambda^i}{i!} x^i e^{-\lambda x}, \qquad x \geq 0,$$

and

$$\bar{B}\left(x\right) = \sum_{i=0}^{m-1} \frac{\mu^i}{i!} x^i e^{-\mu x}, \qquad x \geq 0,$$

with probability density functions (p.d.f.)

$$a\left(x\right) = \frac{\lambda^l x^{l-1}}{(l-1)!} e^{-\lambda x}, \qquad x > 0,$$

and

$$b\left(x\right) = \frac{\mu^m x^{m-1}}{(m-1)!} e^{-\mu x}, \qquad x > 0,$$

respectively.

Consider non-negative random variables

$$N^{+}=\begin{cases}k \text{ if } \frac{Y-X}{T}\in(k-1,k)\,,k\geq 1\\ \qquad 0, \text{otherwise}\end{cases}$$

and

$$N^{-}=\begin{cases}k \text{ if } \frac{X-Y}{T}\in(k,k+1)\,,k\geq 0.\\ \qquad 0, \text{otherwise}\end{cases}$$

Further, consider the difference

$$N=N^{+}-N^{-}.$$

Then obviously

$$EN=EN^{+}-EN^{-}.$$

It is easily to show that the system is stable iff

$$EN<0.$$

We have the equations

$$EN^{+}=\sum_{k=0}^{\infty}P\{Y-X>kT\},$$

$$EN^{-}=\sum_{k=1}^{\infty}P\{X-Y>kT\}.$$

For any $x>0$, we get equations

$$P\{Y-X>x\}=\int_{0}^{\infty}a(t)\bar{B}(t+x)\,dt$$

and

$$P\{X-Y>x\}=\int_{0}^{\infty}b(t)\bar{A}(t+x)\,dt.$$

Taking into account the equations for cumulative d.f. and p.d.f. of Erlangian laws, one obtains the equations

$$P\{Y-X>x\}=\left(\frac{\lambda}{\lambda+\mu}\right)^{l}\sum_{j=0}^{m-1}\frac{\mu^{j}x^{j}}{j!}e^{-\mu x}\sum_{r=0}^{m-j-1}C_{l+r-1}^{r}\left(\frac{\mu}{\lambda+\mu}\right)^{r}$$

and

$$P\{X-Y>x\}=\left(\frac{\mu}{\lambda+\mu}\right)^{m}\sum_{j=0}^{l-1}\frac{\lambda^{j}x^{j}}{j!}e^{-\lambda x}\sum_{r=0}^{l-j-1}C_{m+r-1}^{r}\left(\frac{\lambda}{\lambda+\mu}\right)^{r}$$

for any $x\geq 0$.

To obtain the expressions for EN^{+} and EN^{-}, it suffices to sum up these equations in x for $x=kT$, $k\geq 0$ and $x=kT$, $k\geq 1$, respectively.

It should be noted that infinite series like

$$I_j=\sum_{k=0}^{\infty}(\lambda kT)^{j}e^{-\lambda kT}$$

can be computed through a term-wise differentiation of infinite series. Indeed, obviously

$$\sum_{k=0}^{\infty}e^{-\lambda kTz}=\frac{1}{1-e^{-\lambda Tz}}$$

for any non-negative z. The j-fold term-wise differentiation in z leads to the equation

$$\sum_{k=0}^{\infty}(\lambda kT)^{j}e^{-\lambda kTz}=(-1)^{j}\frac{d^{j}}{dz^{j}}\frac{1}{1-e^{-\lambda Tz}}.$$

Setting $z=1$, one obtains the sum I_j.

Further, we assume that the distributions $A(x)$ and $B(x)$ are hyper-Erlangian. This case can be easily reduced to the above-considered case. Indeed, the p.d.f. $a(x)$ and $b(x)$ are mixtures of some Erlangian p.d.f. $e(x|\lambda_g,l_g)$ and $e(x|\mu_h,m_h)$. It is easy to see that this property implies the corresponding property of the functional N.

4.5. Two models with a combined queueing discipline

A Lakatos-type RQ system reflects the peculiarities of some engineering systems (Koba and Pustova 2012). Meanwhile, a negative aspect of such systems is that long intervals between service times are practically lost for the serving process; thus mean waiting time is rather long. So an idea arises naturally as follows: to combine the FCFS principle with the possibility of a partial servicing of some customers within free intervals. One of the possible disciplines leads to the following model.

Each incoming customer gets its service period on its arrival so that it cannot be changed afterwards. We denote by t_n the nth arrival time provided $t_n < t_{n+1}$ by $Y_n > 0$ the associated service time.

For the definiteness, assume that t_n form a recurrent process; $t_n - t_{n-1}$ are i.i.d.r.v. with a d.f. $A(x) = P\{t_n - t_{n-1} \leq x\}$. Y_n are i.i.d.r.v. with a d.f. $B(x) = P\{Y_n \leq x\}$. Also assume that both $t_n - t_{n-1}$ and Y_n are positive almost surely.

In our model, each customer is serviced either immediately on its arrival, if a required time allows this, or, if not, then the customer is directed to the orbit taking the last place in the queue of "orbited" customers. The rule of reserving time for such customers is controlled by a renewal process. This will be explained later. In such cases, the service time of the customer equals Y'_n, where $Y'_n = Y_n$ (model 1) or Y'_n is a newly generated r.v. with the d.f. $B(x)$ (model 2).

Thus, let $S_n^- = \left(k_n^-; \Delta_{n1}^-, \ldots, \Delta_{n,k_n^-}^-\right)$ denote a number k_n^- of customers who entered before t_n and were not fully serviced before the instant t_n; $\Delta_{ni}^- = \left(X_{ni}^-, Z_{ni}^-\right)$, where $0 \leq X_{n1}^- < Z_{n1}^- < X_{n2}^- < \ldots < X_{n,k_n^-}^- < Z_{n,k_n^-}^-$. These quantities will be interpreted in such a way that time intervals $\left(t_n + X_{ni}^-, t_n + Z_{ni}^-\right)$ being seen from time t_n are complete or residual service times of k_n^- customers who are not fully served up to time t_n. At last, set $\tau_n^- = Z_{n,k_n^-}^-$. In the same way, $S_n^+ = \left(k_n^+; \Delta_{n1}^+, \ldots, \Delta_{nk_n^+}^+\right)$, where $\Delta_{ni}^+ = \left(X_{ni}^+, Z_{ni}^+\right)$. Also set $\tau_n^+ = Z_{n,k_n^+}^+$. Hereinafter, the lower index n relates to an arrival time, whereas the upper signs "-" and "+" relate to prearrival and postarrival moments, respectively.

The variables $X_{ni}^{\pm}$, $Z_{nj}^{\pm}$ are adjusted to the temporary coordinate system with the origin t_n. These symbols mean the variables as X_{ni}^+ relating to time t_n^+ just after the arrival time of the nth customer instead of t_n^-. The dynamics of the servicing process can be described by recurrence equations enabling one to make a transfer from X_n^- to X_n^+ and from X_{n-1}^+ to X_n^-.

Consider the "- to +" case. In any case, $k_n^+ = k_n^- + 1$. If $Y_n < X_{n1}^-$, then $X_{n1}^+ = 0$, $Z_{n1}^+ = Y_n$; $X_{ni}^+ = X_{n,i-1}^-$, $Z_{ni}^+ = Z_{n,i-1}^-$, $2 \leq i \leq k_n^+$. These equations mean that the nth customer is taken for servicing at time t_n (see Figure 4.1 for example).

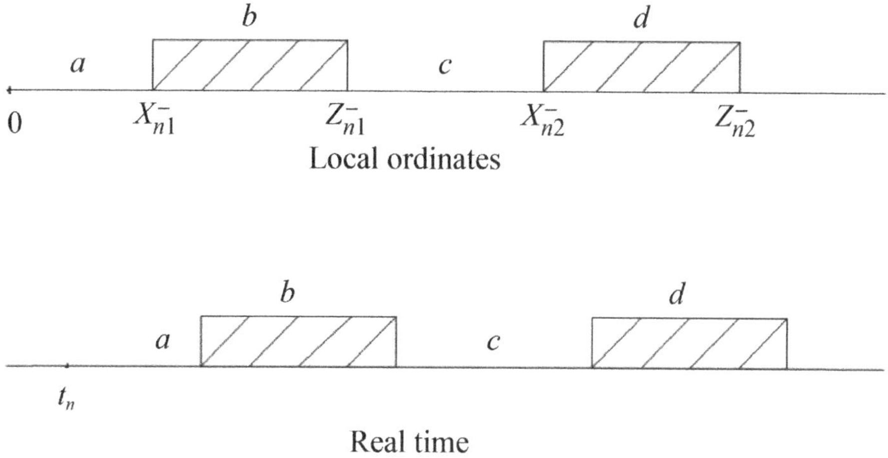

Figure 4.1.

If $X_{n1}^- > 0$ or $Y_n > X_{n1}^-$, then $X_{ni}^+ = X_{ni}^-$, $Z_{ni}^+ = Z_{ni}^-$, $1 \le i \le k_n^-$. To finish the definition, we must use a concept from renewal theory. The variables $X_{n,k_n^+}^+$ and $Z_{n,k_n^+}^+$ are assumed to be defined through independent renewal processes $(U_{nj}, j \ge 0)$ for which $U_{n0} = 0$; $U_{nj} - U_{n,j-1}$ are positive i.i.d.r.v. with a d.f. $C(x) = P\{U_{nj} - U_{n,j-1} \le x\}$, $n \ge 1$. Assume that $\tau_n^- = Z_{n,k_n^-}^-$ is defined. Then $X_{n,k_n^+}^+ = \min\{U_{nj} > \tau_n^-\}$, $\tau_n^+ = Z_{n,k_n^+}^+ = X_{n,k_n^+} + Y'_n$.

One can observe that in the case that a customer cannot be served from the very beginning, then this customer is directed to the end of the queue. Obviously, in the model introduced by Laszlo Lakatos, all the customers follow the FCFS queueing discipline.

At last, consider the t_{n-1}^+ to t_n^- transfer. If $k_{n-1}^+ = 0$, then $k_n^- = 0$. If $t_n - t_{n-1} < X_{n-1,1}^+$, then $k_n^- = k_{n-1}^+$, $X_{nj}^- = X_{n-1,j}^+ - (t_n - t_{n-1})$, $1 \le j \le k_n^-$. If $Z_{n-1,i}^+ < t_n - t_{n-1} < X_{n-1,i+1}^+$ for some i, $1 \le i \le k_{n-1}^+ - 1$, then $k_n^- = k_n^+ - i$, $X_{nj}^- = X_{n-1,j+i}^+ - (t_n - t_{n-1})$, $Z_{nj}^- = Z_{n-1,j+i}^+ - (t_n - t_{n-1})$.

If $X_{n-1,i}^+ < t_n - t_{n-1} < Z_{n-1,i}^+$ for some i, $1 \le i \le k_{n-1}^+$, then $k_n^- = k_{n-1}^+ - i + 1$, $X_{n1}^- = 0$, $Z_{n1}^- = Z_{n-1,i}^+ - (t_n - t_{n-1})$, $X_{nj}^- = X_{n-1,j+i}^+ - (t_n - t_{n-1})$, $Z_{nj}^- = Z_{n-1,j+i}^+ - (t_n - t_{n-1})$, $2 \le j \le k_n^+$.

See also a paper by Koba (2017) for a special queueing retrial discipline on the Markov level. The paper considers the RQ system $M/M/1/0$ with combined discipline of service, namely, a customer from the orbit is served in its turn, but in case of a free channel an arrival from the original flow is serviced immediately. The author obtained the expressions for state probabilities as well as the ergodicity conditions. The system is compared with the Lakatos type system (see also Kovalenko (2015)).

4.6. References

Afanasyeva, L.G., Gnedenko, B.V. (1984). Principles of analytical modeling of a dispatcher in a sector of aircraft movement control. *Appl. Stat. Meth. Manuf. Manag.*, 101–102 (in Russian).

Afanasyeva, L.G., Gnedenko, B.V. (1985). Stochastical and statistical simulation in aircraft movement control. *Appl. Multidim. Stat. Anal. Econ. Qual. Prod. Eval.*, 134–144 (in Russian).

Anisimov, V.V. (1999). Switching stochastic models and applications in retrial queues. *TOP*, 7, 169–186.

Anisimov, V.V., Artalejo, J.R. (2001). Analysis of Markov multiserver retrial queues with negative arrivals. *Queueing Systems,* 39, 157–182

Anisimov, V.V., Kurtulush, M. (2001). Some Markovian queuing retrial systems under light-traffic conditions. *Cybernet. Syst. Anal.*, 37, 876–887.

Artalejo, J.R. (2010). Accessible bibliography on retrial queues: Progress in 2000–2009. *Math. Comp. Model.*, 51, 1071–1081.

Artalejo, J.R., Gomez-Corral, A. (2008). *Retrial Queueing Systems: A Commputational Approach*. Springer, Berlin.

Bocharov, P.P., D'Apice, C., Pechinkin, A.V., Salerno, S. (2004). *Queueing Theory*. Brill Academic Publishers, Utrech.

Falin, G.I., Templeton, J.G.C. (1997). *Retrial Queues*. Chapman & Hall, London.

Falin, G.I. (2008). The $M/M/1$ retrial queue with retrials due to server failures. *Queueing Systems,* 58, 155–160.

Falin, G.I. (2010). A single-server batch arrival queue with returning customers. *Europ. J. Oper. Res.*, 201, 786–790.

Kárász, P., Farcas, G. (2005). Exact solution for a two-type customers retrial system. *Comp. Math. Appl.*, 49, 95–102.

Koba, E.V. (2002). On a $GI/G/1$ retrial queueing with a FIFO queueing discipline. *Theory Stoch. Proc.*, 8, 201–207.

Koba, E.V. (2004). An ergodicity condition for a generalized queueing model of the L.Lakatos type. *Dopovidi NAN Ukrainy*, 11, 70–74 (in Russian).

Koba, E.V., Pustova S.V. (2012). Lakatos queuing systems, their generalization and application. *Cybernet. Syst. Anal.*, 48, 387–396.

Koba, E.V. (2017). Retrial queueing system $M/M/1/0$ with combined service discipline. *Cybernet. Syst. Anal.*, 53, 387–391.

Kovalenko, I.N. (2015). A two-cyclic queuing system. *Cybernet. Syst. Anal.*, 51, 51–55.

Kovalenko, I.N. (2018). The FCFS-RQ system by Laslo Lakatos and its modifications. *Reliab. Theory Appl.*, 2, 52–54.

Lakatos, L. (1994). On a simple continuous cyclic-waiting problem. *Annales Univ. Sci. Budapest, Sect. Comp.*, 14, 105–113.

Lakatos, L. (2002). A retrial system with time-limited tasks. *Theory Stoch. Proc.*, 8, 249–256.

Lakatos, L. (2006). A retrial queueing system with urgent customers. *J. Math. Sci.*, 138, 5405–5409.

Lakatos, L. (2018). Some aspects of the discrete $Geo/G/1$ type cycling waiting systems. In *DCCN 2018, Commun. Comp. Inf. Sci. (CCIS)*, Vishnevskiy, V., Kozyrev, D. (eds.), Springer, Berlin, 919, 288–301.

Lakatos, L., Szeidl, L., Telek, M. (2013). *Introduction to Queueing Systems with Telecommunication Applications*. Springer, Boston.

Langaris, C. (1999). Markovian polling systems with mixed service disciplines and retrial customers. *TOP*, 7, 305–322.

Rogiest, W., Laevens, K., Fiems, D., Bruneel, H. (2006). Analysis of a Lakatos-type queueing system with general service times. *Proc. ORBEL*, 20, 288–301.

Rogiest, W., Laevens, K., Walraevens, J., Bruneel, H. (2007). Analyzing a degenerate buffer with general inter-arrival and service times in discrete time. *Queueing Systems*, 56, 203–212.

5

Parameter Mixing in Infinite-server Queues

Lucas VAN KREVELD[1] and Onno BOXMA[2]

[1]*Korteweg-de Vries Institute, University of Amsterdam, The Netherlands*

[2]*Eindhoven University of Technology, The Netherlands*

We consider two infinite-server queueing models with a so-called mixed arrival process. The arrival process is Poisson, but the arrival intensity is resampled from some distribution at exponentially distributed time intervals. First, we study the case of Coxian service times. For this infinite-server model, we show how to compute all joint moments of the arrival rate and the number of customers, both in transient and steady state. Second, we consider a Markov-modulated infinite-server queue with general service times. The arrival intensity is resampled at the state change epochs of the Markov background process. We develop a procedure to obtain all transient moments of the number of customers given the initial state and the initial arrival rate.

5.1. Introduction

In queueing theory, it is often assumed that the arrival process is a Poisson process with a constant rate. This is not always a realistic assumption, and hence various adaptations have been suggested to better reflect reality. We mention four types of adaptation, in which the arrival process still maintains essential elements and properties of the Poisson process. First, there have been studies of queues with time-inhomogeneous Poisson arrivals (Eick *et al.* 1993), and of queues with Markov-modulated arrivals in which the arrival process is Poisson with rate λ_i when some underlying Markov process is in state i (see Asmussen (2003, Ch. XI), and the pioneering work of Neuts (1981, 1989)). Second, there is a growing interest in

Queueing Theory 1,
coordinated by Vladimir ANISIMOV, Nikolaos LIMNIOS.

queues with Cox arrival processes. These are Poisson processes in which the time-dependent arrival intensity, say $\Lambda(t)$, itself is a stochastic process. The variance of the number of arrivals of a Cox process in a given interval is larger than the mean (whereas they are equal for Poisson); this phenomenon is usually called overdispersion. In a few papers, the process $\{\Lambda(t), t \geq 0\}$ was taken to be a shot-noise process. We refer to (Albrecher and Asmussen 2006) for an example in insurance mathematics and to (Koops *et al.* 2017) for a queueing example. Third, the so-called *mixture* models have been suggested. In such models, the arrival rate is itself a random variable with some distribution. In the queueing literature, it is not so common to consider mixing distributions (Jongbloed and Koole 2001), but this is different in the finance and insurance literature (e.g. Bühlmann (1972) already considered them in 1972 in the context of credibility-based dynamic premium rules). Recent studies of single-server queues with mixing include that of Raaijmakers *et al.* (2019) (in which the random parameter is sampled once and for all) and Klaasse (2017) (in which the random parameter is resampled in each new busy period). Fourth, we would like to mention the theory of *switching processes*, as developed by Anisimov. Such processes describe the operation of stochastic systems with the property that their development in time switches at some random points in time, which may depend on the past system trajectory. In (Anisimov 2008), Anisimov studies asymptotic results for large classes of switching processes.

One class of queueing models for which generalizations of the classical Poisson arrival process have been studied is the class of infinite-server models. Websites provide interesting applications of such systems (but there are many other applications). The arrival rate of website visits typically is not constant in time and could depend on an environment (Markov modulation), or vary according to some stochastic process that takes external events into account (e.g. shot-noise driven), or take a new random value after some time (mixing). Numerous variants of stochastic background processes in infinite-server queues have been studied after the pioneering paper of O'Cinneide and Purdue (1986). Jansen *et al.* (2016) have considered a general càdlàg stochastic process as background process, whereas D'Auria (2008), Anderson *et al.* (2016) and Blom *et al.* (2014) were restricted to the case where the background process is Markov. Next to the already mentioned case of an infinite-server model with a Coxian, shot-noise driven, arrival process (Koops *et al.* 2017), infinite-server systems with the more general Hawkes, or self-exciting, input process have recently been analyzed (Daw and Pender 2018; Koops *et al.* 2018).

This chapter is devoted to the study of infinite-server queues with a mixed Poisson arrival process. We assume that, at exponential intervals, a new value for the arrival rate is sampled. We focus on two different models. The first model is an $M_\Lambda/Cox_n/\infty$ queue. Here, the notation M_Λ is used to emphasize that the arrival process is a Poisson process with a *random* arrival rate. Cox_n indicates that the service times have a Coxian distribution with n phases. The second model allows service times to have a fully general distribution, and furthermore allows Markov modulation: when a Markov background process enters some state i, then a new arrival rate is sampled from some distribution (which is clearly more general than the ordinary Markov modulation, in which one always has rate λ_i if the background state is i). For both models, we consider transient and steady-state moments of the queue length process; in the slightly simpler first model, we even present a procedure to obtain all joint moments of the arrival rate and the numbers of customers in each Coxian service phase.

This chapter is organized as follows. Section 5.2 is devoted to an analysis of the $M_\Lambda/Cox_n/\infty$ queue. We show how all joint moments of the arrival rate and the number of customers in each of the n Coxian phases can be computed. In section 5.3, we consider a Markov-modulated infinite-server queue, with generally distributed service times and with mixing of the arrival rate at modulation epochs. We demonstrate how successive queue length moments can be determined iteratively. Section 5.4 contains some suggestions for further research.

5.2. The $M_\Lambda/\text{Cox}_n/\infty$ queue

In this section, we consider an infinite-server queue where the arrival parameter repeatedly resamples after i.i.d. (independent, identically distributed) exponential amounts of time. We shall analyze the behavior of this queue and make comparisons to "standard" infinite-server queues with a fixed deterministic arrival parameter.

Let us first describe our model in detail. We consider an $M/\text{Cox}_n/\infty$-queue where the arrival intensity $\Lambda(t)$ at any time $t \geq 0$ is a random variable. The arrival intensity is resampled at random times generated by a Poisson process with rate γ, and is drawn according to some distribution G_Λ.

The service times are i.i.d. random variables. The service time distribution we consider is a Coxian with n phases (see (Asmussen 2003, p. 85)). Any customer spends an exp(μ_1) amount of time in phase 1 (i.e. an exponentially distributed amount of time with mean $1/\mu_1$). With probability $1 - p_1$, the service is thereafter completed. Otherwise, the service continues with the next phase, which takes exp(μ_2). Again, with probability $1 - p_2$ the service is completed, and otherwise the service goes on. This procedure continues up to a maximum of n phases.

The Coxian distribution possesses useful exponential properties, while still remaining relatively general. The parameters of a Coxian distribution can be chosen such that it can approximate any distribution D on $[0, \infty)$ arbitrarily closely; for any such D, there exists a sequence of Coxian distributions weakly converging to D. See section III.4 of (Asmussen 2003) for a more detailed discussion.

Our main object of study is the distribution of the number of customers $X(t)$. Throughout the chapter, we assume that $X(0) = 0$ (empty system at time 0), which (as shown later) extends to any initial condition on $X(0)$.

More specifically, we are interested in the distribution of the random vector $(\Lambda(t), X_1(t), ..., X_n(t))$, where $X_i(t)$ represents the number of customers in phase i of their service at time t. To uniquely identify this distribution, we consider here the joint transform $E\left(e^{-s\Lambda(t)} \prod_{i=1}^{n} z_i^{X_i(t)}\right)$. For the analysis of our model, we use a differential equation method introduced in (Koops *et al.* 2018). We observe $(\Lambda(t), X_1(t), ..., X_n(t))$ during a small interval and thus derive a partial differential equation (PDE) for its joint transform. Unfortunately, we are not able to solve the PDE; however, we are able to extract from this PDE a recursive equation in the various moments of $\Lambda(t)$ and the numbers of customers $X_1(t), \ldots, X_n(t)$. Below, we first derive the PDE. We then discuss its solvability and look at special cases. Finally, we manipulate the equation to find a recursion that allows one to iteratively retrieve all moments of the vector $(\Lambda(t), X_1(t), ..., X_n(t))$. Special attention will be given to the steady-state vector $(\Lambda, X_1, ..., X_n)$ and the special Cox_1 case of exponential service times.

5.2.1. *The differential equation*

Let z be a shorthand notation for the vector $(z_1, ..., z_n)^T$. The following theorem describes a PDE for $f(s, z, t) := E\left(e^{-s\Lambda(t)} \prod_{i=1}^{n} z_i^{X_i(t)}\right)$.

THEOREM 5.1.– The joint transform $f(s, z, t)$ satisfies the following PDE with boundary condition $f(0, 1, t) = 1$:

$$\begin{aligned} -\gamma f(s, z, t) &- \frac{\partial}{\partial t} f(s, z, t) + (1 - z_1)\frac{\partial}{\partial s} f(s, z, t) \\ &+ \sum_{i=1}^{n-1} \mu_i(p_i z_{i+1} - z_i + 1 - p_i)\frac{\partial}{\partial z_i} f(s, z, t) \\ &+ \mu_n(1 - z_n)\frac{\partial}{\partial z_n} f(s, z, t) + \gamma f(s, 1, t) f(0, z, t) = 0. \end{aligned} \qquad [5.1]$$

PROOF.– We exploit the fact that $\{(\Lambda(t), X_1(t), \ldots, X_n(t)), t \geq 0\}$ is a Markov process. At some given time t, let us assume $X_i(t) = k_i$ for each i, and $\Lambda(t) = \lambda$. In $[t, t + h), h \downarrow 0$, three things can occur: a customer completes phase i of its service at rate $k_i\mu_i$; a customer arrives at rate λ; the arrival parameter resamples at rate γ. Hence

$$E\Big(e^{-s\Lambda(t+h)}\prod_{i=1}^{n} z_i^{X_i(t+h)}\Big|\Lambda(t)=\lambda, X_1(t)=k_1,...,X_n(t)=k_n\Big)$$

$$= (1-(\lambda+\gamma+\sum_{i=1}^{n} k_i\mu_i)h)e^{-s\lambda}\prod_{i=1}^{n} z_i^{k_i}$$

$$+\sum_{i=1}^{n-1} k_i\mu_i h\left(p_i e^{-s\lambda}\frac{z_{i+1}}{z_i}\prod_{i=1}^{n} z_i^{k_i}+(1-p_i)e^{-s\lambda}\frac{1}{z_i}\prod_{i=1}^{n} z_i^{k_i}\right)$$

$$+ k_n\mu_n h e^{-s\lambda}\frac{1}{z_n}\prod_{i=1}^{n} z_i^{k_i} + \lambda h e^{-s\lambda} z_1\prod_{i=1}^{n} z_i^{k_i}$$

$$+\gamma h E\left(e^{-s\Lambda}\right)\prod_{i=1}^{n} z_i^{k_i} + o(h), \quad h\downarrow 0.$$

Straightforward calculations now yield

$$f(s,z,t+h)-f(s,z,t) = h\Big(-\gamma f(s,z,t)+(1-z_1)\frac{\partial}{\partial s}f(s,z,t)$$

$$+\sum_{i=1}^{n-1}\mu_i(p_i z_{i+1}-z_i+1-p_i)\frac{\partial}{\partial z_i}f(s,z,t)$$

$$+\mu_n(1-z_n)\frac{\partial}{\partial z_n}f(s,z,t)+\gamma f(s,1,t)f(0,z,t)\Big)+o(h), \quad h\downarrow 0.$$

Dividing by h and letting $h\downarrow 0$ leads to the PDE [5.1]. □

Let the vector $(\Lambda, X_1,...,X_n)$ be the steady-state version of the time-dependent $(\Lambda(t), X_1(t),...,X_n(t))$ and write $f(s,z) := E(e^{-s\Lambda}\prod_{i=1}^{n} z_i^{X_i})$. The following corollary then immediately follows from Theorem 5.1.

COROLLARY 5.1.–

$$\begin{aligned}-\gamma f(s,z) &+(1-z_1)\tfrac{\partial}{\partial s}f(s,z)+\sum_{i=1}^{n-1}\mu_i(p_i z_{i+1}-z_i+1-p_i)\tfrac{\partial}{\partial z_i}f(s,z)\\ &+\mu_n(1-z_n)\tfrac{\partial}{\partial z_n}f(s,z)+\gamma f(s,1)f(0,z)=0,\end{aligned} \qquad [5.2]$$

with $f(0,1)=1$.

Solving either of the equations [5.1] or [5.2] is no easy task. Both are in the class of semilinear first-order PDEs. The usual solution approach would therefore be the method of characteristics (see, for example, (Salih 2016)). However, the appearance of $f(0, z, t)$ (and $f(0, z)$) also makes it a delay equation. Hence, we cannot use standard techniques to compute an exact solution.

Numerical approaches for solving partial delay differential equations have been considered, though most often for very specific types. Finding a suitable method seems hard, and it remains a problem to solve [5.1] and [5.2] either analytically or numerically.

NOTE 5.1.– Taking $z_1 = ... = z_n = 1$, equation [5.1] becomes $\frac{\partial}{\partial t} f(s, 1, t) = 0$. Therefore, the marginal transform $E\left(e^{-s\Lambda(t)}\right)$ is independent of t – and hence so is the distribution of $\Lambda(t)$. For each point in time t, $\Lambda(t)$ has distribution G_Λ. Of course, considering $\Lambda(t)$ is still valuable for analyzing correlations with $\Lambda(t')$ or $X_i(t')$ for some $t' \in \mathbb{R}$. From now on, when we are not concerned with such correlations, we simply write Λ instead of $\Lambda(t)$.

NOTE 5.2.– Taking $s = 0$ in [5.2] yields

$$(1 - z_1)\frac{\partial}{\partial s} f(s, z)\bigg|_{s=0} + \sum_{i=1}^{n-1} \mu_i(p_i z_{i+1} - z_i + 1 - p_i)\frac{\partial}{\partial z_i} f(0, z) + \mu_n(1 - z_n)\frac{\partial}{\partial z_n} f(0, z) = 0.$$

Although this equality does not seem to have a direct interpretation, it does when we take $z_1 = ... = z_n$ and subsequently take $z = 1$. We then get

$$E(\Lambda) = \sum_{i=1}^{n-1} \mu_i(1 - p_i)E(X_i) + \mu_n E(X_n). \qquad [5.3]$$

with S representing the sojourn time, one could compare this to Little's formula $\lambda = E(X)/E(S)$ for any standard queue. Since the sojourn time equals the service time in our infinite-server setting, $\frac{1}{E(S)}$ is the rate at which a service is completed. Note that

$$\frac{E(X)}{E(S)} = E(X) \sum_{i=1}^{n} P(\text{customer is in phase } i) \times (\text{completion rate in state } i)$$

$$= E(X) \sum_{i=1}^{n-1} \frac{E(X_i)}{E(X)} \cdot \mu_i (1 - p_i) + E(X) \frac{E(X_n)}{E(X)} \cdot \mu_n$$

$$= \sum_{i=1}^{n-1} \mu_i (1 - p_i) E(X_i) + \mu_n E(X_n),$$

so that Little's formula indeed holds with λ replaced by $E(\Lambda)$.

5.2.2. *Calculating moments*

One important feature of a generating function or Laplace–Stieltjes transform is the ability to quickly extract any moments for the corresponding random variable. In this section, we show that the PDE [5.1] provides enough information to do just that. More specifically, it allows us to calculate any moment of $(\Lambda(t), X_1(t), ..., X_n(t))$ by solving a recursion.

Let k denote the vector $(k_1, ..., k_n)^T$. If we would have an expression for $f(s, z, t)$, then the $(l, k)^{\text{th}}$ moment could be calculated by

$$E\left(\Lambda(t)^l \prod_{i=1}^{n} \frac{X_i(t)!}{(X_i(t) - k_i)!}\right) = (-1)^l \left[\frac{d^l}{ds^l} \prod_{i=1}^{n} \frac{d^{k_i}}{dz_i^{k_i}} f(s, z, t)\right]_{s=0, z=1}. \quad [5.4]$$

NOTE 5.3.– Formally the moment should be defined as $E\left(\Lambda(t)^l \prod_{i=1}^{n} X_i(t)^{k_i}\right)$ rather than $E\left(\Lambda(t)^l \prod_{i=1}^{n} \frac{X_i(t)!}{(X_i(t)-k_i)!}\right)$. However, the former can easily be obtained from the latter.

Applying these same differentiations to PDE [5.1] yields, with d_x^m denoting the mth derivative w.r.t. x:

$$\begin{aligned} -\gamma d_s^l d_{z_1}^{k_1} \cdots d_{z_n}^{k_n} f(s, z, t) &- \frac{\partial}{\partial t} d_s^l d_{z_1}^{k_1} \cdots d_{z_n}^{k_n} f(s, z, t) \\ &+ (1 - z_1) d_s^{l+1} d_{z_1}^{k_1} \cdots d_{z_n}^{k_n} f(s, z, t) - k_1 d_s^{l+1} d_{z_1}^{k_1 - 1} d_{z_2}^{k_2} \cdots d_{z_n}^{k_n} f(s, z, t) \\ &+ \sum_{i=1}^{n-1} \mu_i (p_i z_{i+1} - z_i + 1 - p_i) \frac{\partial}{\partial z_i} d_s^l d_{z_1}^{k_1} \cdots d_{z_n}^{k_n} f(s, z, t) \\ &- \sum_{i=1}^{n-1} k_i \mu_i d_s^l d_{z_1}^{k_1} \cdots d_{z_n}^{k_n} f(s, z, t) \end{aligned}$$

$$+\sum_{i=1}^{n-1} k_{i+1}p_i\mu_i d_s^l d_{z_1}^{k_1}\cdots d_{z_{i-1}}^{k_{i-1}} d_{z_i}^{k_i+1} d_{z_{i+1}}^{k_{i+1}-1} d_{z_{i+2}}^{k_{i+2}}\cdots d_{z_n}^{k_n} f(s,z,t)$$

$$+\mu_n(1-z_n)\frac{\partial}{\partial z_n} d_s^l d_{z_1}^{k_1}\cdots d_{z_n}^{k_n} f(s,z,t) - k_n\mu_n d_s^l d_{z_1}^{k_1}\cdots d_{z_n}^{k_n} f(s,z,t)$$

$$+\gamma d_s^l f(s,1,t) d_{z_1}^{k_1}\cdots d_{z_n}^{k_n} f(0,z,t) = 0.$$

Now define $E(l,k_1,...,k_n,t) := E\Big(\Lambda(t)^l \prod_{i=1}^{n} \frac{X_i(t)!}{(X_i(t)-k_i)!}\Big)$ as a shorthand notation for the joint moment. When we take $s=0$, $z=1$ and use [5.4], we obtain

$$\begin{aligned}\frac{d}{dt}E(l,k_1,...,k_n,t) &= -\left(\gamma+\sum_{i=1}^{n}k_i\mu_i\right)E(l,k_1,...,k_n,t)\\ &\quad + k_1 E(l+1,k_1-1,k_2,...,k_n,t)\\ &\quad + \sum_{j=1}^{n-1} p_j\mu_j k_{j+1} E(l,k_1,...,k_{j-1},k_j+1,k_{j+1}\\ &\quad -1,k_{j+2},...,k_n,t)\\ &\quad + \gamma E(l,0,...,0,t)E(0,k_1,...,k_n,t).\end{aligned} \qquad [5.5]$$

Our aim is to show that the recursive formula [5.5] allows one to iteratively compute all moments of $(\Lambda(t), X_1(t), ..., X_n(t))$, assuming all moments of G_Λ are known. Let us first make the assumption that the system is empty at $t=0$. If not, we split $X_j(t)$ into the customers who arrived before or after $t=0$. The contribution to $X_j(t)$ of arrivals before time 0 can be calculated if we know $(X_1(0), ..., X_n(0))$, by finding the probabilities $p_{ij}(t)$ that a customer who was in phase i at time 0 is in phase j at time t. Conditioning on the case that it takes exactly u time to move from phase i to phase j,

$$\begin{aligned}p_{ij}(t) &= p_i\cdots p_{j-1}\int_0^t \sum_{k=i}^{j-1} C_{k,j-1}\mu_k e^{-\mu_k u} e^{-\mu_j(t-u)}du\\ &= p_i\cdots p_{j-1}\sum_{k=i}^{j-1} C_{k,j-1}\frac{\mu_k}{\mu_k-\mu_j}\left(e^{-\mu_j t}-e^{-\mu_k t}\right).\end{aligned}$$

Here, $C_{k,j-1} = \prod_{m\in\{i,...,j-1\}\setminus\{k\}} \frac{\mu_m}{\mu_m-\mu_k}$ and $\sum_{k=i}^{j-1} C_{k,j-1}\mu_k e^{-\mu_k u}$ is the density of a hypoexponential distribution (see (Ross 2010, p. 310)). Recall that a hypoexponential random variable is a sum of independent exponentials (in our case $j-i$ exponentials with rates μ_i,...,μ_{j-1}, respectively). It can also be seen as a Coxian random variable for which all continuation probabilities $p_k=1$.

Let $Y_{ij}(t) \sim \text{Bernoulli}(p_{ij}(t))$ for $i \leq j$, and let $Y_{ij,m}(t)$, $m = 1, 2, \dots$ be i.i.d. copies of $Y_{ij}(t)$. Then the fraction of customers from $X_j(t)$ that arrived before $t = 0$ equals

$$\sum_{i=1}^{j} \sum_{m=1}^{X_i(0)} Y_{ij,m}(t).$$

Now that we have found the fraction of customers from $X_j(t)$ that arrived before $t = 0$, let us assume for the remainder of the section that the system is empty at time 0.

THEOREM 5.2.– All moments of $(\Lambda(t), X_1(t), ..., X_n(t))$ can be computed using [5.5], and have the form

$$E\Big(\Lambda(t)^l \prod_{i=1}^{n} \frac{X_i(t)!}{(X_i(t) - k_i)!}\Big) = \sum_{j=1}^{N} a_j e^{-b_j t}, \qquad [5.6]$$

for some constants $a_1, ..., a_N$ and non-negative constants $N, b_1, ..., b_N$.

PROOF.– We provide an inductive argument. In remark 5.1, we have seen that the distribution of $\Lambda(t)$ is independent of t. In particular, $E(\Lambda^l(t))$ is constant in t, so the statement is true when taking $k_1 = ... = k_n = 0$.

Let us take a look at the structure of the recursion [5.5]. The derivative of the moment with $(l, k_1, ..., k_n)$ equals a linear function consisting of four unknown components: (1) the moment itself, (2) the same moment with $k_1 - 1$ and $l + 1$, (3) the same moment with $k_j - 1$ and $k_{j-1} + 1$ for some $j = 1, ..., n - 1$, and (4) the same moment with $l = 0$.

Let $K = \sum_{i=1}^{n} k_i$. We have shown above that the statement is true for $K = 0$. The inductive step is split into two parts. First, assume we know all moments of order $(l, K - 1)$ and we want to calculate all moments of order $(0, K)$. Note that when substituting $(0, K)$, component 4 coincides with component 1, and component 2 is known by assumption. From $E(1, K - 1, 0, ..., 0, t)$, we can calculate an arbitrary moment $E(0, k_1, ..., k_n, t)$ of order $(0, K)$ with the following scheme:

$$(1, K-1, 0, ..., 0) \to (0, K, 0, ..., 0) \to (0, K-1, 1, 0, ..., 0)$$
$$\to (0, K-2, 2, 0, ..., 0) \to ... \to (0, k_1, K - k_1, 0, ..., 0) \to ... \to (0, k_1, ..., k_n).$$

Note that for each arrow, a differential equation needs to be solved.

For the second part of the induction step, we assume that all moments of order $(l', K-1)$ and $(0, K)$ are known, and want to calculate all moments of order (l, K). In this case, both components 2 and 4 are known by assumption. The arbitrary moment $E(l, k_1, ..., k_n, t)$ of order (l, K) can now be calculated using the following scheme:

$$(l+1, K-1, ..., 0) \to (l, K, 0, ..., 0) \to (l, K-1, 1, 0, ..., 0)$$
$$\to (l, K-2, 2, 0, ..., 0) \to ... \to (l, k_1, K-k_1, 0, ..., 0) \to ... \to (l, k_1, ..., k_n).$$

Combining both schemes, we can iteratively find an expression for any moment $E(l, k_1, ..., k_n, t)$. All that remains is to show that the moments take the form of [5.6]. This will be done by induction on the number of iterations.

Assume all moments calculated thus far take the form of [5.6]. Note that each iteration requires the differential equation [5.5] to be solved. This differential equation has the form

$$\frac{d}{dt}g(t) = -Cg(t) + \sum_{m=1}^{M} \tilde{a}_m g_m(t),$$

where $C, M, \tilde{a}_m$ are constants of which C, M are non-negative and $g_m(t)$ are known functions from previous iterations. Since the $g_m(t)$ have the required form by assumption, it holds that

$$\frac{d}{dt}g(t) = -Cg(t) + \sum_{m=1}^{M} \tilde{a}_m \sum_{j=1}^{N_m} a_{j,m} e^{-b_{j,m}t} = -Cg(t) + \sum_{m=1}^{N'} a'_m e^{-b'_m t}, \qquad [5.7]$$

for some constants C, N', a'_m, b'_m of which C, N', b'_m are non-negative. Solving this differential equation quickly leads to the result of the theorem. □

NOTE 5.4.– It is an interesting question how many different terms a moment consists of, given its order $(l, k_1, ..., k_n)$. In other words, how large is N in [5.6]? To answer this, note that when we solve the differential equation [5.7] the only exponent we have not seen in previous iterations is $-Ct$. So N is at most the number of possible values of C. Checking with [5.5], we conclude that $N = O\left(\left(\sum_{i=1}^{n} k_i\right)^n\right)$.

In terms of computational complexity, observe that [5.5] has $O(n)$ terms, each one consisting of one moment. Since a moment has $O\left(\left(\sum_{i=1}^{n} k_i\right)^n\right)$ terms, one iteration requires $O\left(n \cdot \left(\sum_{i=1}^{n} k_i\right)^n\right)$ operations. Note also that we need one iteration for each moment of lesser order. Therefore, the calculation of a moment of order $(l, k_1, ..., k_n)$ requires $O\left(n \cdot \left(\sum_{i=1}^{n} k_i\right)^{2n}\right)$ operations.

Now that we have seen that all moments can be retrieved, and which form they have, let us calculate some of these moments. This gives an idea how the system behaves in general and presents a quantitative way to compare our queue to other models.

One can check that expressions from higher moments become very large and complicated, requiring much computational effort and making analysis difficult. However, for some lower moments and special cases, we can derive "nicer" expressions. This enables us to give intuitive explanations and perform proper analyses on the formulas we used.

We start with an expression for the mean number of customers.

THEOREM 5.3.– Let $C_{i,j} = \prod_{m \neq i}^{j} \frac{\mu_m}{\mu_m - \mu_i}$. Given that the system is empty at $t = 0$, the mean number of customers in phase j at time t equals

$$E(X_j(t)) = \left(\prod_{i=1}^{j-1} p_i\right) \frac{E(\Lambda)}{\mu_j} \left(1 - \sum_{i=1}^{j} C_{i,j} e^{-\mu_i t}\right), \qquad j = 1, ..., n. \quad [5.8]$$

PROOF.– Let us first consider the case $j = 1$. Substituting $k_1 = 1$ and $l = k_2 = ... = k_n = 0$ gives the differential equation

$$\begin{aligned} \frac{d}{dt} E(X_1(t)) &= -(\gamma + \mu_1) E(X_1(t)) + E(\Lambda(t)) + \gamma E(X_1(t)) \\ &= -\mu_1 E(X_1(t)) + E(\Lambda). \end{aligned}$$

With $E(X_1(0)) = 0$, its solution is

$$E(X_1(t)) = \frac{E(\Lambda)}{\mu_1} \left(1 - e^{-\mu_1 t}\right). \qquad [5.9]$$

Therefore, [5.8] holds for $j = 1$.

Now let $j \in \{2, ..., n\}$. The differential equation corresponding to $k_j = 1$ is

$$\begin{aligned} \frac{d}{dt} E(X_j(t)) &= -(\gamma + \mu_j) E(X_j(t)) + \mu_{j-1} p_{j-1} E(X_{j-1}(t)) + \gamma E(X_j(t)) \\ &= -\mu_j E(X_j(t)) + \mu_{j-1} p_{j-1} E(X_{j-1}(t)). \end{aligned} \qquad [5.10]$$

To show that [5.8] is indeed a solution to this equation, we differentiate [5.8] and rearrange terms, using that $\frac{\mu_j-\mu_i}{\mu_j}C_{i,j} = C_{i,j-1}$:

$$\begin{aligned}\frac{d}{dt}E(X_j(t)) &= \left(\prod_{i=1}^{j-1} p_i\right) E(\Lambda) \sum_{i=1}^{j} \frac{\mu_i}{\mu_j} C_{i,j} e^{-\mu_i t} \\ &= \left(\prod_{i=1}^{j-1} p_i\right) E(\Lambda) \left(\sum_{i=1}^{j-1} \left(1 - \frac{\mu_j - \mu_i}{\mu_j}\right) C_{i,j} e^{-\mu_i t} + C_{j,j} e^{-\mu_j t}\right) \\ &= \left(\prod_{i=1}^{j-1} p_i\right) E(\Lambda) \left(-\left(1 - \sum_{i=1}^{j} C_{i,j} e^{-\mu_i t}\right) + \left(1 - \sum_{i=1}^{j-1} C_{i,j-1} e^{-\mu_i t}\right)\right) \\ &= -\mu_j E(X_j(t)) + \mu_{j-1} p_{j-1} E(X_{j-1}(t)).\end{aligned}$$

To verify that solution [5.8] to differential equation [5.10] is the desired solution, we need to show that [5.8] satisfies the boundary condition $E(X_j(0)) = 0$. So what remains is to check that $\sum_{i=1}^{j} C_{i,j} = 1$. We can do this by observing that

$$\sum_{i=1}^{j} C_{i,j} = \sum_{i=1}^{j} C_{i,j} \cdot \int_0^{\infty} \mu_i e^{-\mu_i t} dt = \int_0^{\infty} \sum_{i=1}^{j} \mu_i C_{i,j} e^{-\mu_i t} dt.$$

As has been proven (Ross 2010, pp. 309–310), the integrand is a density function of a random variable on $[0, \infty)$. It follows that $\sum_{i=1}^{j} C_{i,j} = 1$, which concludes the proof. □

For further moments, it is necessary to find $E\left(\Lambda(t)X_1(t)\right)$ first. So with $l = k_1 = 1$ and $k_2 = ... = k_n = 0$, [5.5] gives

$$\begin{aligned}\frac{d}{dt}E(\Lambda(t)X_1(t)) &= -(\gamma+\mu_1)E\left(\Lambda(t)X_1(t)\right) + E(\Lambda^2) + \gamma E(\Lambda)E(X_1(t)) \\ &= -(\gamma+\mu_1)E\left(\Lambda(t)X_1(t)\right) + \frac{\mu_1 E(\Lambda^2) + \gamma E(\Lambda)^2}{\mu_1} - \frac{\gamma E(\Lambda)^2}{\mu_1} e^{-\mu_1 t}.\end{aligned}$$

Here, we used the $E(X_1(t))$ we found in [5.9]. Combined with the boundary condition $E(\Lambda(0)X_1(0)) = 0$, the solution equals

$$\begin{aligned}E(\Lambda(t)X_1(t)) &= \frac{\mathrm{Var}(\Lambda)}{\gamma+\mu_1} + \frac{E(\Lambda)^2}{\mu_1} - \frac{E(\Lambda)^2}{\mu_1} e^{-\mu_1 t} - \frac{\mathrm{Var}(\Lambda)}{\gamma+\mu_1} e^{-(\gamma+\mu_1)t} \\ &= \frac{E(\Lambda)^2}{\mu_1}\left(1 - e^{-\mu_1 t}\right) + \frac{\mathrm{Var}(\Lambda)}{\gamma+\mu_1}\left(1 - e^{-(\gamma+\mu_1)t}\right).\end{aligned} \quad [5.11]$$

Next, we use the same trick for $E\left(X_1(t)\left(X_1(t)-1\right)\right)$. The differential equation for $k_1 = 2$ and $l = k_2 = \ldots = k_n = 0$ is

$$\frac{d}{dt}E\left(X_1(t)\left(X_1(t)-1\right)\right) = -2\mu_1 E\left(X_1(t)\left(X_1(t)-1\right)\right) + 2E(\Lambda(t)X_1(t)),$$

with $E\left(X_1(0)\left(X_1(0)-1\right)\right) = 0$. This has the solution

$$E\left(X_1(t)\left(X_1(t)-1\right)\right) = \frac{\mu_1 E(\Lambda^2) + \gamma E(\Lambda)^2}{\mu_1^2(\gamma+\mu_1)} - \frac{2E(\Lambda)^2}{\mu_1^2}e^{-\mu_1 t}$$
$$+ \frac{2\text{Var}(\Lambda)}{(\gamma+\mu_1)(\gamma-\mu_1)}e^{-(\gamma+\mu_1)t} + \frac{\gamma E(\Lambda)^2 - \mu_1 E(\Lambda^2)}{\mu_1^2(\gamma-\mu_1)}e^{-2\mu_1 t}, \qquad [5.12]$$

and it follows that

$$\text{Var}(X_1(t)) = \frac{E(\Lambda)}{\mu_1} + \frac{\text{Var}(\Lambda)}{\mu_1(\gamma+\mu_1)} - \frac{E(\Lambda)}{\mu_1}e^{-\mu_1 t}$$
$$+ \frac{2\text{Var}(\Lambda)}{(\gamma+\mu_1)(\gamma-\mu_1)}e^{-(\gamma+\mu_1)t} - \frac{\text{Var}(\Lambda)}{\mu_1(\gamma-\mu_1)}e^{-2\mu_1 t}. \qquad [5.13]$$

It is easily verified that $\gamma = \mu_1$ is not a singularity.

The correlation between $\Lambda(t)$ and $X_1(t)$ follows from [5.11] and [5.13]:

$$\text{Cor}(\Lambda(t), X_1(t)) = \frac{\text{Cov}(\Lambda(t), X_1(t))}{\sqrt{\text{Var}(\Lambda)\text{Var}(X_1(t))}} \qquad [5.14]$$

$$= \frac{\left(1 - e^{-(\gamma+\mu_1)t}\right)\sqrt{D_\Lambda \mu_1(\gamma-\mu_1)}}{\sqrt{(\gamma+\mu_1)^2(\gamma-\mu_1)(1-e^{-\mu_1 t}) + D_\Lambda(\gamma+\mu_1)\left(\gamma - \mu_1 + 2\mu_1 e^{-(\gamma+\mu_1)t} - (\gamma+\mu_1)e^{-2\mu_1 t}\right)}},$$

where $D_\Lambda = \frac{\text{Var}(\Lambda)}{E(\Lambda)}$ is the dispersion index. Interestingly, the arrival process manifests itself only through D_Λ. The positive correlation that the formula indicates makes sense, since correlation between $\Lambda(t)$ and $X_1(t)$ can only occur when $\Lambda(t)$ shows variability.

As it is visible in Figure 5.1, there appears to be a time where the correlation is maximal. This will be the time t_{max} that enough customers could have arrived to show correlation, but a large fraction of $X_1(t_{\text{max}})$ still originates from the current Λ (and not from previous ones).

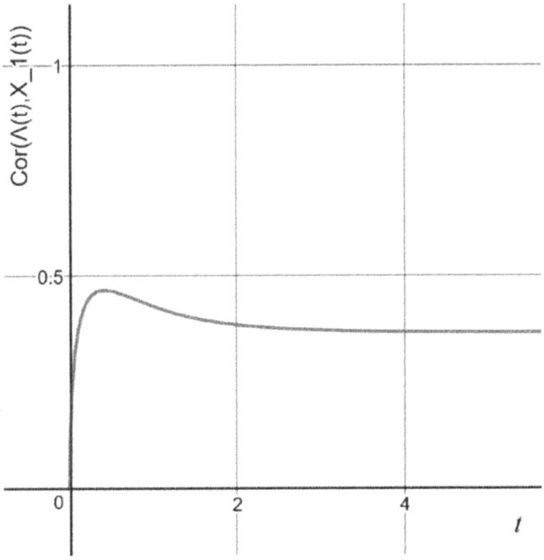

Figure 5.1. $\mathrm{Cor}(\Lambda(t), X_1(t))$ *for* $\mu_1 = 1$, $\gamma = 2$ *and* $D_\Lambda = 2$. *For a color version of this figure, see www.iste.co.uk/anisimov/queueing1.zip*

5.2.3. *Steady state*

In this section, we consider the steady-state behavior of the infinite-server queue, again focusing on a recursion for the joint moments of arrival rate and numbers of customers in phases $1, \ldots, n$. Let $E\left(l, k_1, ..., k_n\right) := E\Big(\Lambda^l \prod\limits_{i=1}^{n} \frac{X_i!}{(X_i-k_i)!}\Big)$. Letting $t \to \infty$ in [5.5] yields the following steady-state recursion:

$$E\left(l, k_1, ..., k_n\right) = \frac{1}{\gamma + \sum\limits_{i=1}^{n} k_i \mu_i} \Bigg(k_1 E\left(l+1, k_1-1, k_2, ..., k_n\right) \tag{5.15}$$

$$+ \sum_{j=1}^{n-1} p_j \mu_j k_{j+1} E\left(l, k_1, ..., k_{j-1}, k_j+1, k_{j+1}-1, k_{j+2}, ..., k_n\right)$$

$$+ \gamma E(l, 0, ..., 0) E\left(0, k_1, ..., k_n\right) \Bigg).$$

Note that an iteration step only requires some basic operations, whereas in the transient case each iteration required a differential equation to be solved.

Let us find the steady-state mean number of customers in phase j. Taking $k_j = 1$ (and everything else 0) yields

$$E(X_j) = \frac{p_{j-1}\mu_{j-1}}{\mu_j} E(X_{j-1}), \quad j = 2, 3, \ldots; \quad E(X_1) = \frac{E(\Lambda)}{\mu_1}.$$

This simple recursion results in

$$E(X_j) = \frac{E(\Lambda)}{\mu_j} \prod_{i=1}^{j-1} p_i, \qquad j = 1, ..., n, \qquad [5.16]$$

which of course we could also have found from [5.8] letting $t \to \infty$.

NOTE 5.5.– Recall that the mean number of customers in an ordinary $M/M/\infty$-queue is $\frac{\lambda}{\mu}$, the arrival rate divided by the departure rate. Formula [5.16] represents something similar: the expected arrival rate over the departure rate in phase j. To see this, note that with $\prod_{i=1}^{j-1} p_i$ being the probability that a customer reaches phase j, $E(\Lambda) \prod_{i=1}^{j-1} p_i$ is the expected arrival rate of phase j.

With the steady-state mean at hand, formula [5.8] has a nice interpretation. We consider the steady-state number of customers who are in phase j (given in [5.16]), and distinguish the fraction that has been in the system for at least t time, and the fraction that has not. Where the steady-state mean consists of both fractions, the transient mean [5.8] only consists of the second (assuming an empty system at time 0). Now note that a customer in phase j has gone through j exponential phases (1 up to $j - 1$ completely and the past part of an exponential phase j). So the second fraction is proportional to the probability that j exponential phases are traversed before time t. According to formula [5.8], this probability should be $\left(1 - \sum_{i=1}^{j} \prod_{m \neq i}^{j} \frac{\mu_m}{\mu_m - \mu_i} e^{-\mu_i t}\right)$. Indeed, this is precisely the distribution function of a hypoexponential random variable.

Another quantity that is greatly simplified by considering steady state is $\text{Cov}(\Lambda, X_j)$. For this, we take $l = k_j = 1$ in [5.15] and find

$$E(\Lambda X_j) = \frac{1}{\gamma + \mu_j} \left(p_{i-1}\mu_{i-1} E(\Lambda X_{j-1}) + \gamma E(\Lambda) E(X_j)\right),$$

such that

$$\begin{aligned}\mathrm{Cov}(\Lambda, X_j) &= \frac{1}{\gamma + \mu_j}(p_{i-1}\mu_{i-1}\mathrm{Cov}(\Lambda, X_{j-1}) + p_{i-1}\mu_{i-1}E(\Lambda)E(X_{j-1}) \\ &- \mu_i E(\Lambda)E(X_j)) \\ &= \frac{p_{i-1}\mu_{i-1}}{\gamma + \mu_j}\mathrm{Cov}(\Lambda, X_{j-1}), \quad j = 2, 3, \ldots,\end{aligned}$$

returns a simple recursion. We also observe that

$$\begin{aligned}&\mathrm{Cov}(\Lambda, X_1) = E(\Lambda X_1) - E(\Lambda)E(X_1) \\ &= \frac{E(\Lambda^2)}{\gamma + \mu_1} + \frac{\gamma}{\gamma + \mu_1}E(\Lambda)E(X_1) - E(\Lambda)E(X_1) = \frac{\mathrm{Var}(\Lambda)}{\gamma + \mu_1},\end{aligned}$$

hence the solution is

$$\mathrm{Cov}(\Lambda, X_j) = \frac{\mathrm{Var}(\Lambda)}{\gamma + \mu_j}\prod_{i=1}^{j-1}\frac{p_i\mu_i}{\gamma + \mu_i}, \qquad j = 1, ..., n. \qquad [5.17]$$

This formula also provides an interesting interpretation. Recall that $\frac{\mu}{\gamma+\mu}$ is the probability that an exponential time with rate μ is shorter than an exponential time with rate γ. Therefore, $\prod_{i=1}^{j-1}\frac{p_i\mu_i}{\gamma+\mu_i}$ is the probability that a customer traverses $j-1$ service phases before the arrival parameter resamples. The appearance of this probability is intuitive, since the current Λ can only influence the current X_j if customers generated from this Λ arrive in phase j before a new Λ is drawn.

Next, we check the variance of the number of customers in phase 2. To do this, we substitute $k_2 = 2$ in [5.15]. In the calculation, we also use [5.17] for $j = 2$ and [5.12] for $t \to \infty$:

$$\begin{aligned}E(X_2(X_2-1)) &= \frac{p_1\mu_1}{\mu_2}E(X_1X_2) \\ &= \frac{p_1\mu_1}{\mu_2(\mu_1+\mu_2)}E(\Lambda X_2) + \frac{p_1^2\mu_1^2}{\mu_2(\mu_1+\mu_2)}E(X_1(X_1-1)) \\ &= \frac{p_1\mu_1}{\mu_2(\mu_1+\mu_2)}\left(\frac{p_1\mu_1\mathrm{Var}(\Lambda)}{(\gamma+\mu_1)(\gamma+\mu_2)} + \frac{p_1E(\Lambda)^2}{\mu_2}\right) \\ &+ \frac{p_1^2\mu_1^2}{\mu_2(\mu_1+\mu_2)}\frac{\mu_1E(\Lambda^2)+\gamma E(\Lambda)^2}{\mu_1^2(\gamma+\mu_1)} \\ &= \frac{p_1^2\mu_1(\gamma+\mu_1+\mu_2)\mathrm{Var}(\Lambda)}{\mu_2(\mu_1+\mu_2)(\gamma+\mu_1)(\gamma+\mu_2)} + \frac{p_1^2E(\Lambda)^2}{\mu_2^2},\end{aligned}$$

so that

$$\mathrm{Var}(X_2) = \frac{p_1^2 \mu_1(\gamma + \mu_1 + \mu_2)\mathrm{Var}(\Lambda)}{\mu_2(\mu_1 + \mu_2)(\gamma + \mu_1)(\gamma + \mu_2)} + \frac{p_1 E(\Lambda)}{\mu_2}. \qquad [5.18]$$

Let us state the variance of the steady-state number of customers in phase 1:

$$\mathrm{Var}(X_1) = \frac{\mathrm{Var}(\Lambda)}{\mu_1(\gamma + \mu_1)} + \frac{E(\Lambda)}{\mu_1} \qquad [5.19]$$

($t \to \infty$ in [5.13]). It is easy to check that $E(X_2) = E(X_1)$ and $\mathrm{Var}(X_2) \leq \mathrm{Var}(X_1)$ when $p_1 = 1$ and $\mu_1 = \mu_2 = \mu$. A numerical result shown next will elaborate on this last remark.

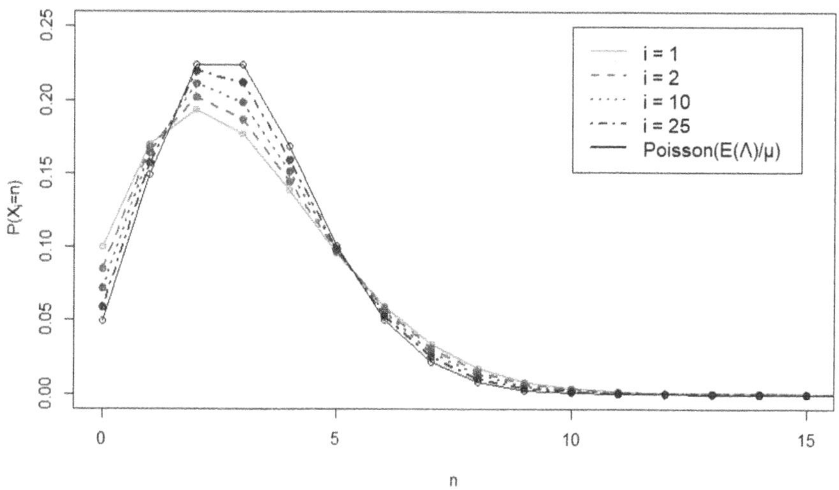

Figure 5.2. *Estimated distribution of X_i for several phases i, with $\Lambda \sim U(0,6)$ and $\gamma = \mu = 1$. For a color version of this figure, see www.iste.co.uk/anisimov/queueing1.zip*

Figure 5.2 visualizes the results of a simulation, where the service time distribution is taken to be Erlang with a large number of phases. The results give an impression of how the distribution of X_i evolves for growing i. Each line represents the distribution of the number of customers in the service phase corresponding to its color. Clearly, higher phases show less variability (while the mean remains unchanged due to $p_1 =$

... $= p_n = 1$ and $\mu_1 = ... = \mu_n = \mu$, see [5.16]). It should also be noted that the distribution of the number of customers in phase i tends to a Poisson distribution for large i. This can be explained by the fact that consecutive customers spread out over different phases. The effect of the mixed arrival rate then slowly decays for increasing phase numbers, and eventually the stream of customers from phase $i-1$ to i barely differs from a Poisson process. Therefore, it is intuitive that X_i converges to a Poisson random variable with mean $\frac{E(\Lambda)}{\mu}$.

Using the variance of X_2, we now calculate the correlation

$$\mathrm{Cor}(\Lambda, X_2) = \frac{\frac{p_1\mu_1 \mathrm{Var}(\Lambda)}{(\gamma+\mu_1)(\gamma+\mu_2)}}{\sqrt{\frac{p_1 \mathrm{Var}(\Lambda)}{\mu_2}\left(E(\Lambda) + \frac{p_1\mu_1(\gamma+\mu_1+\mu_2)\mathrm{Var}(\Lambda)}{(\mu_1+\mu_2)(\gamma+\mu_1)(\gamma+\mu_2)}\right)}}$$

$$= \mu_1\sqrt{\frac{p_1\mu_2(\mu_1+\mu_2)D_\Lambda}{(\gamma+\mu_1)^2(\gamma+\mu_2)^2(\mu_1+\mu_2) + p_1\mu_1(\gamma+\mu_1)(\gamma+\mu_2)(\gamma+\mu_1+\mu_2)D_\Lambda}}.$$

In the special case that $\mu_1 = \mu_2 = \mu$, we have

$$\mathrm{Cor}(\Lambda, X_2) = \frac{\mu}{\gamma+\mu}\sqrt{\frac{2p_1\mu D_\Lambda}{2(\gamma+\mu)^2 + p_1(\gamma+2\mu)D_\Lambda}}.$$

To compare, we let $t \to \infty$ in [5.14] to get $\mathrm{Cor}(\Lambda, X_1)$:

$$\mathrm{Cor}(\Lambda, X_1) = \sqrt{\frac{D_\Lambda \mu_1}{(\gamma+\mu_1)^2 + D_\Lambda(\gamma+\mu_1)}}. \qquad [5.20]$$

In cases where the service rates are equal (i.e. $\mu_1 = \mu_2 = \mu$), it holds that

$$\frac{\mathrm{Cor}(\Lambda, X_1)}{\mathrm{Cor}(\Lambda, X_2)} = \frac{\gamma+\mu}{\mu}\sqrt{\frac{2(\gamma+\mu)^2 + p_1(\gamma+2\mu)D_\Lambda}{2p_1\left((\gamma+\mu)^2 + D_\Lambda(\gamma+\mu)\right)}}$$

$$= \sqrt{\frac{2(\gamma+\mu)^3 + 2p_1\mu^2 D_\Lambda + p_1(\gamma^2+3\gamma\mu)D_\Lambda}{2p_1\mu^2(\gamma+\mu) + 2p_1\mu^2 D_\Lambda}} \geq 1,$$

because $p_1 \leq 1$. An immediate consequence from this calculation is that $\mathrm{Cor}(\Lambda, X_1) \geq \mathrm{Cor}(\Lambda, X_2)$ when the service rates are equal. This is not surprising, since customers "generated" by the current arrival rate Λ immediately arrive in phase 1, but take some time before entering phase 2 (if they remain in the system after phase 1). So earlier phases have a more direct connection with the current arrival rate.

The inequality does not necessarily hold for different service rates. To see this, note that when p_1 is high and μ_1 is high compared to μ_2 and γ, customers spend more time in phase 2 than in phase 1 before the arrival parameter resamples.

Let us finally take a look at $X_1 + X_2$, which is the total number of customers for $n = 2$. Its mean and variance can easily be computed, but its correlation with Λ is less straightforward. To find it, we first have to determine $\text{Cov}(X_1, X_2)$. This is done by taking $k_1 = k_2 = 1$ in [5.15], and using the moments given in [5.17], [5.12] and [5.16]:

$$\begin{aligned}\text{Cov}(X_1, X_2) &= E(X_1X_2) - E(X_1)E(X_2)\\ &= \frac{1}{\mu_1+\mu_2}E(\Lambda X_2) + \frac{p_1\mu_1}{\mu_1+\mu_2}E\left(X_1(X_1-1)\right) - E(X_1)E(X_2)\\ &= \frac{1}{\mu_1+\mu_2}\left(\text{Cov}(\Lambda, X_2) + E(\Lambda)E(X_2)\right)\\ &+ \lim_{t\to\infty}\frac{p_1\mu_1}{\mu_1+\mu_2}E\left(X_1(t)(X_1(t)-1)\right) - E(X_1)E(X_2)\\ &= \frac{p_1(\gamma+\mu_1+\mu_2)\text{Var}(\Lambda)}{(\gamma+\mu_1)(\gamma+\mu_2)(\mu_1+\mu_2)}.\end{aligned} \quad [5.21]$$

Then with [5.17], [5.13], [5.18] and [5.21], we find

$$\text{Cor}(\Lambda, X_1+X_2) = \frac{\text{Cov}(\Lambda, X_1) + \text{Cov}(\Lambda, X_2)}{\sqrt{\text{Var}(\Lambda)\left(\text{Var}(X_1) + \text{Var}(X_2) + 2\text{Cov}(X_1, X_2)\right)}}$$

$$= (\gamma + p_1\mu_1 + \mu_2) \quad [5.22]$$

$$\times\sqrt{\frac{\mu_1\mu_2(\mu_1+\mu_2)D_\Lambda}{(\gamma+\mu_1)(\gamma+\mu_2)\left((p_1\mu_1+\mu_2)(\gamma+\mu_1)(\gamma+\mu_2)(\mu_1+\mu_2) + \mu_1\mu_2\gamma D_\Lambda + (p_1\mu_1+\mu_2)^2(\gamma+\mu_1+\mu_2)D_\Lambda\right)}}.$$

Note that when $p_1 = 0$, customers never reach phase 2. It is easy to check that in that case, it indeed holds that $\text{Cor}(\Lambda, X_1 + X_2) = \text{Cor}(\Lambda, X_1)$.

5.2.4. $M_\Lambda/M/\infty$

In this section, we give a few results for the special Coxian case where the service time distribution is exponential with parameter μ. An easy way to recover the main results is by observing that this service time distribution can be obtained by taking $p_1 = ... = p_{n-1} = 0$ and $\mu_1 = \mu$. We then find

$$-\gamma f(s,z) + (1-z)\frac{\partial}{\partial s}f(s,z) + \mu(1-z)\frac{\partial}{\partial z}f(s,z) + \gamma f(s,1)f(0,z) = 0 \quad [5.23]$$

as the PDE for the joint transform of (Λ, X) and

$$E\left(\Lambda^l \frac{X!}{(X-k)!}\right) = \frac{k}{\gamma + k\mu} E\left(\Lambda^{l+1} \frac{X!}{(X-k+1)!}\right) + \frac{\gamma}{\gamma + k\mu} E\left(\Lambda^l\right) E\left(\frac{X!}{(X-k)!}\right) \quad [5.24]$$

as the recursion for moments. The observant reader may have noticed that all p_i, $i > 1$ are irrelevant when $p_1 = 0$, because customers can only be in phase 1. The extended assumption is made to more easily copy results from Coxian service times.

If $\Lambda \sim \exp(\theta)$, we can derive an expression for the form of any moment.

THEOREM 5.4.– Let $\Lambda \sim \exp(\theta)$. Then the recursion [5.24] has solution

$$E\left(\Lambda^l \frac{X!}{(X-k)!}\right) = \frac{l!}{\theta^{l+k} \prod_{i=1}^{k} (\gamma + i\mu)} \sum_{j=0}^{k} c_j(l,k) \left(\frac{\gamma}{\mu}\right)^{k-j}, \quad [5.25]$$

where $c_0(l,k) = 1$, $c_k(l,k) = \frac{k!(l+k)!}{l!}$ and $c_j(l,k)$ satisfies the recursion

$$c_j(l,k) = k(l+1)c_{j-1}(l+1, k-1) + c_j(1, k-1), \qquad j = 1, ..., k-1. \quad [5.26]$$

For a proof by induction, we refer to (van Kreveld 2018). Let us here only check that [5.26] has a solution. The key observation is that the second argument is lowered in each step of the recursion. Note that when we express $c_j(l,k)$ into coefficients of the form $c_{j'}(l', k-1)$ by using [5.26], we have either $j' = 0$, $j' \in \{1, ..., k-2\}$ or $j' = k-1$. In the first and third case, the value is given, and in the second case we can use [5.26] again. Repeating this procedure $k-1$ times leaves us with only unknown terms with $k = 1$. Since $j \leq k$, the remaining coefficients are of the form $c_0(l', 1)$ and $c_1(l', 1)$. These are both known, so the recursion ends after $k-1$ steps.

As we have seen, the mean number of customers is similar in $M/M/\infty$ and $M_\Lambda/M/\infty$, the only difference being that the fixed arrival rate λ is replaced by the expected arrival rate $E(\Lambda)$. Mixing turns out to have a larger effect on the variance. Without mixing, the variance of X is equal to the mean, but our model has

$$\text{Var}(X) = \frac{E(\Lambda)}{\mu} + \frac{\text{Var}(\Lambda)}{\mu(\gamma + \mu)}, \quad [5.27]$$

as a consequence of [5.13] by taking $\mu_1 = \mu$ and $t \to \infty$. The first term is the variance without mixing, and the second is caused by the variability of the arrival rate. It is therefore no surprise that the second term is linear in $\mathrm{Var}(\Lambda)$.

Heemskerk *et al.* (2017) provide another interesting comparison for the variance. They consider the same model, the only difference being *deterministic* resample intervals with common length Δ. For the variance of the number of customers, Heemskerk *et al.* (2017) obtain

$$\mathrm{Var}(X_H) = \frac{E(\Lambda)}{\mu} + \frac{\left(1 - e^{-\mu\Delta}\right)\mathrm{Var}(\Lambda)}{(1 + e^{-\mu\Delta})\mu^2}, \qquad [5.28]$$

where X_H denotes the steady-state number of customers in this model.

For a fair comparison between [5.27] and [5.28], we take $\gamma = \frac{1}{\Delta}$, such that the mean resample interval lengths of both models agree. It clearly holds that $\mathrm{Var}(X) > \mathrm{Var}(X_H)$ if and only if $\frac{1}{\gamma+\mu} > \frac{1-e^{-\mu\Delta}}{\mu(1+e^{-\mu\Delta})}$. Hence

$$\mathrm{Var}(X) > \mathrm{Var}(X_H) \quad \text{if and only if} \quad 1 + 2\mu\Delta > e^{\mu\Delta} \quad (\text{so } \mu\Delta < 1.25.....). \qquad [5.29]$$

One might expect that randomness of resample interval lengths leads to more variability in X. Surprisingly, this result shows that is not always true.

For a good image of the distribution of X, we use a simple simulation, of which the details can be found in (van Kreveld 2018). Keeping track of the number of customers in the $M_\Lambda/M/\infty$, the results for simulating up to $t = 10^6$ can be seen in Figure 5.3.

Note that with $\Lambda \sim U(0, 6)$ and $\gamma = \mu = 1$, we have $E(X) = 3$ and $\mathrm{Var}(X) = 4.5$. On a single run, the simulation rarely returns an error of more than 0.01 for each quantity.

To investigate the role of the resample rate γ for the distribution of X, we make a plot for several values of γ (Figure 5.4). Although the mean number of customers is the same for each value, we see that the variance decreases when γ increases. This observation is in line with formula [5.27].

A particularly noteworthy case is $\gamma \to \infty$. It then follows from [5.27] that $\mathrm{Var}(X) \to E(X)$. In fact, looking at Figure 5.4 we may even suspect that X converges to a Poisson random variable. This is intuitively explained as follows: when the number of resamples between two arrivals grows to infinity, the arrival process behaves like a Poisson process with rate $E(\Lambda)$.

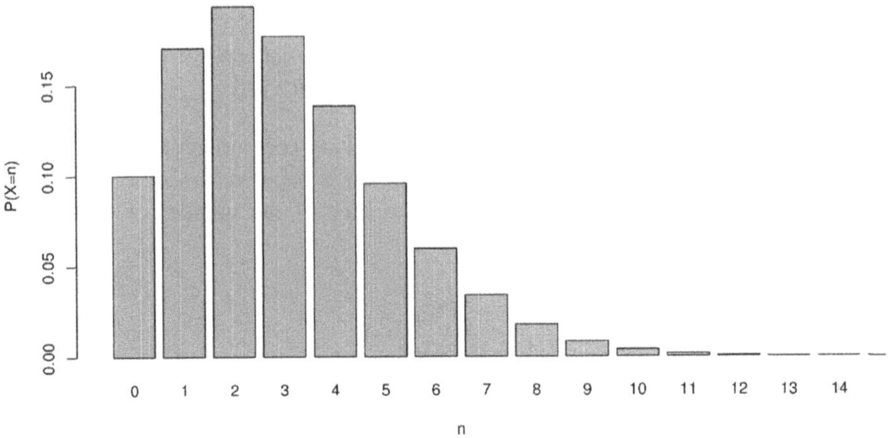

Figure 5.3. *Histogram estimating the distribution of X, with $\Lambda \sim U(0, 6)$ and $\gamma = \mu = 1$*

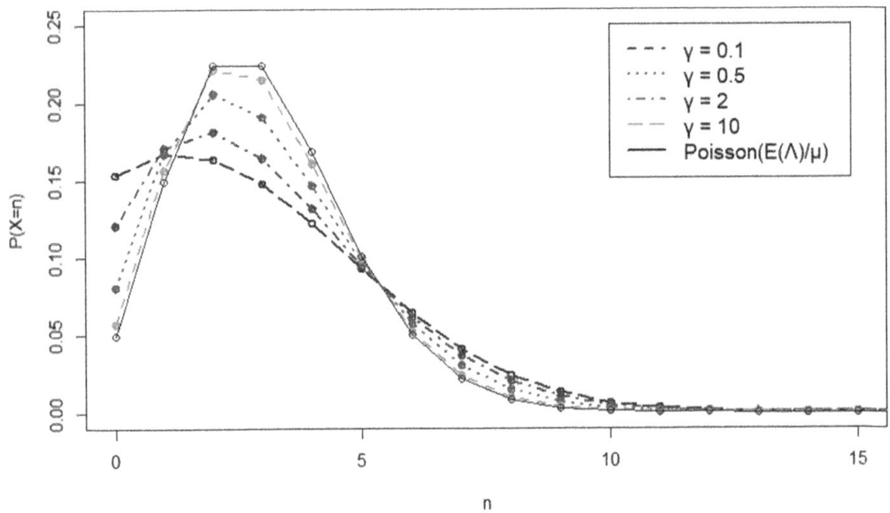

Figure 5.4. *Estimated distribution of X for different values of γ, with $\Lambda \sim U(0, 6)$ and $\mu = 1$. For a color version of this figure, see www.iste.co.uk/anisimov/queueing1.zip*

The last quantity we analyze regarding the steady-state vector (Λ, X) is the correlation between Λ and X. It is given by [5.20], replacing μ_1 by μ as is explained at the start of this section.

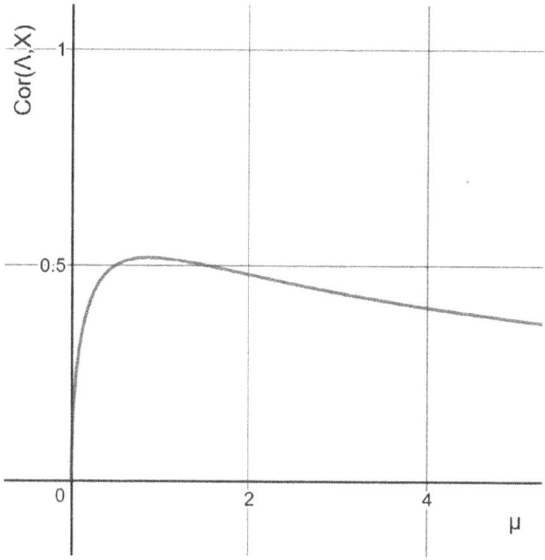

Figure 5.5. $\mathrm{Cor}(X, \Lambda)$ *for* $D_\Lambda = 1$ *and* $\gamma = 0.5$. *For a color version of this figure, see www.iste.co.uk/anisimov/queueing1.zip*

In Figure 5.5, we see that the correlation is low both when the service rate is very low or very high. This can be explained as follows: for the correlation to be high, on the one hand the service rate should not be too high since we want customers generated by the current Λ to stay as long as possible. On the other hand, the service rate should not be so low that "old" customers from previously sampled Λ stay in the system for too long. The correlation turns out to be maximal for $\mu = \sqrt{\gamma^2 + \frac{\gamma}{D_\Lambda}}$, attaining the value

$$\mathrm{Cor}(X, \Lambda) = \sqrt{\frac{\sqrt{\frac{\gamma(1+D_\Lambda\gamma)}{D_\Lambda}}}{D_\Lambda\left(2\gamma^2 + \frac{\gamma}{D_\Lambda} + 2\gamma\sqrt{\gamma^2 + \frac{\gamma}{D_\Lambda}}\right) + \left(\gamma + \sqrt{\gamma^2 + \frac{\gamma}{D_\Lambda}}\right)}}$$

$$= \sqrt{\frac{1}{2\sqrt{D_\Lambda\gamma(1 + D_\Lambda\gamma)} + 2D_\Lambda\gamma + 1}}.$$

Notice its dependence only on $D_\Lambda\gamma$.

We conclude this section with a brief discussion of scaling limits. We consider a scaling of the arrival rate $\Lambda(t)$ with a factor N together with a scaling of the resample rate γ with factor N^α. The value of α determines which rate speeds up faster. If $\alpha > 1$, the arrival rate changes very often, while if $\alpha < 1$, the arrival rate rarely changes relative to the number of arrivals. We check the behavior of the $M_\Lambda/M/\infty$ under this scaling and discuss the effect of the value of α. We follow the idea of Heemskerk *et al.* (2017, pp. 8–12), in which they consider a model with the arrival parameter resampling after a fixed time Δ.

Let $X^{(N)}$ be the steady-state number of customers under the corresponding scaling. Then we easily see that

$$E\left(X^{(N)}\right) = \frac{NE(\Lambda)}{\mu},$$

and from [5.27] that

$$\begin{aligned}\operatorname{Var}\left(X^{(N)}\right) &= \frac{NE(\Lambda)}{\mu} + \frac{N^2\operatorname{Var}(\Lambda)}{\mu\left(N^\alpha\gamma + \mu\right)}\\ &\sim \frac{NE(\Lambda)}{\mu} + \frac{N^{2-\alpha}\operatorname{Var}(\Lambda)}{\mu\gamma}\end{aligned} \qquad [5.30]$$

when N tends to infinity. This can be compared to the asymptotic variance of Heemskerk *et al.* (2017). Denoting by $X_H^{(N)}$ the number of customers in their model, they find

$$\operatorname{Var}\left(X_H^{(N)}\right) \sim \frac{NE(\Lambda)}{\mu} + \frac{\Delta N^{2-\alpha}\operatorname{Var}(\Lambda)}{2\mu}. \qquad [5.31]$$

In both models, the variance increases as the average length of a resample interval ($\frac{1}{\gamma}$ and Δ) increases. However, for the variances to be equal, it must be that $\frac{1}{\gamma} = \frac{\Delta}{2}$. In other words, our more random model needs twice as many resamples to obtain the same asymptotic variance.

From [5.30], the limiting behavior of the variance is

$$\operatorname{Var}\left(X^{(N)}\right) \sim \frac{NE(\Lambda)}{\mu}\mathbf{1}_{\{\alpha\geq 1\}} + \frac{N^{2-\alpha}\operatorname{Var}(\Lambda)}{\mu\gamma}\mathbf{1}_{\{\alpha\leq 1\}}. \qquad [5.32]$$

Let $\beta = \max\{1, 2-\alpha\}$. The asymptotic behavior suggests the following central limit theorem.

CONJECTURE 5.1.– Assume all moments of Λ are finite. If $N \to \infty$, then

$$\frac{X^{(N)} - E\left(X^{(N)}\right)}{\sqrt{N^\gamma}} \to N(0, \sigma^2), \qquad [5.33]$$

with $\sigma^2 = \frac{E(\Lambda)}{\mu}\mathbf{1}_{\{\alpha\geq 1\}} + \frac{\operatorname{Var}(\Lambda)}{\mu\gamma}\mathbf{1}_{\{\alpha\leq 1\}}$.

The statement of the conjecture is very similar to theorem 2.1 from Heemskerk *et al.* (2017). However, in the proof they use they remark that the resample interval lengths are constant. This property is critical in the proof, making it hard to prove our conjecture using the same idea. Nonetheless, it is expected that the statement holds in our model as well.

5.3. Mixing in Markov-modulated infinite-server queues

In this section, we consider a Markov-modulated infinite-server queue. The Markov background process moves through a finite number of states $\mathcal{N} = \{1, ..., n\}$; be aware that this n has nothing to do with the number of phases in the Coxian distribution considered in the previous section. While the process is in state $i \in \mathcal{N}$, it takes an exponential amount of time with parameter γ_i before the state changes. When it does, it has probability p_{ij} to move to state j. Of course, $p_{ij} \geq 0$ for each $(i, j) \in \mathcal{N} \times \mathcal{N}$ and $\sum_{j=1}^{n} p_{ij} = 1$ for each $i \in \mathcal{N}$. Note that the Markov process behaves independently of the queue and acts only as a generator.

In state $i \in \mathcal{N}$, customers arrive according to a Poisson process with random rate Λ_i, drawn according to a probability density function $g_{\Lambda_i}(\cdot)$. This rate is drawn at the start of the current state, and resampled once the state changes again; the arrival parameter also resamples if the Markov process moves to the *same* state. B_i is the service time of a customer arriving when the background process is in state i.

The difference with other Markov-modulated queueing models lies within the random parameter Λ_i. In standard Markov-modulated queues, the interarrival times have a fixed distribution depending on i; most commonly an exponential distribution with rate λ_i. In our case, when we apply mixing, the interarrival time distribution is exponential with a *random* rate Λ_i.

NOTE 5.6.– Observe that the workload of a customer is only dependent on the state in which it *enters* the system, not necessarily on the current state. In case the state of the background process changes before the customer ends its service, service continues with the service time drawn from the previous state.

In this section, we present the idea of Blom *et al.* (2014) applied to our concept of mixing the arrival parameter. First, we develop a differential equation that shows similarities with [5.1] from the $M_\Lambda/\text{Cox}_n/\infty$ queue. Like in the previous section, we indicate how this differential equation can be used to obtain queue length moments.

5.3.1. *The differential equation*

We follow a similar approach as in Blom *et al.* (2014, section 3). Let $S(t) \in \mathcal{N}$ be the state of the Markovian background process at time t. Also, let $\bar{X}_{i,\lambda}(t)$ be the number of customers at time t given $X(0) = 0$, $S(0) = i$ and $\Lambda(0) = \lambda$. Assuming an empty system at time zero is, for the same reason as in section 5.2.2, not very restrictive. Say for example that there are m customers at time zero. Each of those can be characterized by the state it arrived in and its residual service time. With this information, we can calculate the probabilities $q_1, ..., q_m$ that the corresponding customer is still in the system at time t. Let $Z_j(t) \sim \text{Bernoulli}(q_j)$ for $j = 1, ..., m$. Then the total number of customers at time t equals $\bar{X}_{i,\lambda}(t) + \sum_{j=1}^{m} Z_j(t)$.

Consider the small time interval $(0, h)$. Define $f_{i,\lambda}(z,t) := E(z^{\bar{X}_{i,\lambda}(t)})$. To remove the condition $\Lambda(0) = \lambda$, we define $\bar{X}_i(t)$ as the number of customers at time t given $X(0) = 0$ and $S(0) = i$. Moreover, let $f_i(z,t) = E(z^{\bar{X}_i(t)}) = \int_0^\infty g_{\Lambda_i}(\lambda) E(z^{\bar{X}_{i,\lambda}(t)}) d\lambda$ be the corresponding generating function. It is easily seen that

$$f_{i,\lambda}(z,t) = \Big(1 + \lambda h(z-1)P(B_i \geq t)\Big) \times \left(h\gamma_i \sum_{j=1}^{n} p_{ij} f_j(z, t-h) + (1 - h\gamma_i) f_{i,\lambda}(z, t-h)\right) + o(h),$$

which yields the differential equation

$$\gamma_i \sum_{j=1}^{n} p_{ij} f_j(z,t) - \gamma_i f_{i,\lambda}(z,t) - \frac{\partial}{\partial t} f_{i,\lambda}(z,t) + \lambda(z-1)P(B_i \geq t) f_{i,\lambda}(z,t) = 0. \quad [5.34]$$

Removing the condition $\Lambda(0) = \lambda$ then gives

$$\gamma_i \sum_{j=1}^{n} p_{ij} f_j(z,t) - \gamma_i f_i(z,t) - \frac{\partial}{\partial t} f_i(z,t) + (z-1)P(B_i \geq t) E\Big(\Lambda_i f_{i,\Lambda_i}(z,t)\Big) = 0. \quad [5.35]$$

This last formula has some interesting consequences. For example, by letting $t \to \infty$, we find the steady-state formula

$$E(z^{\bar{X}_i(\infty)}) = \sum_{j=1}^{n} p_{ij} E(z^{\bar{X}_j(\infty)}).$$

Note that after an infinite amount of time, the state of the background process at $t = 0$ should be irrelevant. Since $\sum_{j=1}^{n} p_{ij} = 1$, the above equality indeed holds.

Another interesting situation occurs when $n = 1$, so that the system is always in state 1. Note that this is a generalization of our previous $M_\Lambda/\text{Cox}_n/\infty$ model, the service times now having a fully general distribution. Equation [5.35] in this case becomes

$$\frac{\partial}{\partial t} f_1(z,t) = (z-1)P(B_1 \geq t)E\Big(\Lambda_1 f_{1,\Lambda_1}(z,t)\Big).$$

Also taking the derivative with respect to z and taking $z = 1$ gives

$$\frac{d}{dt} E(\bar{X}_1(t)) = P(B_1 \geq t)E(\Lambda_1),$$

which, with the condition that $E(\bar{X}_1(0)) = 0$, yields

$$E(\bar{X}_1(t)) = E(\Lambda_1) \int_0^t P(B_1 \geq u)du.$$

Taking $B_1 \sim \exp(\mu_1)$ here, for example, gives

$$E(\bar{X}_1(t)) = \frac{E(\Lambda_1)}{\mu_1}\left(1 - e^{-\mu_1 t}\right). \qquad [5.36]$$

This formula of course agrees with formula [5.9] for the mean number of customers in phase 1 of the $M_\Lambda/\text{Cox}_n/\infty$ queue. When only considering customers in the first phase of a Coxian distribution, successive phases are irrelevant since there are infinitely many servers. Therefore, both models describe the $M_\Lambda/M/\infty$, and $\bar{X}_1(t) \stackrel{d}{=} X_1(t)$ when the background process has only one state and $B_1 \sim \exp(\mu_1)$.

5.3.2. *Calculating moments*

In the $M_\Lambda/\text{Cox}_n/\infty$ from section 5.2, we saw that despite being unable to find the generating function, we could use the differential equation to compute all moments. We will do the same here with the Markov-modulated queue, both with and without the initial condition on $\Lambda(0)$.

To obtain factorial moments, we take from [5.34] the kth derivative with respect to z in $z = 1$:

$$\frac{d}{dt}E\left(\frac{\bar{X}_{i,\lambda}(t)!}{(\bar{X}_{i,\lambda}(t)-k)!}\right) = -\gamma_i E\left(\frac{\bar{X}_{i,\lambda}(t)!}{(\bar{X}_{i,\lambda}(t)-k)!}\right)$$
$$+\gamma_i \sum_{j=1}^{n} p_{ij} E\left(\frac{\bar{X}_j(t)!}{(\bar{X}_j(t)-k)!}\right)$$
$$+k\lambda P\left(B_i \geq t\right) E\left(\frac{\bar{X}_{i,\lambda}(t)!}{(\bar{X}_{i,\lambda}(t)-k+1)!}\right). \qquad [5.37]$$

Doing the same with [5.35] gives

$$\frac{d}{dt}E\left(\frac{\bar{X}_{i}(t)!}{(\bar{X}_{i}(t)-k)!}\right) = -\gamma_i E\left(\frac{\bar{X}_{i}(t)!}{(\bar{X}_{i}(t)-k)!}\right)$$
$$+\gamma_i \sum_{j=1}^{n} p_{ij} E\left(\frac{\bar{X}_j(t)!}{(\bar{X}_j(t)-k)!}\right)$$
$$+kP\left(B_i \geq t\right) \int_0^\infty g_{\Lambda_i}(\lambda)\lambda E\left(\frac{\bar{X}_{i,\lambda}(t)!}{(\bar{X}_{i,\lambda}(t)-k+1)!}\right) d\lambda. \qquad [5.38]$$

THEOREM 5.5.– All moments $E\left(\frac{\bar{X}_{i,\lambda}(t)!}{(\bar{X}_{i,\lambda}(t)-k)!}\right)$ and $E\left(\frac{\bar{X}_{i}(t)!}{(\bar{X}_{i}(t)-k)!}\right)$ can be iteratively computed for all $k \in \mathbb{N}$.

PROOF.– Note that equation [5.38] can be seen as a system of linear differential equations

$$\frac{d}{dt}\overrightarrow{E(k,t)} = A\overrightarrow{E(k,t)} + \overrightarrow{c(t)}, \qquad [5.39]$$

where

- $\overrightarrow{E(k,t)}$ is the n-dimensional vector consisting of $E\left(\frac{\bar{X}_{1}(t)!}{(\bar{X}_{1}(t)-k)!}\right)$ up to $E\left(\frac{\bar{X}_{n}(t)!}{(\bar{X}_{n}(t)-k)!}\right)$;
- A is a matrix with entries $a_{ij} = \gamma_i p_{ij}$ for $i \neq j$ and $a_{ii} = \gamma_i(p_{ii} - 1)$;

- $c(t)_i = kP(B_i \geq t) \int\limits_0^{\infty} g_{\Lambda_i}(\lambda)\lambda E\left(\frac{\bar{X}_{i,\lambda}(t)!}{(\bar{X}_{i,\lambda}(t)-k+1)!}\right) d\lambda.$

In the same way, [5.37] can be viewed as n separate differential equations

$$\frac{d}{dt}E\left(\frac{\bar{X}_{i,\lambda}(t)!}{(\bar{X}_{i,\lambda}(t)-k)!}\right) = -\gamma_i E\left(\frac{\bar{X}_{i,\lambda}(t)!}{(\bar{X}_{i,\lambda}(t)-k)!}\right) + c(\lambda,t)_i, \quad i = 1,...,n, \qquad [5.40]$$

with

$$c(\lambda,t)_i = \gamma_i \sum_{j=1}^{n} p_{ij} E\left(\frac{\bar{X}_j(t)!}{(\bar{X}_j(t)-k)!}\right) + k\lambda P(B_i \geq t)\, E\left(\frac{\bar{X}_{i,\lambda}(t)!}{(\bar{X}_{i,\lambda}(t)-k+1)!}\right).$$

The main idea is that when we solve the equations in the right order, the $c(t)_i$ and $c(\lambda,t)_i$ are known from previous iterations. If these quantities are known, we can find any moment by solving equations [5.39] and [5.40] via ordinary differential equation methods.

We now proceed to show a sufficient calculation order to complete the proof, by induction. For $k = 1$, we have $c(t)_i = P(B_i \geq t)E(\Lambda_i)$, enabling us to calculate $E(\bar{X}_i(t))$ for $i = 1,...,n$ with [5.39]. Now, $c(\lambda,t)_i = \gamma_i \sum\limits_{j=1}^{n} p_{ij} E(\bar{X}_j(t)) + P(B_i \geq t)\lambda$ is known such that with [5.40] we can also find $E(\bar{X}_{i,\lambda}(t))$.

Suppose we know all moments of order $k-1$. Then $c(t)_i$ can be found by assumption, so we can solve the system [5.39] to find $E\left(\frac{\bar{X}_i(t)!}{(\bar{X}_i(t)-k)!}\right)$ for $i = 1,...,n$. With these quantities found and the induction hypothesis, note that now also $c(\lambda,t)_i$ are known. By solving [5.40], the proof is complete. □

The next step is to calculate some specific moments with equations [5.37] and [5.38]. Specifying the service time distributions allows for expressions without integrals, so let us assume from now that $B_i \sim \exp(\mu_i)$ for $i = 1,...,n$. The exponential distribution is a natural choice and gives simple expressions.

Note that we already found $E(\bar{X}_1(t))$ for $n = 1$ in [5.36]. We proceed with finding the mean, conditioned on $\Lambda(0) = \lambda$, and the variance for $n = 1$. After that we move on to $n = 2$.

With $k = 1$ and [5.36], differential equation [5.37] reads

$$\begin{aligned}\frac{d}{dt}E(\bar{X}_{1,\lambda}(t)) &= -\gamma_1 E(\bar{X}_{1,\lambda}(t)) + \gamma_1 E(\bar{X}_1(t)) + \lambda P(B_1 \geq t)\\ &= -\gamma_1 E(\bar{X}_{1,\lambda}(t)) + \frac{\gamma_1 E(\Lambda_1)}{\mu_1} + \left(\lambda - \frac{\gamma_1 E(\Lambda_1)}{\mu_1}\right)e^{-\mu_1 t}.\end{aligned}$$

One can easily check that, with boundary condition $E\left(\bar{X}_{1,\lambda}(0)\right) = 0$,

$$E\left(\bar{X}_{1,\lambda}(t)\right) = \frac{E(\Lambda_1)}{\mu_1} + \frac{\mu_1\lambda - \gamma_1 E(\Lambda_1)}{\mu_1(\gamma_1 - \mu_1)} e^{-\mu_1 t} - \frac{\lambda - E(\Lambda_1)}{\gamma_1 - \mu_1} e^{-\gamma_1 t}.$$

Some interpretation of this formula can be obtained when we write it as

$$E\left(\bar{X}_{1,\lambda}(t)\right) = \frac{E(\Lambda_1)}{\mu_1}\left(1 - e^{-\mu_1 t}\right) + \frac{e^{-\mu_1 t} - e^{-\gamma_1 t}}{\gamma_1 - \mu_1}(\lambda - E(\Lambda_1)). \qquad [5.41]$$

We recognize the first term as the formula for $E\left(\bar{X}_1(t)\right)$. The second term is a non-negative number times the difference between the conditioned starting arrival parameter λ and its expectation $E(\Lambda_1)$. In other words, the difference in the mean queue size conditioned on $\Lambda(0) = \lambda$ or $\Lambda(0) = \Lambda_1$, is linear in $\lambda - E(\Lambda_1)$. It is evident that when $t > 0$, $E\left(\bar{X}_{1,\lambda}(t)\right) \geq E\left(\bar{X}_1(t)\right)$ if and only if $\lambda \geq E(\Lambda_1)$, with equality if and only $\lambda = E(\Lambda_1)$.

It also follows from [5.41] that the difference between $E\left(\bar{X}_{1,\lambda}(t)\right)$ and $E\left(\bar{X}_1(t)\right)$ is largest at $t = \frac{\ln(\gamma_1) - \ln(\mu_1)}{\gamma_1 - \mu_1}$. The existence of a maximum is intuitive: at $t = 0$, the system still has to set up, and when $t \to \infty$, the value of $\Lambda(0)$ has become irrelevant. In both of these cases, it holds that $E\left(\bar{X}_{1,\lambda}(t)\right) = E\left(\bar{X}_1(t)\right)$.

Formula [5.41] also holds for the $M_\Lambda/\text{Cox}_n/\infty$ setting, in the sense that $E(X_1(t)|\Lambda(0) = \lambda) = E\left(\bar{X}_{1,\lambda}(t)\right)$ (see the end of section 5.3.1). The approach of section 5.2 does not enable us to find moments conditioned on $\Lambda(0)$, so this is an interesting observation.

Moving on to moments of second order, we take $k = 2$ in [5.38] and find

$$\frac{d}{dt}E\left(\bar{X}_1(t)\left(\bar{X}_1(t) - 1\right)\right) = 2P\left(B_1 \geq t\right)\int_0^\infty g_{\Lambda_i}(\lambda)\lambda E\left(\bar{X}_{1,\lambda}(t)\right)d\lambda$$

$$= 2e^{-\mu_1 t}\int_0^\infty g_{\Lambda_1}(\lambda)\lambda\left(\frac{E(\Lambda_1)}{\mu_1} + \frac{\mu_1\lambda - \gamma_1 E(\Lambda_1)}{\mu_1(\gamma_1 - \mu_1)}e^{-\mu_1 t} - \frac{\lambda - E(\Lambda_1)}{\gamma_1 - \mu_1}e^{-\gamma_1 t}\right)d\lambda$$

$$= \frac{2E(\Lambda_1)^2}{\mu_1}e^{-\mu_1 t} + \frac{2\mu_1 E(\Lambda_1^2) - 2\gamma_1 E(\Lambda_1)^2}{\mu_1(\gamma_1 - \mu_1)}e^{-2\mu_1 t} - \frac{2\text{Var}(\Lambda_1)}{\gamma_1 - \mu_1}e^{-(\gamma_1 + \mu_1)t}.$$

The only solution with $E\left(\bar{X}_1(0)\left(\bar{X}_1(0)-1\right)\right)=0$ is

$$E\big(\bar{X}_1(t)\left(\bar{X}_1(t)-1\right)\big)=\frac{\mu_1 E(\Lambda_1^2)+\gamma_1 E(\Lambda_1)^2}{\mu_1^2(\gamma_1+\mu_1)}-\frac{2E(\Lambda_1)^2}{\mu_1^2}e^{-\mu_1 t}$$
$$+\frac{\gamma_1 E(\Lambda_1)^2-\mu_1 E(\Lambda_1^2)}{\mu_1^2(\gamma_1-\mu_1)}e^{-2\mu_1 t}+\frac{2\mathrm{Var}(\Lambda_1)}{(\gamma_1-\mu_1)(\gamma_1+\mu_1)}e^{-(\gamma_1+\mu_1)t}, \qquad [5.42]$$

and hence

$$\mathrm{Var}(\bar{X}_1(t))=\frac{E(\Lambda_1)}{\mu_1}+\frac{\mathrm{Var}(\Lambda_1)}{\mu_1(\gamma_1+\mu_1)}-\frac{E(\Lambda_1)}{\mu_1}e^{-\mu_1 t}$$
$$-\frac{\mathrm{Var}(\Lambda_1)}{\mu_1(\gamma_1-\mu_1)}e^{-2\mu_1 t}+\frac{2\mathrm{Var}(\Lambda_1)}{(\gamma_1-\mu_1)(\gamma_1+\mu_1)}e^{-(\gamma_1+\mu_1)t}. \qquad [5.43]$$

This is in agreement with [5.13]; see also the end of section 5.3.1.

We proceed by studying the variance when the starting arrival rate is fixed, i.e. $\Lambda(0)=\lambda$. Using formula [5.38] for $k=2$ yields

$$\frac{d}{dt}E\big(\bar{X}_{1,\lambda}(t)(\bar{X}_{1,\lambda}(t)-1)\big)=-\gamma_1 E\left(\bar{X}_{1,\lambda}(t)\left(\bar{X}_{1,\lambda}(t)-1\right)\right)$$
$$+\gamma_1 E\big(\bar{X}_1(t)\left(\bar{X}_1(t)-1\right)\big)+2\lambda e^{-\mu_1 t}E\left(\bar{X}_{1,\lambda}(t)\right).$$

Observe that the latter two terms are known from [5.42] and [5.41], hence we have a solvable linear differential equation. As usual, we set the boundary condition at $E\big(\bar{X}_{1,\lambda}(0)(\bar{X}_{1,\lambda}(0)-1)\big)=0$, leading to $E\big(\bar{X}_{1,\lambda}(t)(\bar{X}_{1,\lambda}(t)-1)\big)$ and subsequently to

$$\mathrm{Var}\big(\bar{X}_{1,\lambda}(t)\big)=E\big(\bar{X}_{1,\lambda}(t)\big)+\frac{\mathrm{Var}(\Lambda_1)}{\mu_1(\gamma_1+\mu_1)}$$
$$+\left(\frac{2\mathrm{Var}(\Lambda_1)}{\mu_1(\gamma_1-2\mu_1)}-\frac{2\left(\lambda-E(\Lambda_1)\right)^2}{\mu_1(\gamma_1-2\mu_1)}\right)e^{-\gamma_1 t}$$
$$+\left(-\frac{\gamma_1\mathrm{Var}(\Lambda_1)}{\mu_1(\gamma_1-\mu_1)(\gamma_1-2\mu_1)}+\frac{\gamma_1\left(\lambda-E(\Lambda_1)\right)^2}{(\gamma_1-\mu_1)^2(\gamma_1-2\mu_1)}\right)e^{-2\mu_1 t}$$
$$+\left(-\frac{2\gamma_1\mathrm{Var}(\Lambda_1)}{\mu_1(\gamma_1-\mu_1)(\gamma_1+\mu_1)}+\frac{2\gamma_1\left(\lambda-E(\Lambda_1)\right)^2}{\mu_1(\gamma_1-\mu_1)^2}\right)e^{-(\gamma_1+\mu_1)t}$$
$$-\frac{\left(\lambda-E(\Lambda_1)\right)^2}{(\gamma_1-\mu_1)^2}e^{-2\gamma_1 t}. \qquad [5.44]$$

Note here that the initial arrival intensity only appears as $\lambda - E(\Lambda_1)$, and that $\gamma = \mu_1$ and $\gamma = 2\mu_1$ are removable singularities.

So far, we have only obtained explicit expressions for moments in the Markov-modulated queue with the background process having only one state. In order to better analyze the effect of the background process on the queue, we now consider an example with $n = 2$. This should give an impression of the behavior of the Markov-modulated queue with multiple states.

For $n = 2$, $k = 1$, [5.39] takes the form

$$\frac{d}{dt}\begin{pmatrix} E\left(\bar{X}_1(t)\right) \\ E\left(\bar{X}_2(t)\right) \end{pmatrix} = \begin{pmatrix} -\gamma_1 p_{12} & \gamma_1 p_{12} \\ \gamma_2 p_{21} & -\gamma_2 p_{21} \end{pmatrix} \begin{pmatrix} E\left(\bar{X}_1(t)\right) \\ E\left(\bar{X}_2(t)\right) \end{pmatrix} + \begin{pmatrix} E(\Lambda_1)e^{-\mu_1 t} \\ E(\Lambda_2)e^{-\mu_2 t} \end{pmatrix},$$

since $p_{11} - 1 = -p_{12}$ and $p_{22} - 1 = -p_{21}$. We can solve this system with the eigenvalue method (see, for example, (Shen 2013)). Define $\pi := \gamma_1 p_{12} + \gamma_2 p_{21}$ as the sum of the state transition rates. It is easily seen that the eigenvectors of the system are $\begin{pmatrix} 1 \\ 1 \end{pmatrix}$ and $\begin{pmatrix} \gamma_1 p_{12} \\ -\gamma_2 p_{21} \end{pmatrix}$ with eigenvalues 0 and $-\pi$, respectively. Therefore, the solution is

$$\begin{pmatrix} E\left(\bar{X}_1(t)\right) \\ E\left(\bar{X}_2(t)\right) \end{pmatrix} = \begin{pmatrix} 1 & \gamma_1 p_{12} \\ 1 & -\gamma_2 p_{21} \end{pmatrix} \begin{pmatrix} y_1(t) \\ y_2(t) \end{pmatrix}$$

with

$$\frac{d}{dt}\begin{pmatrix} y_1(t) \\ y_2(t) \end{pmatrix} = \begin{pmatrix} 0 & 0 \\ 0 & -\pi \end{pmatrix} \begin{pmatrix} y_1(t) \\ y_2(t) \end{pmatrix} + \begin{pmatrix} 1 & \gamma_1 p_{12} \\ 1 & -\gamma_2 p_{21} \end{pmatrix} \begin{pmatrix} E(\Lambda_1)e^{-\mu_1 t} \\ E(\Lambda_2)e^{-\mu_2 t} \end{pmatrix}.$$

The latter is a pair of separate linear differential equations. Solving these, substituting in the above solution and again using the boundary condition $E(\bar{X}_1(0)) = E(\bar{X}_2(0)) = 0$, yields

$$\begin{aligned} E(\bar{X}_1(t)) &= \tfrac{\gamma_2 p_{21} E(\Lambda_1)}{\pi\mu_1}\left(1 - e^{-\mu_1 t}\right) + \tfrac{\gamma_1 p_{12} E(\Lambda_2)}{\pi\mu_2}\left(1 - e^{-\mu_2 t}\right) \\ &\quad + \tfrac{\gamma_1 p_{12} E(\Lambda_1)}{\pi} \cdot \tfrac{e^{-\mu_1 t} - e^{-\pi t}}{\pi - \mu_1} - \tfrac{\gamma_1 p_{12} E(\Lambda_2)}{\pi} \cdot \tfrac{e^{-\mu_2 t} - e^{-\pi t}}{\pi - \mu_2}, \\ E(\bar{X}_2(t)) &= \tfrac{\gamma_2 p_{21} E(\Lambda_1)}{\pi\mu_1}\left(1 - e^{-\mu_1 t}\right) + \tfrac{\gamma_1 p_{12} E(\Lambda_2)}{\pi\mu_2}\left(1 - e^{-\mu_2 t}\right) \\ &\quad - \tfrac{\gamma_2 p_{21} E(\Lambda_1)}{\pi} \cdot \tfrac{e^{-\mu_1 t} - e^{-\pi t}}{\pi - \mu_1} + \tfrac{\gamma_2 p_{21} E(\Lambda_2)}{\pi} \cdot \tfrac{e^{-\mu_2 t} - e^{-\pi t}}{\pi - \mu_2}. \end{aligned} \quad [5.45]$$

Quantities in [5.45] that should be recognized are

- $\frac{\gamma_2 p_{21}}{\pi}$ and $\frac{\gamma_1 p_{12}}{\pi}$, the long-term fractions of finding the Markov process in states 1 and 2, respectively;

• $\frac{E(\Lambda_i)}{\mu_i}(1-e^{-\mu_i t})$, the mean number of customers when the Markov process is always in state i ($i=1,2$);

• $\frac{e^{-\mu_i t}-e^{-\pi t}}{\pi-\mu_i}$, a non-negative quantity that increases whenever π or μ_i increases ($i=1,2$).

With these equations we can also find some other relevant quantities. For example, suppose that the state at time 0 is not known, but the background process is in steady state. Then $P(S(0)=1)=\frac{\gamma_2 p_{21}}{\pi}$ and $P(S(0)=2)=\frac{\gamma_1 p_{12}}{\pi}$, so it follows that the mean number of customers at time t is

$$\begin{aligned}&\frac{\gamma_2 p_{21}}{\pi}E(\bar{X}_1(t))+\frac{\gamma_1 p_{12}}{\pi}E(\bar{X}_2(t))\\&\qquad=\frac{\gamma_2 p_{21}E(\Lambda_1)}{\pi\mu_1}\left(1-e^{-\mu_1 t}\right)+\frac{\gamma_1 p_{12}E(\Lambda_2)}{\pi\mu_2}\left(1-e^{-\mu_2 t}\right).\end{aligned} \qquad [5.46]$$

Another consequence of [5.45] is the steady-state mean

$$E(\bar{X})=\frac{\gamma_2 p_{21}E(\Lambda_1)}{\pi\mu_1}+\frac{\gamma_1 p_{12}E(\Lambda_2)}{\pi\mu_2}. \qquad [5.47]$$

Keep in mind that there is no subscript needed, since in steady state it does not matter in which state we started. In the formula, we see the steady-state mean of each individual state, multiplied by the fraction that the corresponding state was active. The fact that the steady-state mean is just the sum of these parts underlines that customers from different states do not interfere with each other.

Let us now consider $E(\bar{X}|S=1)$: the steady-state expected number of customers given the background process is in state 1. Before we can give an expression, we have to define $\tilde{X}_i$ as the portion of the current customers who arrived when the state was i. Distinction between $\tilde{X}_1$ and $\tilde{X}_2$ is relevant because customers who arrived in a different state will have a different service rate. For instance, the first term of [5.47] is the mean number of customers with service rate μ_1.

Now define T_1 as an arbitrary time in steady state when the background process moves from state 1 to state 2. Also, let $T_{2\to 1}$ be the last time before T_1 that the state changed from 2 to 1. We split $\tilde{X}_1(T_1)$ into the customers who arrived before or after

$T_{2\to1}$, and then condition on the value of $T_1 - T_{2\to1}$. It holds that

$$E\left(\tilde{X}_1(T_1)\right) = E\left(\tilde{X}_1(T_{2\to1})\right) \cdot P(B_1 \geq T_1 - T_{2\to1}) + E\left(\bar{X}_1(T_1 - T_{2\to1})\right)$$

$$= \int_0^\infty \gamma_1 p_{12} e^{-\gamma_1 p_{12} t} \left(E\left(\tilde{X}_1(T_{2\to1})\right) e^{-\mu_1 t} + E\left(\bar{X}_1(t) | S(u)\right.\right.$$

$$\left.\left. = 1 \text{ for all } u \in [0,t]\right)\right) dt$$

$$= \int_0^\infty \gamma_1 p_{12} e^{-\gamma_1 p_{12} t} \left(E\left(\tilde{X}_1(T_{2\to1})\right) e^{-\mu_1 t} + \frac{E(\Lambda_1)}{\mu_1}\left(1 - e^{-\mu_1 t}\right)\right) dt$$

$$= \frac{\gamma_1 p_{12}}{\gamma_1 p_{12} + \mu_1} E\left(\tilde{X}_1(T_{2\to1})\right) + \frac{E(\Lambda_1)}{\mu_1}\left(1 - \frac{\gamma_1 p_{12}}{\gamma_1 p_{12} + \mu_1}\right).$$

Let T_2 and $T_{1\to2}$ be the symmetric versions of T_1 and $T_{2\to1}$. We then have

$$E\left(\tilde{X}_1(T_2)\right) = E\left(\tilde{X}_1(T_{1\to2})\right) \cdot P(B_1 \geq T_2 - T_{1\to2})$$

$$= \int_0^\infty \gamma_2 p_{21} e^{-\gamma_2 p_{21} t} E\left(\tilde{X}_1(T_{1\to2})\right) e^{-\mu_1 t} dt$$

$$= \frac{\gamma_2 p_{21}}{\gamma_2 p_{21} + \mu_1} E\left(\tilde{X}_1(T_{1\to2})\right).$$

A useful observation here is that $T_1 \stackrel{d}{=} T_{1\to2}$ and $T_2 \stackrel{d}{=} T_{2\to1}$, since in each case, both times represent a steady-state time at the end of a period of one state. As a result, we have a system of two linear equations. The solution is

$$E\left(\tilde{X}_1(T_1)\right) = \frac{(\gamma_2 p_{21} + \mu_1) E(\Lambda_1)}{(\pi + \mu_1)\mu_1}, \quad E\left(\tilde{X}_1(T_2)\right) = \frac{\gamma_2 p_{21} E(\Lambda_1)}{(\pi + \mu_1)\mu_1}. \tag{5.48}$$

Note the absence of $E(\Lambda_2)$ and μ_2 caused by the fact that customers from different states do not influence each other.

Let $T_{S=1}$ be an arbitrary steady-state time with $S(T_{S=1}) = 1$. We will show that $E\left(\tilde{X}_1(T_1)\right) = E\left(\tilde{X}_1(T_{S=1})\right) = E(\tilde{X}_1 | S = 1)$. Note that we can view the Markov background process as a Poisson process with rate $\gamma_1 p_{12}$, and we can view its events as periods where the state is 2. We assume those periods take 0 time, so that the rate at which an event happens is always $\gamma_1 p_{12}$. A nice property of the Poisson process is that when we pick an arbitrary point in time ($T_{S=1}$), the amount of time since the last event is exponentially distributed with parameter $\gamma_1 p_{12}$. The same holds for T_1, so $\tilde{X}_1(T_{S=1}) \stackrel{d}{=} \tilde{X}_1(T_1)$, hence in particular $E(\tilde{X}_1|S=1) = E\left(\tilde{X}_1(T_1)\right) = \frac{(\gamma_2 p_{21}+\mu_1)E(\Lambda_1)}{(\pi+\mu_1)\mu_1}$.

As a quick sanity check, we calculate $E(\tilde{X}_1)$, the steady-state mean number of customers who arrived when the background process was in state 1 (i.e. the customers who have service rate μ_1). As mentioned, this should give the first term of [5.47]. The calculation below verifies this:

$$E(\tilde{X}_1) = E(\tilde{X}_1|S=1) \cdot P(S=1) + E(\tilde{X}_1|S=2) \cdot P(S=2)$$
$$= \frac{(\gamma_2 p_{21} + \mu_1)\, E(\Lambda_1)}{(\pi + \mu_1)\mu_1} \cdot \frac{\gamma_2 p_{21}}{\pi} + \frac{\gamma_2 p_{21} E(\Lambda_1)}{(\pi + \mu_1)\mu_1} \cdot \frac{\gamma_1 p_{12}}{\pi} = \frac{\gamma_2 p_{21} E(\Lambda_1)}{\pi \mu_1}.$$

For symmetry reasons, [5.48] can be transformed into

$$E\left(\tilde{X}_2(T_1)\right) = \frac{\gamma_1 p_{12} E(\Lambda_2)}{(\pi + \mu_2)\mu_2}, \quad E\left(\tilde{X}_2(T_2)\right) = \frac{(\gamma_1 p_{12} + \mu_2)\, E(\Lambda_2)}{(\pi + \mu_2)\mu_2}. \tag{5.49}$$

Now we are finally ready to find the mean total number of customers conditioned on the current state:

$$E\left(\bar{X}|S=1\right) = E\left(\tilde{X}_1(T_1)\right) + E\left(\tilde{X}_2(T_1)\right)$$
$$= \frac{\gamma_2 p_{21} E(\Lambda_1)}{(\pi + \mu_1)\mu_1} + \frac{\gamma_1 p_{12} E(\Lambda_2)}{(\pi + \mu_2)\mu_2} + \frac{E(\Lambda_1)}{\pi + \mu_1}, \tag{5.50}$$

and analogously,

$$E\left(\bar{X}|S=2\right) = \frac{\gamma_2 p_{21} E(\Lambda_1)}{(\pi + \mu_1)\mu_1} + \frac{\gamma_1 p_{12} E(\Lambda_2)}{(\pi + \mu_2)\mu_2} + \frac{E(\Lambda_2)}{\pi + \mu_2}. \tag{5.51}$$

Given the complexity of calculations with two states, formulas for general n will be costly to derive analytically. However, if one assumes that the Markov process is in steady state at time 0, one can use a shortcut found by Blom *et al.* (2014) for the mean number of customers. It is based on the well-known rate-in = rate-out balance equation $\sum_{i=1}^{n} \pi_i \gamma_i p_{ij} = \pi_j \gamma_j$, π_i being the steady-state probabilities for each state.

To find the mean number of customers in this case, we take $k = 1$ in [5.38], multiply by π_i and sum over i. This results in

$$\frac{d}{dt}\sum_{i=1}^{n}\pi_i E\left(\bar{X}_i(t)\right) = -\sum_{i=1}^{n}\pi_i\gamma_i E\left(\bar{X}_i(t)\right) + \sum_{i=1}^{n}\pi_i\gamma_i\sum_{j=1}^{n}p_{ij}E\left(\bar{X}_j(t)\right)$$
$$+\sum_{i=1}^{n}\pi_i P\left(B_i \geq t\right)E(\Lambda_i) = \sum_{i=1}^{n}\pi_i P\left(B_i \geq t\right)E(\Lambda_i),$$

and hence

$$\sum_{i=1}^{n}\pi_i E\left(\bar{X}_i(t)\right) = \sum_{i=1}^{n}\pi_i E(\Lambda_i)\int_0^t P\left(B_i \geq u\right)du. \qquad [5.52]$$

With relatively easy methods, we found a general formula for the mean. For instance, $n = 2$ and exponential service times yields [5.46], which was more difficult to find as we had to solve a system of differential equations. The disadvantage of the quicker method is that it gives no information about the case where the starting state $S(0)$ is fixed.

5.4. Discussion and future work

We have considered two infinite-server queueing models with a mixed arrival process. For the $M_\Lambda/Cox_n/\infty$ model with exponential resample times, we showed how to compute all joint moments of the arrival rate and the number of customers, both in transient and steady state. For a Markov-modulated queue with general service times, we gave a procedure to obtain all moments of the number of customers given the initial state and the initial arrival rate.

Since moments define a distribution, finding a way to compute them is a significant step toward finding the exact queue size distribution, even though we could not find a closed formula for all moments. Possible future research may include finding the exact distributions of the considered random variables. From our point of view, a way to do this is solving the PDE corresponding to that model.

In related papers about infinite-server queues, a considerable amount of work has been done regarding scaling limits and large deviation principles (Anderson *et al.* 2016; Blom *et al.* 2014; Heemskerk *et al.* 2017; Jansen *et al.* 2016). It is likely that similar results can be proven for our models (see Conjecture 5.1).

Finally, it would be interesting to study the $M/M/\infty$ queue in which not the arrival rate but the *service rate* is described by a stochastic process. That infinite-server queue seems less amenable to a recursive approach.

5.5. References

Adan, I., Resing, J. (2015). Queueing systems [Online]. Available: http://www.win.tue.nl/~iadan/queueing.pdf.

Albrecher, H., Asmussen, S. (2006). Ruin probabilities and aggregate claims distributions for shot noise Cox processes. *Scandinavian Actuarial Journal*, 86–110.

Anderson, D., Blom, J., Mandjes, M., Thorsdottir, H., de Turck, K. (2016). A functional central limit theorem for a Markov-modulated infinite-server queue. *Methodology and Computing in Applied Probability*, 18, 153–168.

Anisimov, V. (2008). *Switching Processes in Queueing Models*. ISTE Ltd, London, and John Wiley & Sons, New York.

Asmussen, S. (2003). *Applied Probability and Queues*, 2nd ed. Springer-Verlag, New York.

Blom, J., Kella, O., Mandjes, M., Thorsdottir, H. (2014). Markov-modulated infinite-server queues with general service times. *Queueing Systems*, 76, 403–424.

Bühlmann, H. (1972). Ruinwahrscheinlichkeit bei erfahrungstarifiertem portefeuille. *Bulletin de l'Association des Actuaires Suisses*, 2, 131–140.

D'Auria, B. (2008). $M/M/\infty$ queues in semi-Markovian random environment. *Queueing Systems*, 58, 221–237.

Daw, A., Pender, J. (2018). Queues driven by Hawkes processes. *Stochastic Systems*, 8, 192–229.

Eick, S., Massey, W., Whitt, W. (1993). The physics of the $M_t/G/\infty$ queue. *Management Science*, 39, 241–252.

Heemskerk, M., van Leeuwaarden, J., Mandjes, M. (2017). Scaling limits for infinite-server systems in a random environment. *Stochastic Systems*, 7, 1–31.

Jansen, H.M., Mandjes, M.R.H., De Turck, K., Wittevrongel, S. (2016). A large deviations principle for infinite-server queues in a random environment. *Queueing Systems*, 82, 199–235.

Jongbloed, G., Koole, G. (2001). Managing uncertainty in call centres using Poisson mixtures. *Applied Stochastic Models in Business and Industry*, 17, 307–318.

Klaasse, B. (2017). Queuing and insurance risk models with mixing dependencies. Report, Eindhoven University of Technology.

Koops, D., Boxma, O., Mandjes, M. (2017). Networks of $./G/\infty$ queues with shot-noise driven arrival intensities. *Queueing Systems*, 86, 301–325.

Koops, D., Saxena, M., Boxma, O., Mandjes, M. (2018). Infinite-server queues with Hawkes input. *Journal of Applied Probability*, 55, 920–943.

van Kreveld, L.R. (2018). Parameter mixing in infinite server queues. MSc Thesis, University of Utrecht.

Neuts, M.F. (1981). *Matrix-Geometric Solutions in Stochastic Models*. Johns Hopkins University Press, Baltimore.

Neuts, M.F. (1989). *Structured Stochastic Matrices of $M/G/1$ Type and their Applications*. Marcel Dekker, New York.

O'Cinneide, C.A., Purdue, P. (1986). The $M/M/\infty$ queue in a random environment. *Journal of Applied Probability*, 23, 175–184.

Raaijmakers, Y., Albrecher, H., Boxma, O. (2019). The single server queue with mixing dependencies. *Methodology and Computing in Applied Probability*, 21(82), forthcoming.

Ross, S. M. (2010). *Introduction to Probability Models*, 10th edition. Elsevier, New York.

Salih, A. (2016). *Method of Characteristics*. Indian Institute of Space Science and Technology, Thiruvananthapuram.

Shen, W. (2013). Introduction to ordinary and partial differential equations [Online]. Available: https://www.math.psu.edu/shen_w/LNDE.pdf

6

Application of Fast Simulation Methods of Queueing Theory for Solving Some High-dimension Combinatorial Problems

Igor KUZNETSOV[1] and Nickolay KUZNETSOV[2]

[1] *National Technical University of Ukraine "Igor Sikorsky Kyiv Polytechnical Institute", Kiev, Ukraine*

[2] *V.M. Glushkov Institute of Cybernetics, National Academy of Sciences of Ukraine, Kiev, Ukraine*

Fast Monte Carlo simulation used in queueing theory to evaluate the systems efficiency indexes is here applied to solve two high-dimension combinatorial problems where other approaches fail.

The first problem is dealing with the estimation of the number of k-dimensional subspaces of an arbitrary weight ω of an n-dimensional vector space over a Galois field containing q elements. Upper and lower bounds are constructed through an analytical-statistical approach.

The second problem is dealing with the evaluation of so-called "good" permutations. A permutation $(s_0, s_1, \ldots, s_{N-1})$ of symbols $(0, 1, \ldots, N-1)$ is called "good" if the set $(t_0, t_1, \ldots, t_{N-1})$ formed by the rule $t_i = i + s_i \mod (N)$, $i = 0, 1, \ldots, N-1$, is also a permutation. The fast simulation method is proposed. Its implementation on the SCIT-4 multiprocessor computer complex enables us to evaluate the number of "good" permutations for any $N \leq 305$ with a relative error less than 1%.

Queueing Theory 1,
coordinated by Vladimir ANISIMOV, Nikolaos LIMNIOS.

Both fast simulation techniques are illustrated by numerical examples.

6.1. Introduction

The Monte Carlo simulation is the most widely used technique in engineering practice to evaluate system probability characteristics. Sometimes, it is the only practical method to obtain numerical estimates when we deal with objects, which are too large and complex to be analyzed by using analytical (explicit or asymptotic) methods. Standard or direct Monte Carlo simulation is based on the statistical reproduction of system behavior with the evaluation of some functionals on the system trajectories (Buslenko *et al.* 1973; Rubinstein 1981; Kleijnen 1987). At the same time, practical experience shows that the application of the standard Monte Carlo simulation to the evaluation of low-probability characteristics is impractical because of the excessive amount of computing time used even with modern computing systems.

During recent decades, essential efforts have been directed to the development of new efficient fast simulation methods known as variance reduction techniques. In fact, we mean the special indirect simulation techniques aimed at reducing the variance of the estimate whilst preserving its unbiasedness or at least the asymptotical unbiasedness. The best algorithms produce estimates with bounded relative errors (or coefficients of variation) while the system characteristic being evaluated tends toward zero.

Among the most widely used approaches aimed at reducing the variance of estimates, we can mention only some of the mainstreams of investigation: analytical-statistical methods (Kovalenko 1980; Kovalenko and Kuznetsov 1988; Kovalenko *et al.* 1997; Kuznetsov 2009; Kuznetsov and Shumskaya 2013), importance sampling (Glasserman 2004; Heidelberger 1995; Li *et al.* 2011; Kuznetsov 2014, 2017), stratified sampling (Fox and Glynn 1990; Plaks 1984; Shumskaya 2003), the method of multilevel splitting (Glasserman *et al.* 1999), etc. There we also some surveys of fast simulation techniques aimed at evaluating the probability of different rare events (see Kovalenko *et al.* (1997); Blanchet and Lam (2011); Lagnoux (2006)).

Monte Carlo simulation proved to be very efficient in solving some high-dimension combinatorial problems, in particular, importance sampling was used to evaluate the permanent of the matrix (Kuznetsov 1996) and the number of Latin rectangles (Kuznetsov 2009) and stratified sampling was applied to solve the knapsack problem (Kuznetsov 1998) (see also Kuznetsov (2016); Masol (2001); Kuznetsov (2010)).

The chapter is organized as follows. In section 6.1, we investigate the number of k-dimensional subspaces of a given weight ω of an n-dimensional vector space over a Galois field containing q elements (we use the notation $\left[\begin{smallmatrix} n \\ k \end{smallmatrix}\middle|\omega\right]$). The estimation of $\left[\begin{smallmatrix} n \\ k \end{smallmatrix}\middle|\omega\right]$ is of interest for encoding problems and for the solution of the problem of information protection against unauthorized access. Fast Monte Carlo simulation used in queueing theory makes it possible to construct the upper and lower bounds for $\left[\begin{smallmatrix} n \\ k \end{smallmatrix}\middle|\omega\right]$ for any values of n, k and ω (for particular cases $k = 1$ and $k = 2$ we give explicit analytical formulae).

In section 6.2, we deal with the evaluation of the number of permutations, which are known as "good" permutations. An exhaustive analysis of the fundamentals of the application of "good" permutations in cryptography is presented in Cooper *et al.* (1999). The principles of using such permutations in the rotor cryptographic systems are described in Konheim (1991).

In order to enumerate all "good" permutations of the dimension N, we need to analyze $N!$ permutations. In this section, we propose a new approach enabling the estimation of the number M_N of "good" permutations for large values of the dimension N. The implementation of fast simulation algorithms on the multiprocessor computer complex SCIT-4 at the V.M. Glushkov Institute of Cybernetics of the NAS of Ukraine makes it possible to calculate M_N for any $N \leq 305$ with the confidence level 0.99 and the relative error of only 1%.

Both fast simulation techniques are illustrated by numerical examples.

6.2. Upper and lower bounds for the number of some k-dimensional subspaces of a given weight over a finite field

Let us consider an n-dimensional vector space V_n over a finite field $GF(q)$ (a Galois field) containing q elements where q is a power of a prime number. It is well known (Andrews 1998) that the total number of different k-dimensional subspaces $V_n^{(k)}$ of the space V_n equals

$$\begin{bmatrix} n \\ k \end{bmatrix} = \prod_{i=0}^{k-1} \frac{q^{n-i} - 1}{q^{k-i} - 1}, \quad 1 \leq k \leq n. \qquad [6.1]$$

As the weight of a vector $v \in V_n$, we take the number of its non-zero components. By the weight of a k-dimensional subspace $V_n^{(k)}$ of the space V_n we mean the number, which equals the minimal weight of non-zero vectors $v \in V_n^{(k)}$. We denote by $\left[\begin{smallmatrix} n \\ k \end{smallmatrix}\middle|\omega\right]$ the number of k-dimensional subspaces $V_n^{(k)}$ of the space V_n, each of which has a weight ω. It follows from the algorithm (Calabi and Wilf 1977) aimed to construct the set of base vectors of a k-dimensional subspace $V_n^{(k)}$ that the weight ω cannot exceed $n-k+1$ and, hence, we have

$$\begin{bmatrix} n \\ k \end{bmatrix} = \sum_{\omega=1}^{n-k+1} \left[\begin{matrix} n \\ k \end{matrix} \middle| \omega \right], \quad 1 \leq k \leq n. \tag{6.2}$$

It is a common practice to characterize the efficiency of information encoding with the help of linear (n, k) codes (it is another name for a subspace $V_n^{(k)}$) by the number k/n (see, for example, MacWilliams and Sloane (1977)). An increase in efficiency leads to a decrease in the weight ω, which is not always desirable for the process of encoding information since it decreases its correcting capabilities and reduces them to zero when $\omega = 1$. Nevertheless, the subspaces $V_n^{(k)}$ of the small weight ω, in particular, $\omega = 1$, are used for encoding (since they make it possible to increase efficiency) with subsequent transmission by highly reliable communication channels, which is discussed in Masol (2001). At the same time, the estimation of the number of subspaces $V_n^{(k)}$ of a given weight ω is of interest for encoding problems and for the solution of the problem of information protection against unauthorized access.

We should mention the total lack of results that allow one to compute or estimate $\left[\begin{matrix} n \\ k \end{matrix} \middle| \omega \right]$ for arbitrary values of n, k and ω. Results were obtained only in special cases. In particular, in Masol (1989, 1996) recurrent formulae for the evaluation of $\left[\begin{matrix} n \\ k \end{matrix} \middle| \omega \right]$ are presented for $\omega = 1$ and $\omega = 2$.

In this chapter, a new approach based on fast simulation of small probabilities is proposed. It is based on the fundamental idea of analytical-statistical methods (Kovalenko 1975, 1980). A method can be called analytical-statistical if it uses both analytical formulae and statistical simulation to increase the accuracy of estimates. The first analytical-statistical method was based on the small parameter method: among the system characteristics we choose a value, which can be considered as a small parameter ε (e.g. the rate of the input flow). Then the system characteristic is represented as a series in powers of ε (analytical part), and the coefficients of such expansion are evaluated by the Monte Carlo method (statistical part). It is often enough to take only a few first terms to reach a good approximation. Such an approach proved to be very efficient in queueing theory and reliability analysis (Kovalenko *et al.* 1997). In contrast to queueing theory, we propose to expand $\left[\begin{matrix} n \\ k \end{matrix} \middle| \omega \right]$ in terms of powers of a large parameter q (typical values are $q = 2^8$, $q = 2^{16}$ and larger). The fast simulation method proposed above makes it possible to construct the upper and lower bounds for $\left[\begin{matrix} n \\ k \end{matrix} \middle| \omega \right]$ for any values of n, k and ω (for particular cases $k = 1$ and $k = 2$, we give explicit analytical formulae). The accuracy of the estimates is illustrated by numerical examples.

6.2.1. *A general fast simulation algorithm*

In enumerating all k-dimensional subspaces $V_n^{(k)}$ of a vector space V_n, the key role is played by the algorithm (Calabi and Wilf 1977) for the construction of the set of base vectors of this subspace. This algorithm amounts to the construction of the matrix A of the size $k \times n$ (see Figure 6.1). Let $\overline{r} = (r_1, \ldots, r_k)$ be a k-dimensional vector whose components are ordered positive integers, $U = \{\overline{r} : 1 \leq r_1 < r_2 < \ldots < r_k \leq n\}$.

1) Choose an arbitrary vector $\overline{r} \in U$.

2) In the matrix A with k rows and n columns, form the unit matrix of the size $k \times k$ with columns having the numbers specified by the vector $\overline{r}$.

3) Write zeros into all positions $j > r_i$ in the ith row of the matrix A $(1 \leq i \leq k)$.

4) Fill the remaining positions in the matrix A with arbitrary elements of the field.

$$\begin{array}{ccc} r_1 & r_2 & r_k \\ \downarrow & \downarrow & \downarrow \end{array}$$

$$A = \begin{pmatrix} a_{11}\ldots a_{1r_1-1}1 & 0 & \ldots & 0 & 0 & 0 & \ldots & 0 & 00\ldots 0 \\ a_{21}\ldots a_{2r_1-1}0 a_{2r_1+1}\ldots a_{2r_2-1}1 & 0 & \ldots & 0 & 00\ldots 0 \\ \vdots \ \ldots & \vdots & \vdots & \vdots & \ldots & \vdots & \vdots & \vdots & \ldots & \vdots & \vdots \vdots \ldots \vdots \\ a_{k1}\ldots a_{kr_1-1}0 a_{kr_1+1}\ldots a_{kr_2-1}0 a_{kr_2+1}\ldots a_{kr_k-1}1 0\ldots 0 \end{pmatrix}$$

Figure 6.1. *A matrix determining the base vectors of some subspace*

The rows of the matrix constructed are base vectors of some subspace $V_n^{(k)} \subset V_n$. Therefore, the problem of determining the number of all k-dimensional subspaces of the weight ω is reduced to the evaluation of the number of matrices A with base vectors whose linear combination makes it possible to construct some vector containing exactly ω non-zero components and, at the same time, there is no linear combination of base vectors with a smaller number of non-zero components.

We use the following notations later on:

$\left[\begin{matrix} n \\ k \end{matrix} \middle| \omega; \overline{r}\right]$ is the number of k-dimensional subspaces of the weight ω for a fixed vector $\overline{r} \in U$;

$$U_\omega(L) = \{\overline{r} \in U : \ r_1 \geq \omega, \ (r_2 - 2) + (r_3 - 3) + \ldots + (r_k - k) = L\},$$

$(k-1)\,(\omega-1) \le L \le (k-1)\,(n-k);$

$|U_\omega(L)|$ is the number of components in the set $U_\omega(L)$.

Taking into account that $\left[\begin{smallmatrix} n \\ k \end{smallmatrix}\middle|\, \omega;\, \overline{r}\right] = 0$ for $\overline{r} \notin \bigcup_{L=(k-1)\,(\omega-1)}^{(k-1)\,(n-k)} U_\omega(L)$, the following relation holds true for any $\omega \in \{1, \ldots, n-k+1\}$:

$$\left[\begin{matrix} n \\ k \end{matrix}\middle|\, \omega\right] = \sum_{\overline{r}\in U} \left[\begin{matrix} n \\ k \end{matrix}\middle|\, \omega;\, \overline{r}\right] = \sum_{L=(k-1)\,(\omega-1)}^{(k-1)\,(n-k)} \sum_{\overline{r}\in U_\omega(L)} \left[\begin{matrix} n \\ k \end{matrix}\middle|\, \omega;\, \overline{r}\right]$$

$$= \sum_{L=(k-1)\,(\omega-1)}^{(k-1)\,(n-k)} |U_\omega(L)|\; q^{L+\omega-1} \sum_{\overline{r}\in U_\omega(L)} \frac{1}{q^{L+\omega-1}} \left[\begin{matrix} n \\ k \end{matrix}\middle|\, \omega;\, \overline{r}\right] \frac{1}{|U_\omega(L)|}$$

$$= Z_\omega \sum_{L=(k-1)\,(\omega-1)}^{(k-1)\,(n-k)} \frac{|U_\omega(L)|\, q^{L+\omega-1}}{Z_\omega} \sum_{\overline{r}\in U_\omega(L)} c_\omega(\overline{r};\, L)\, \frac{1}{|U_\omega(L)|}$$

$$= Z_\omega\, \mathbf{E}_\nu\, \mathbf{E}\,\{c_\omega(\overline{\gamma}_\nu;\, \nu)\,|\,\nu\}, \qquad [6.3]$$

where

$$Z_\omega = \sum_{L=(k-1)\,(\omega-1)}^{(k-1)\,(n-k)} |U_\omega(L)|\, q^{L+\omega-1}, \quad c_\omega(\overline{r};\, L) = \frac{1}{q^{L+\omega-1}} \left[\begin{matrix} n \\ k \end{matrix}\middle|\, \omega;\, \overline{r}\right], \qquad [6.4]$$

and the expectation $\mathbf{E}_\nu$ is taken over the distribution of the random value ν, which takes the value $L \in \{(k-1)\,(\omega-1),\, (k-1)\,(\omega-1)+1,\, \ldots,\, (k-1)\,(n-1)\}$ with the probability $|U_\omega(L)|\, q^{L+\omega-1}/Z_\omega$; for a fixed $\nu = L$, a random vector $\overline{\gamma}_L$ is uniformly distributed over the set $U_\omega(L)$.

The formula [2.3] is based on the idea of importance sampling. An expansion in powers of q (see the simulation algorithm of the random variable ν) makes it possible to rank the summands according to their "weights". Note that the coefficient $c_\omega(\overline{r};\, L)$ is uniformly bounded for any values of $\overline{r}$, ω, L and q.

The relations [2.3] and [2.4] contain unknown values $|U_\omega(L)|$ and $c_\omega(\overline{r};\, L)$. The value of $|U_\omega(L)|$ can be evaluated due to following simple recurrence relations. Denote $\varrho_\omega(n,k,L) = |U_\omega(L)|$. For $2 \le i \le k$, $(i-1)\,(\omega-1) \le l \le (i-1)\,(m-i)$ the following relation

$$\varrho_\omega(m,i,l) = \sum_{j=i+\left]\frac{l}{i-1}\right[}^{\min\{m,\; l+i-(\omega-1)\,(i-2)\}} \varrho_\omega(j-1,i-1,l-(j-i)) \qquad [6.5]$$

holds true where $\left]\frac{l}{i-1}\right[= \min\left\{j:\ j \geq \frac{l}{i-1}\right\}$. The relation [2.5] is complemented by boundary conditions: $\varrho_\omega(m,1,0) = m - \omega + 1$ for any $m \geq \omega$, and $\varrho_\omega(m,i,(\omega-1)(i-1)) = 1$ for any $m \geq \omega+1, 2 \leq i \leq m-\omega+1$.

The formula [2.5] follows from the following considerations. Determine the values, which can take r_i if we know that $r_1 \geq \omega$ and

$$(r_2 - 2) + (r_3 - 3) + \ldots + (r_i - i) = l. \qquad [6.6]$$

Since $r_s - s \geq \omega - 1$ for any $s = 2, \ldots, i-1$, it follows from [2.6] that $(\omega-1)(i-2) + r_i - i \leq l$, i.e. $r_i \leq l + i - (\omega-1)(i-2)$. At the same time, $r_i \leq m$. Therefore, $r_i \leq \min\{m,\ l + i - (\omega-1)(i-2)\}$. It follows from [2.6] that

$$r_i = l - (r_{i-1} + r_{i-2} + \ldots + r_2) + \frac{(i-1)(i+2)}{2} \geq l - [(r_i - 1) + (r_i - 2)$$

$$+ \ldots + (r_i - i + 2)] + \frac{(i-1)(i+2)}{2} = l - (i-2)\, r_i + i\,(i-1),$$

i.e. $r_i \geq i + \left]\frac{l}{i-1}\right[$. If $r_i = j$, then $\omega \leq r_1 < r_2 < \ldots < r_{i-1} \leq j-1$; moreover, the value l is decreasing on the value $j - i$ (it is the number of places in the row i, which should be filled with elements of the field – see the algorithm below).

The problem of estimation of $c_\omega(\overline{r};\ L)$ for arbitrary values of n, k, $\overline{r}$, L and ω is considerably more complicated. We will focus our attention only on this problem in the following sections.

Suppose now we possess some method for the evaluation of the coefficient $c_\omega(\overline{r};\ L)$ for any values of parameters n, k, $\overline{r}$, L and ω. Then equations [2.3] and [2.4] imply a simple algorithm for the construction of unbiased estimates for $\left[\begin{array}{c|c} n & \omega \\ k & \end{array}\right]$.

1) Compute $|U_\omega(L)| = \varrho_\omega(n,k,L)$ for all $L = (k-1)(\omega-1), \ldots, (k-1)(n-k)$ (see the recurrence formula [2.5]).

2) Compute Z_ω due to [2.4].

3) Simulate the random value ν, which equals $L \in \{(k-1)(\omega-1), \ldots, (k-1)(n-k)\}$ with the probability $|U_\omega(L)|\, q^{L+\omega-1} / Z_\omega$. Let $\nu = L$.

4) Simulate the random vector $\overline{\gamma}_L$ uniformly distributed over the set $U_\omega(L)$. Let $\overline{\gamma}_L = \overline{r}$.

5) Compute $c_\omega(\overline{r};\, L)$.

6) As the estimate for $\left[\begin{smallmatrix} n \\ k \end{smallmatrix} \middle| \omega \right]$ constructed in one trial take $\hat{\beta}_\omega(n,k) = Z_\omega\, c_\omega(\overline{r};\, L)$.

The relations [2.3] and [2.4] imply the following statement.

THEOREM 6.1.– The estimate $\hat{\beta}_\omega(n,k)$ is unbiased, i.e. $\mathbf{E}\, \hat{\beta}_\omega(n,k) = \left[\begin{smallmatrix} n \\ k \end{smallmatrix} \middle| \omega \right]$.

The algorithm formulated above can be applied under the condition that the coefficient $c_\omega(\overline{r};\, L)$ can be calculated for any values of the parameters n, k, $\overline{r}$, L and ω. Let us concentrate our attention on the evaluation of $c_\omega(\overline{r};\, L)$.

Let $\overline{r} = (r_1, \ldots, r_k)$ be some fixed vector determining the number of columns forming the unit matrix A (see Figure 6.1). The number of positions in the ith row ($i = 1, \ldots, k$), which have to be filled with elements of the field, is equal to $r_i - 1 - (i-1) = r_i - i$. It is obvious that $r_i - i \le r_j - j$ for $i < j$. Hereafter, the columns of the matrix A with the numbers specified by the vector $\overline{r}$ and also with the numbers $j > r_k$ will be excluded from the consideration. Thus, we obtain a rectangular matrix $B = (b_{ij})$ of the size $k \times (r_k - k)$ with $b_{ij} = 0$ for $j > r_i - i$; at all other positions (i,j) one of q elements of the field can be located. If $N(\overline{r})$ denotes the total number of the disposition variants of the field elements in the matrix B, then

$$N(\overline{r}) = \prod_{i=1}^{k} q^{r_i - i} = q^{\sum\limits_{i=1}^{k} (r_i - i)}. \qquad [6.7]$$

Let $Y_\omega(\overline{r})$ denote the event: "rows of the matrix A with randomly chosen components a_{ij}, $j \le r_i - i$, form the basis of the k-dimensional vector subspace of the weight ω". Then

$$\left[\begin{matrix} n \\ k \end{matrix} \middle| \omega;\, \overline{r} \right] = N(\overline{r})\, \mathbf{P}\{Y_\omega(\overline{r})\}. \qquad [6.8]$$

Taking into account that $L = \sum\limits_{i=2}^{k} (r_i - i)$, the relations [2.4], [2.7] and [2.8] imply that

$$c_\omega(\overline{r};\, L) = \frac{1}{q^{L+\omega-1}} \left[\begin{matrix} n \\ k \end{matrix} \middle| \omega;\, \overline{r} \right] = \frac{N(\overline{r})}{q^{L+\omega-1}}\, \mathbf{P}\{Y_\omega(\overline{r})\} = q^{r_1-\omega}\, \mathbf{P}\{Y_\omega(\overline{r})\}. \qquad [6.9]$$

In the following, we concentrate our attention on the evaluation of the probability $\mathbf{P}\{Y_\omega(\overline{r})\}$.

6.2.2. *An auxiliary algorithm*

In this section, an auxiliary algorithm underlying the construction of the upper and lower estimates of the probability $\mathbf{P}\{Y_\omega(\overline{r})\}$ is described.

We consider a Galois field with s non-zero elements. Let $\overline{x}$ be a fixed m-dimensional vector whose components are the non-zero elements of the field. Denote by $\varphi_m^{(1)}(\sigma, s)$ the number of vectors $\overline{y}$ possessing the following two properties: (a) there is such an element of the Galois field $\beta \neq 0$ that the vector $\overline{x} + \beta\overline{y}$ contains exactly σ non-zero components (the weight of the linear combination of $\overline{x}$ and $\overline{y}$ equals σ); (b) for any non-zero element of the Galois field $\gamma \neq \beta$, the vector $\overline{x} + \gamma\overline{y}$ contains at least σ non-zero components. In other words, the minimal weight of vectors $\overline{x} + \beta\overline{y}$ equals σ. Moreover, we denote

$$\varphi_m^{(0)}(\sigma, s) = \varphi_m^{(1)}(\sigma, s) + \varphi_m^{(1)}(\sigma+1, s) + \ldots + \varphi_m^{(1)}(m-1, s),\ 0 \le \sigma \le m-1. \quad [6.10]$$

LEMMA 6.1.– For $\sigma < m - 1$, the following recurrence relations hold true:

$$\varphi_m^{(0)}(\sigma, s) = \varphi_m^{(1)}(\sigma+1, s) + \sum_{j=1}^{\left[\frac{m}{m-\sigma}\right]} \frac{m!}{[(m-\sigma)!]^j\,[m - j\,(m-\sigma)]!}\, C_s^j\, c_j(m, \sigma, s), \quad [6.11]$$

$$\varphi_m^{(1)}(\sigma, s) = \sum_{j=1}^{\left[\frac{m}{m-\sigma}\right]} \frac{m!}{[(m-\sigma)!]^j\,[m - j\,(m-\sigma)]!}\, C_s^j\, c_j(m, \sigma, s), \quad [6.12]$$

where

$$c_j(m, \sigma, s) = \begin{cases} 1 & \text{if } m'(j) = 0, \\ (s-j)^{m'(j)} & \text{if } 0 < m'(j) < m - \sigma, \\ (s-j)^{m-\sigma} - (s-j) & \text{if } m'(j) = m - \sigma, \\ \varphi_{m'(j)}^{(0)}(m'(j) - (m-\sigma) + 1, s - j) & \text{if } m'(j) > m - \sigma, \end{cases}$$

with $m'(j) = m - j\,(m - \sigma)$. The initial conditions are as follows:

$$\varphi_m^{(0)}(\sigma, s) = 0 \text{ if } \sigma \ge m \text{ or } s = 0; \quad \varphi_m^{(0)}(m-1, s) = 0 \text{ if } s < m;$$

$$\varphi_m^{(0)}(m-1, s) = \varphi_m^{(1)}(m-1, s) = \prod_{j=1}^{m} (s - j + 1) \text{ if } s \ge m. \quad [6.13]$$

PROOF.– The case when $\sigma = m - 1$ (see formula [2.13]) is the simplest. In fact, the first component of the vector $\overline{y}$ can take any of s values, for obvious reasons the second component can take $s - 1$ values (otherwise the weight of the linear combination will be less than $m - 1$). Similarly, the third component can take $s - 2$ values, etc. For $\sigma < m - 1$, it follows from the relation [2.10] that

$$\varphi_m^{(0)}(\sigma, s) = \varphi_m^{(0)}(\sigma + 1, s) + \varphi_m^{(1)}(\sigma, s). \quad [6.14]$$

Therefore, it is enough to prove the relation [2.12]; the relation [2.11] follows immediately from [2.12] and [2.14].

The equality [2.12] follows from the following considerations. The vector $\overline{x} + \beta\,\overline{y}$ contains exactly σ non-zero components only if for some $\beta \neq 0$ there is a set of $m - \sigma$ components of the vector $\overline{y}$, which satisfy the relations $x_{i_l} + \beta\, y_{i_l} = 0$, $l = 1, \ldots, m - \sigma$. The number of such sets cannot exceed $\left[\frac{m}{m-\sigma}\right]$. Let us consider the case when the vector $\overline{y}$ contains exactly j sets of the components satisfying the property mentioned above. The rules of combinatorics imply that there are

$$C_m(m - \sigma, \ldots, m - \sigma, m') = \frac{m!}{[(m - \sigma)!]^j\,[m - j\,(m - \sigma)]!}$$

ways to divide a vector consisting of m components into j sets consisting of $m - \sigma$ components each and one set consisting of $m' = m - j\,(m - \sigma)$ components.

In this case, with j sets should be associated j pairwise different elements of the Galois field which can be done by C_s^j ways. A coefficient $c_j(m, \sigma, s)$ determines the number of such m'-dimensional vectors $\overline{y'}$ that for any non-zero element of Galois field β taking $s - j$ values, the weight of the linear combination $\overline{x'} + \beta\,\overline{y'}$ will be greater than $m' - (m - \sigma)$, where $\overline{x'}$ is an m'-dimensional vector obtained from $\overline{x}$ by eliminating components from j sets mentioned above. If $m' = 0$, then $c_j(m, \sigma, s) = 1$. If $0 < m' < m - \sigma$, then the components of the vector $\overline{y'}$ can take arbitrary values except for j fixed values, i.e. $c_j(m, \sigma, s) = (s - j)^{m'}$. If $m' = m - \sigma$, then the combinations in which all the components of the vector $\overline{y'}$ differ from the corresponding components of the vector $\overline{x'}$ in the same factor should be eliminated from all the possible combinations ($s - j$ inadmissible combinations), i.e. $c_j(m, \sigma, s) = (s - j)^{m'} - (s - j)$. Let $m' > m - \sigma$. In this case, we have to determine the number of such vectors $\overline{y'}$ that the weight of any linear combination $\overline{x'} + \beta\,\overline{y'}$ is not less than $\sigma' = m' - (m - \sigma) + 1$. Thus, $c_j(m, \sigma, s) = \varphi_{m'}^{(0)}(\sigma', s')$, where $s' = s - j$. Summing over all the possible values of j (the corresponding sets of vectors $\overline{y}$ do not intersect), we obtain the relation [2.12]. The lemma is proved.

6.2.3. *Exact analytical formulae for the cases* $k = 1$ *and* $k = 2$

For the simplest cases when $k = 1$ and $k = 2$, the value of $\left[\begin{smallmatrix} n \\ k \end{smallmatrix}\middle| \omega\right]$ can be evaluated due to exact analytical formulae.

THEOREM 6.2.– The following equalities hold:

$$\left[\begin{matrix} n \\ 1 \end{matrix}\middle| \omega\right] = (q-1)^{\omega-1} \sum_{r=\omega}^{n} C_{r-1}^{\omega-1}, \qquad [6.15]$$

$$\left[\begin{matrix} n \\ 2 \end{matrix}\middle| \omega\right] = \sum_{r_1=\omega}^{n-1} \sum_{r_2=r_1+1}^{n} \left\{ \sum_{l=\omega-1}^{r_1-1} C_{r_1-1}^{l} (q-1)^l \right.$$

$$\times \sum_{(s_1,s_2)\in S(l)} C_l^{s_1} C_{r_2-2-l}^{s_2} (q-1)^{s_2} \psi(l, s_1, s_2) \qquad [6.16]$$

$$\left. + \sum_{l=\omega}^{r_1-1} C_{r_1-1}^{l} (q-1)^l \sum_{(s_1,s_2)\in U(l)} C_l^{s_1} C_{r_2-2-l}^{s_2} (q-1)^{s_2} \pi(l, s_1, s_2) \right], \qquad [6.17]$$

where

$$S(\omega-1) = \{(s_1, s_2) : \ 0 \le s_1 \le \omega - 1, \ 0 \le s_2 \le r_2 - \omega - 1,$$
$$s_1 + s_2 \ge \omega - 1\}, \qquad [6.18]$$

$$S(l) = \{(s_1, s_2) : s_1 + s_2 = \omega - 1, 0 \le s_2 \le \min\{\omega - 1, r_2 - 2 - l\}\},$$
$$l = \omega, \ldots, r_1 - 1, \qquad [6.19]$$

$$U(l) = \{(s_1, s_2) : \ s_1 + s_2 \ge \omega, \ 0 \le s_1 \le l, 0 \le s_2 \le r_2 - 2 - l\},$$
$$l = \omega, \ldots, r_1 - 1, \qquad [6.20]$$

$$\psi(l, s_1, s_2) = \sum_{\sigma=s_1-s_2-l+\omega-2}^{s_1-1} \varphi_{s_1}^{(1)}(\sigma, q-1) \ \text{ if } s_1 - s_2 - l + \omega - 2 > 0, [6.21]$$

$$\psi(l, s_1, s_2) = (q-1)^{s_1} \ \text{ if } s_1 - s_2 - l + \omega - 2 \le 0, \qquad [6.22]$$

$$\pi(l, s_1, s_2) = \varphi_{s_1}^{(1)}(s_1 - s_2 - l + \omega - 2, q - 1) \ \text{ if } 0 \le s_1 - s_2 - l + \omega - 2 < s_1, [6.23]$$

$$\pi(l, s_1, s_2) = 0 \quad \text{otherwise,} \tag{6.24}$$

$\{\varphi_{s_1}^{(1)}(\sigma, q-1)\}$ are determined by recurrent formulae [2.11]–[2.13].

PROOF.– Let $k = 1$. The event $Y_\omega(\overline{r})$ occurs if and only if a single row of the matrix B contains exactly $\omega - 1$ non-zero elements. Therefore, we have

$$\mathbf{P}\{Y_\omega(\overline{r})\} = C_{r_1-1}^{\omega-1} \left(\frac{q-1}{q}\right)^{\omega-1} \left(\frac{1}{q}\right)^{r_1-\omega}. \tag{6.25}$$

The relations [2.7], [2.8] and [2.24] imply the formula [2.15]:

$$\left[\begin{array}{c|c} n & \omega \\ 1 & \end{array}\right] = \sum_{r_1=\omega}^{n} \left[\begin{array}{c|c} n & \omega;\ r_1 \\ 1 & \end{array}\right] = (q-1)^{\omega-1} \sum_{r=\omega}^{n} C_{r-1}^{\omega-1}.$$

The case $k = 2$ is more complicated. We have

$$\left[\begin{array}{c|c} n & \omega \\ 2 & \end{array}\right] = \sum_{r_1=\omega}^{n-1} \sum_{r_2=r_1+1}^{n} \left[\begin{array}{c|c} n & \omega;\ r_1 r_2 \\ 2 & \end{array}\right]. \tag{6.26}$$

Our aim is to evaluate the probability of the event $Y_\omega(\overline{r})$, and then to take advantage of the formulae [2.7] and [2.8]. Introduce the following events:

$Z_1 = \{$the first row of the matrix B contains $\omega - 1$ non-zero elements, and the second one contains not less than $\omega - 1$ non-zero elements; any linear combination of the first two rows of the matrix B has at least $\omega - 2$ non-zero elements$\}$;

$Z_2 = \{$the first row of the matrix B contains more than $\omega - 1$ non-zero elements, and the second one contains $\omega - 1$ non-zero elements; any linear combination of the first two rows of the matrix B has at least $\omega - 2$ non-zero elements$\}$;

$Z_{1,2} = \{$the first two rows of the matrix B contain more than $\omega - 1$ non-zero elements; there is a linear combination of the first two rows of this matrix containing $\omega - 2$ non-zero elements, and any other linear combination contains not less than $\omega - 2$ non-zero elements$\}$.

Since the events are disjoint, we have

$$\mathbf{P}\{Y_\omega(\overline{r})\} = \mathbf{P}\{Z_1\} + \mathbf{P}\{Z_2\} + \mathbf{P}\{Z_{1,2}\}. \tag{6.27}$$

The probability of the event Z_1 can be evaluated as follows. Among r_1-1 positions of the first row, we choose $\omega-1$ positions where non-zero elements are to be disposed (the corresponding probability equals $C_{r_1-1}^{\omega-1}\left(\frac{q-1}{q}\right)^{\omega-1}\left(\frac{1}{q}\right)^{r_1-\omega}$). Denote the set of these positions by E. Let us consider the disposition of elements in the second row. Introduce the following event: $H(s_1,s_2)=\{$the second row contains s_1 non-zero elements located at the positions from the set E, and s_2 non-zero elements located at the other positions available for filling$\}$. It is obvious that $(s_1,s_2)\in S(\omega-1)$ (see [2.17]). The probability of this event is equal to

$$p(l,s_1,s_2)=C_l^{s_1}\left(\frac{q-1}{q}\right)^{s_1}\left(\frac{1}{q}\right)^{l-s_1}C_{r_2-2-l}^{s_2}\left(\frac{q-1}{q}\right)^{s_2}\left(\frac{1}{q}\right)^{r_2-2-l-s_2} \quad [6.28]$$

where $l=\omega-1$. Taking into account the disposition of the zero and non-zero elements in the first two rows, their linear combination contains at least $(\omega-1-s_1)+s_2$ non-zero elements. If $s_2\geq s_1-1$, then the third condition of the event Z_1 is fulfilled. Set $\theta(\omega-1,s_1,s_2)=1$ for this case. If $s_2<s_1-1$, then the lemma from the previous section implies that the probability of fulfillment of the third condition in the event Z_1 satisfies the following relation:

$$\theta(\omega-1,s_1,s_2)=\frac{\psi(\omega-1,s_1,s_2)}{(q-1)^{s_1}},$$

where $\psi(\omega-1,s_1,s_2)$ is determined by [2.20] and [2.21]. Finally, we have

$$\mathbf{P}\{Z_1\}=C_{r_1-1}^{\omega-1}\left(\frac{q-1}{q}\right)^{\omega-1}\left(\frac{1}{q}\right)^{r_1-\omega} \times\sum_{(s_1,s_2)\in S(\omega-1)}p(\omega-1,s_1,s_2)\theta(\omega-1,s_1,s_2). \quad [6.29]$$

A similar reasoning yields the following formulae for the probabilities of the events Z_2 and $Z_{1,2}$:

$$\mathbf{P}\{Z_2\}=\sum_{l=\omega}^{r_1-1}C_{r_1-1}^{l}\left(\frac{q-1}{q}\right)^{l}\left(\frac{1}{q}\right)^{r_1-1-l}\sum_{(s_1,s_2)\in S(l)}p(l,s_1,s_2)\frac{\psi(l,s_1,s_2)}{(q-1)^{s_1}}, \quad [6.30]$$

$$\mathbf{P}\{Z_{1,2}\}=\sum_{l=\omega}^{r_1-1}C_{r_1-1}^{l}\left(\frac{q-1}{q}\right)^{l}\left(\frac{1}{q}\right)^{r_1-1-l}\sum_{(s_1,s_2)\in U(l)}p(l,s_1,s_2)\frac{\pi(l,s_1,s_2)}{(q-1)^{s_1}}, \quad [6.31]$$

where $S(l), U(l), \{\psi(l, s_1, s_2)\}$ and $\{\pi(l, s_1, s_2)\}$ are determined by formulae [2.18]–[2.23]. And finally, formulae [2.7], [2.8] and [2.25]–[2.30] imply the formula [2.16]. The theorem is proved.

6.2.4. *The upper and lower bounds for the probability* $\mathrm{P}\{Y_\omega(\overline{r})\}$

Let $\overline{r}$ be a fixed vector (it is generated by the algorithm for the construction of the unbiased estimates for $\left[\begin{matrix} n \\ k \end{matrix} \middle| \omega \right]$). Assume that at every position of the matrix $B = (b_{ij})$ accessible for filling an element of Galois field can be located with probability $1/q$. The event $Y_\omega(\overline{r}) = \{$rows of the matrix A form the basis of a k-dimensional vector subspace of the weight $\omega\}$ occurs if and only if one of k disjoint events $D_\omega^{(i)}(\overline{r}),\ i = 1, \ldots, k$, takes place. An event $D_\omega^{(i)}(\overline{r})$ is stipulated by three conditions:

a) the first $i-1$ rows of the matrix A form the basis of an $(i-1)$-dimensional vector space with the weight not less than $\omega + 1$;

b) the ith row of the matrix A contains exactly ω non-zero elements or, for some $m \in \{1, \ldots, i-1\}$, $1 \le j_1 < j_2 < \ldots < j_m \le i-1$, there is a linear combination of the ith row with the rows $j_1, \ldots, j_m$ containing exactly ω non-zero elements; at the same time, it is not possible to find a linear combination of the ith row with the preceding ones determining a vector with a smaller number of non-zero elements;

c) for all $j = i+1, \ldots, k$, the first j rows of the matrix A form the basis of a j-dimensional vector space of the weight ω.

It is obvious that

$$\mathbf{P}\{Y_\omega(\overline{r})\} = \sum_{i=1}^{k} \mathbf{P}\{D_\omega^{(i)}(\overline{r})\}. \qquad [6.32]$$

In order to evaluate $\mathbf{P}\{D_\omega^{(i)}(\overline{r})\}$, we have to evaluate the probability of occurrence of one of the s dependent events. The most universal formula is based on the inclusions–exclusions principle. For the large values of s, the problem of finding the probabilities of all the pairwise intersections of the events mentioned above becomes unreal. Therefore, we use the sum of probabilities of the corresponding events as an upper bound; the lower bound is trivial. Denote:

$$\nu_{jl} = \begin{cases} 0 \text{ if zero element is located at } l\text{th position of } j\text{th row of the matrix } B, \\ 1 \text{ otherwise,} \end{cases}$$

$l = 1, \dots, r_k - k, \quad j = 1, \dots, k, \quad \overline{\nu} = (\nu_{jl});$

$$p_j(l; \overline{r}) = C^l_{r_j - j} \left(1 - \frac{1}{q}\right)^l \left(\frac{1}{q}\right)^{r_j - j - l}$$

$$= C^l_{r_j - j} \frac{(q-1)^l}{q^{r_j - j}}, \quad j = 1, \dots, k, \quad l = 0, \dots, r_j - j, \qquad [6.33]$$

is the probability that the jth row of the matrix B contains l non-zero elements (the corresponding row of the matrix A contains $l + 1$ non-zero elements). The simultaneous fulfillment of the conditions (a)–(c) means that each of the first $i - 1$ rows of the matrix A contains at least $\omega + 1$ non-zero elements, and each of the rows $i, \dots, k$ contains at least ω non-zero elements. Therefore, we have

$$\mathbf{P}\{D^{(i)}_\omega(\overline{r})\} = \prod_{j=1}^{i-1} \sum_{l=\omega}^{r_j - j} p_j(l; \overline{r})$$

$$\times \prod_{j=i+1}^{k} \sum_{l=\omega-1}^{r_j - j} p_j(l; \overline{r}) \sum_{s=\omega-1}^{r_i - i} p_i(s; \overline{r}) \; \mathbf{E}_{\overline{\nu}(s)} \, \mathbf{P}\{D^{(i)}_\omega(\overline{r}) \,|\, \overline{\nu}(s)\}, \qquad [6.34]$$

where an expectation $\mathbf{E}_{\overline{\nu}(s)}$ is taken over the distribution of the random vector $\overline{\nu}(s) = \{\nu_{jl}(s)\}$. Each of the components $\nu_{jl}(s), \; l = 1, \dots, r_j - j, \; j = 1, \dots, k$, takes the value 0 with the probability $\frac{1}{q}$ and the value 1 with the probability $\frac{q-1}{q}$ provided that $\nu_{j1}(s) + \dots + \nu_{jr_j - j}(s) \geq \omega, \; j = 1, \dots, i-1, \; \nu_{i1}(s) + \dots + \nu_{ir_i - i}(s) = s, \; \nu_{j1}(s) + \dots + \nu_{jr_j - j}(s) \geq \omega - 1, \; j = i+1, \dots, k$.

We first construct an upper bound for the probability $\mathbf{P}\{D^{(i)}_\omega(\overline{r}) \,|\, \overline{\nu}(s)\}$, supposing that the components of the vector $\overline{\nu}(s)$ are known and fixed. Disregarding the conditions (a) and (b), we obtain the following inequalities:

$$\mathbf{P}\{D^{(i)}_\omega(\overline{r}) \,|\, \overline{\nu}(s)\} \leq 1 \text{ when } s = \omega - 1, \qquad [6.35]$$

$$\mathbf{P}\{D^{(i)}_\omega(\overline{r}) | \overline{\nu}(s)\} \leq \sum_{m=1}^{i-1} \sum_{1 \leq j_1 < \dots < j_m \leq i-1} \mathbf{P}\{E^{(i)}_{\omega,s,m}(\overline{r}, \overline{j}) | \overline{\nu}(s)\}$$
$$\text{when } s = \omega, \dots, r_i - i, \qquad [6.36]$$

where $E^{(i)}_{\omega,s,m}(\overline{r}, \overline{j}) = \{$there is a linear combination (containing exactly ω non-zero elements) of the ith row of the matrix A with s non-zero elements and rows specified by the vector $\overline{j} = (j_1, \ldots, j_m)$; at the same time, there are no such linear combinations determining a vector with a smaller number of non-zero elements$\}$. Construct an upper bound for the probability $\mathbf{P}\{E^{(i)}_{\omega,s,m}(\overline{r},\overline{j})|\overline{\nu}(s)\}$. Denote

$$\mu_l(s) = \begin{cases} 0 \text{ if } \nu_{j_z l}(s) = 0 \text{ for all } z = 1, \ldots, m, \\ 1 \text{ otherwise}, \end{cases} \qquad l = 1, \ldots, r_i - i.$$

Let w be the number of such indices l such that $\mu_l(s) = 1$ and $\nu_{il}(s) = 1$, and δ be the number of such indices l such that $\mu_l(s) + \nu_{il}(s) = 1$. It is obvious that any linear combination of the ith row with the rows specified by the vector $\overline{j}$ contains at least $m+1+\delta$ non-zero elements. If $m+1+\delta > \omega$, then $\mathbf{P}\{E^{(i)}_{\omega,s,m}(\overline{r},\overline{j})|\overline{\nu}(s)\} = 0$. Let $\sigma = \omega - (m+1+\delta) \geq 0$. Then

$$\begin{aligned} \mathbf{P}\{E^{(i)}_{\omega,s,m}(\overline{r},\overline{j})|\overline{\nu}(s)\} &\leq (q-1)^{m-1}\frac{\varphi^{(1)}_w(\sigma, q-1)}{(q-1)^w} \\ &= (q-1)^{m-w-1}\varphi^{(1)}_w(\sigma, q-1). \end{aligned} \qquad [6.37]$$

In fact, $(q-1)^{m-1}$ is the number of possible ways of constructing different (up to a factor) linear combinations of rows specified by the vector $\overline{j}$. Let $\overline{\mu}$ be a fixed linear combination of m rows. Then $(q-1)^w$ is the number of all the possible ways of filling w positions with non-zero elements in the ith row, $\varphi^{(1)}_w(\sigma, q-1)$ is the number of ways of filling w positions in the ith row for which there is a linear combination of the ith row and a vector $\overline{\mu}$ containing exactly σ non-zero elements, and there are no linear combinations with a smaller number of non-zero elements. The relations [2.31]–[2.36] make it possible to formulate an algorithm for the construction of an upper bound for the probability $\mathbf{P}\{Y_\omega(\overline{r})\}$.

THEOREM 6.3.– The following inequality holds true:

$$\mathbf{P}\{Y_\omega(\overline{r})\} \leq \mathbf{E}\alpha^{(up)}_\omega(\overline{r}).$$

The algorithm for the construction of independent trials of the random variable $\alpha^{(up)}_\omega(\overline{r})$ is formulated as follows.

1) Evaluate the probabilities $\{p_j(l;\overline{r})\}$ by the formula [2.32].

2) For each $i = 1, \ldots, k$ perform the following steps:

- evaluate (see [2.33]):

$$d^{(i)}_\omega(\overline{r}) = p_i(\omega-1;\overline{r}) \prod_{j=1}^{i-1} \sum_{l=\omega}^{r_j-j} p_j(l;\overline{r}) \prod_{j=i+1}^{k} \sum_{l=\omega-1}^{r_j-j} p_j(l;\overline{r}); \qquad [6.38]$$

- simulate the state of components of the vector $\overline{\nu} = \{\nu_{jl}\}$; each of its components takes the value 0 with the probability $\frac{1}{q}$ and 1 with the probability $\frac{q-1}{q}$ provided that $\nu_{j1} + \ldots + \nu_{jr_j-j} \geq \omega$, $j = 1, \ldots, i$, $\nu_{j1} + \ldots + \nu_{jr_j-j} \geq \omega - 1$, $j = i+1, \ldots, k$.

3) As the independent trial of the random variable $\alpha_\omega^{(up)}(\overline{r})$ take

$$\alpha_\omega^{(up)}(\overline{r}) = \sum_{i=1}^{k} \left[d_\omega^{(i)}(\overline{r}) + h_\omega^{(i)}(\overline{r}) \sum_{m=1}^{i-1} \right.$$

$$\left. \times \sum_{1 \leq j_1 < \ldots < j_m \leq i-1} (q-1)^{m-w(\overline{j})-1} \varphi_w^{(1)}(\sigma(\overline{j}), q-1) \right], \qquad [6.39]$$

where

$$h_\omega^{(i)}(\overline{r}) = \prod_{j=1}^{i} \sum_{l=\omega}^{r_j-j} p_j(l;\overline{r}) \prod_{j=i+1}^{k} \sum_{l=\omega-1}^{r_j-j} p_j(l;\overline{r}), \qquad [6.40]$$

and the values $w(\overline{j})$ and $\sigma(\overline{j})$ are computed according to the same rules as w and σ in the formula [2.36]; here, as above, $\overline{j} = (j_1, \ldots, j_m)$.

Let us construct a lower bound for the probability $\mathbf{P}\{D_\omega^{(i)}(\overline{r})\}$. The formula [2.33] implies the following lower estimate

$$\mathbf{P}\{D_\omega^{(i)}(\overline{r})\} \geq p_i(\omega-1;\overline{r})$$

$$\times \mathbf{P}\{D_\omega^{(i)}(\overline{r})\} \geq p_i(\omega-1;\overline{r}) \prod_{j=1}^{i-1} \sum_{l=\omega}^{r_j-j} p_j(l;\overline{r})$$

$$\times \prod_{j=i+1}^{k} \sum_{l=\omega-1}^{r_j-j} p_j(l;\overline{r})\, \mathbf{E}_{\overline{\nu}^{(i)}}\, \mathbf{P}\{D_\omega^{(i)}(\overline{r}) | \overline{\nu}^{(i)}\}, \qquad [6.41]$$

where the expectation $\mathbf{E}_{\overline{\nu}^{(i)}}$ is taken over the distribution of the random vector $\overline{\nu}^{(i)} = \{\nu_{jl}^{(i)}\}$. Each of the components $\nu_{jl}^{(i)}$, $l = 1, \ldots, r_j - j$, $j = 1, \ldots, k$, takes the value 0 with the probability $\frac{1}{q}$ and the value 1 with the probability $\frac{q-1}{q}$ provided that $\nu_{j1}^{(i)} + \ldots + \nu_{jr_j-j}^{(i)} \geq \omega$, $j = 1, \ldots, i-1$, $\nu_{i1}^{(i)} + \ldots + \nu_{ir_i-i}^{(i)} = \omega - 1$, $\nu_{j1}^{(i)} + \ldots + \nu_{jr_j-j}^{(i)} \geq \omega - 1$, $j = i+1, \ldots, k$.

Let us construct a lower bound for the probability $\mathbf{P}\{D_\omega^{(i)}(\overline{r}) \,|\, \overline{\nu}^{(i)}\}$ supposing that the components of the vector $\overline{\nu}^{(i)}$ are known and fixed. Introduce the following events: $V_\omega^{(i)}(\overline{r},\gamma) = \{$the first γ rows of the matrix A form the basis of a γ-dimensional vector space with the weight not less than $\omega+1\}$, $\gamma=1,\ldots,k$. It is obvious that $V_\omega^{(i)}(\overline{r},\gamma+1) \subset V_\omega^{(i)}(\overline{r},\gamma)$, $\gamma=1,\ldots,k-1$, and $V_\omega^{(i)}(\overline{r},1)$ is a certain event. Then we have

$$D_\omega^{(i)}(\overline{r}) = \bigcap_{\gamma=2}^{i-1} V_\omega^{(i)}(\overline{r},\gamma) \cap \bigcap_{\gamma=i+1}^{k} V_{\omega-1}^{(i)}(\overline{r},\gamma).$$

Therefore, we obtain

$$\mathbf{P}\{D_\omega^{(i)}(\overline{r})\,|\,\overline{\nu}^{(i)}\} = \prod_{\gamma=2}^{i-1} \mathbf{P}\{V_\omega^{(i)}(\overline{r},\gamma)\,|\,\overline{\nu}^{(i)};\, V_\omega^{(i)}(\overline{r},\gamma-1)\}$$

$$\times \prod_{\gamma=i+1}^{k} \mathbf{P}\{V_{\omega-1}^{(i)}(\overline{r},\gamma)\,|\,\overline{\nu}^{(i)};\, V_{\omega-1}^{(i)}(\overline{r},\gamma-1)\} \qquad [6.42]$$

(by the definition of the vector $\overline{\nu}^{(i)}$, the weight of a γ-dimensional vector space cannot be greater than ω when $\gamma \geq i+1$). At the same time, we have

$$\mathbf{P}\{V_\omega^{(i)}(\overline{r},\gamma)\,|\,\overline{\nu}^{(i)};\, V_\omega^{(i)}(\overline{r},\gamma-1)\} = 1-\mathbf{P}\{\overline{V}_\omega^{(i)}(\overline{r},\gamma)\,|\,\overline{\nu}^{(i)};\, V_\omega^{(i)}(\overline{r},\gamma-1)\}, [6.43]$$

where $\overline{V}_\omega^{(i)}(\overline{r},\gamma)$ denotes a complementary event to $V_\omega^{(i)}(\overline{r},\gamma)$. The occurrence of the event $\overline{V}_\omega^{(i)}(\overline{r},\gamma)$ provided that the event $V_\omega^{(i)}(\overline{r},\gamma-1)$ has occurred means that the elements in the γth row are located in such a way that a linear combination of this row and preceding rows make it possible to obtain a vector in which the number of non-zero elements is less than $\omega+1$. In order to construct an upper bound for the probability of the event $\overline{V}_\omega^{(i)}(\overline{r},\gamma)$, we will take an advantage of the same considerations as above. In fact, when $\gamma \geq 2$, we have

$$\mathbf{P}\{\overline{V}_\omega^{(i)}(\overline{r},\gamma)\,|\,\overline{\nu}^{(i)};\, V_\omega^{(i)}(\overline{r},\gamma-1)\}$$

$$\leq \sum_{m=1}^{\gamma-1} \sum_{1\leq j_1<\ldots<j_m\leq\gamma-1} \mathbf{P}\{W_{\omega,m}^{(i)}(\overline{r},\overline{j},\gamma)|\overline{\nu}^{(i)};\, V_\omega^{(i)}(\overline{r},\gamma-1)\}, \qquad [6.44]$$

where $W^{(i)}_{\omega,m}(\overline{r},\overline{j},\gamma) = \{$there is a linear combination of the γth row of the matrix A and the rows specified by the vector $\overline{j} = (j_1,\dots,j_m)$, and this combination contains no more than ω non-zero elements$\}$. Denote

$$\mu_l = \begin{cases} 0 \text{ if } \nu^{(i)}_{j_z l} = 0 \text{ for all } z = 1,\dots,m, \\ 1 \text{ otherwise,} \end{cases} \qquad l = 1,\dots,r_\gamma - \gamma.$$

Let w be the number of such indices l such that $\mu_l = 1$ and $\nu^{(i)}_{\gamma l} = 1$, and δ be the number of such indices l such that $\mu_l + \nu^{(i)}_{\gamma l} = 1$. It is obvious that any linear combination of the γth row with the rows specified by the $\overline{j}$th vector contains at least $m+1+\delta$ non-zero elements. If $m+1+\delta > \omega$, then $\mathbf{P}\{W^{(i)}_{\omega,m}(\overline{r},\overline{j},\gamma) \,|\, \overline{\nu}^{(i)}; V^{(i)}_{\omega}(\overline{r},\gamma-1)\} = 0$. Let $\sigma = \omega - (m+1+\delta) \geq 0$. Then

$$\mathbf{P}\{W^{(i)}_{\omega,m}(\overline{r},\overline{j},\gamma) \,|\, \overline{\nu}^{(i)}; V^{(i)}_{\omega}(\overline{r},\gamma-1)\} \;\leq\; (q-1)^{m-1} \frac{\sum\limits_{\tau=0}^{\sigma} \varphi^{(1)}_{w}(\tau,q-1)}{(q-1)^{w}}$$

$$= (q-1)^{m-w-1} \sum_{\tau=0}^{\sigma} \varphi^{(1)}_{w}(\tau,q-1). \qquad [6.45]$$

In fact, $(q-1)^{m-1}$ is the number of possible ways of constructing different (up to a factor) linear combinations of the rows specified by the vector $\overline{j}$. Let $\overline{\mu}$ be a fixed linear combination of m rows. Then $(q-1)^{w}$ is the number of all the possible ways of filling w positions with non-zero elements in the γth row, and $\sum\limits_{\tau=0}^{\sigma} \varphi^{(1)}_{w}(\tau,q-1)$ is the number of ways of filling w positions with non-zero elements in the γth row for which there is a linear combination of the γth row and the vector $\overline{\mu}$ containing no more than σ non-zero elements.

The probabilities $\mathbf{P}\{V^{(i)}_{\omega-1}(\overline{r},\gamma) \,|\, \overline{\nu}^{(i)}; V^{(i)}_{\omega-1}(\overline{r},\gamma-1)\}$, $\gamma = i+1,\dots,k$, can be evaluated by similar formulae.

The relations [2.31], [2.32] and [2.40]–[2.44] make it possible to formulate an algorithm for the construction of a lower bound for the probability $\mathbf{P}\{Y_\omega(\overline{r})\}$.

THEOREM 6.4.– The following inequality holds true:

$$\mathbf{P}\{Y_\omega(\overline{r})\} \;\geq\; \mathbf{E}\alpha^{(low)}_{\omega}(\overline{r}).$$

The algorithm for the construction of independent trials of the random variable $\alpha^{(low)}_{\omega}(\overline{r})$ is formulated as follows:

1) evaluate the probabilities $\{p_j(l;\overline{r})\}$ by the formula [2.32];

2) for each $i = 1, \ldots, k$ perform the following steps:
- evaluate $d_\omega^{(i)}(\overline{r})$ by the formula [2.37];

- simulate the state of components of the vector $\overline{\nu}^{(i)} = \{\nu_{jl}^{(i)}\}$; each of its components takes the value 0 with the probability $\frac{1}{q}$ and the value 1 with the probability $\frac{q-1}{q}$ provided that $\nu_{j1}^{(i)} + \ldots + \nu_{jr_j-j}^{(i)} \geq \omega,\ j = 1, \ldots, i-1,\ \nu_{i1}^{(i)} + \ldots + \nu_{ir_i-i}^{(i)} = \omega - 1,\ \ \nu_{j1} + \ldots + \nu_{jr_j-j} \geq \omega - 1,\ j = i+1, \ldots, k$;

- construct the lower bounds $\{g_\omega^{(i)}(\overline{r},\gamma)\}$ for the probabilities $\mathbf{P}\{V_\omega^{(i)}(\overline{r},\gamma)\,|\,\overline{\nu}^{(i)}; V_\omega^{(i)}(\overline{r},\gamma-1)\},\ \gamma = 2, \ldots, i-1$, and the lower bounds $\{g_{\omega-1}^{(i)}(\overline{r},\gamma)\}$ for the probabilities $\mathbf{P}\{V_{\omega-1}^{(i)}(\overline{r},\gamma)\,|\,\overline{\nu}^{(i)};\ V_{\omega-1}^{(i)}(\overline{r},\gamma-1)\},\ \gamma = i+1, \ldots, k$ (see [2.42]–[2.44]).

3) as the independent trial of the random variable $\alpha_\omega^{(low)}(\overline{r})$ take

$$\alpha_\omega^{(low)}(\overline{r}) = \sum_{i=1}^{k}\left[d_\omega^{(i)}(\overline{r})\prod_{\gamma=2}^{i-1} g_\omega^{(i)}(\overline{r},\gamma)\prod_{\gamma=i+1}^{k} g_{\omega-1}^{(i)}(\overline{r},\gamma)\right]. \qquad [6.46]$$

The algorithms presented above in Theorems 2.3 and 2.4 make it possible to construct upper and lower bounds for $\left[\begin{array}{c|c} n & \omega \\ k & \end{array}\right]$. At the same time, as numerous examples demonstrate, the approximation based on the combination of upper and lower bounds (see the formulae [2.38] and [2.45]) is the most exact, namely,

$$\begin{aligned}\alpha_\omega^{(appr)}(\overline{r}) &= \sum_{i=1}^{k}\left[d_\omega^{(i)}(\overline{r})\prod_{\gamma=2}^{i-1} g_\omega^{(i)}(\overline{r},\gamma)\prod_{\gamma=i+1}^{k} g_{\omega-1}^{(i)}(\overline{r},\gamma)\right.\\ &\left.+\, h_\omega^{(i)}(\overline{r})\sum_{m=1}^{i-1}\ \sum_{1\leq j_1<\ldots<j_m\leq i-1}(q-1)^{m-w(\overline{j})-1}\varphi_w^{(1)}(\sigma(\overline{j}),q-1)\right], \end{aligned} \qquad [6.47]$$

where $\{d_\omega^{(i)}(\overline{r})\}$ and $\{h_\omega^{(i)}(\overline{r})\}$ are evaluated due to the formulae [2.37] and [2.39].

6.2.5. *Numerical results*

We illustrate the accuracy of the upper and lower bounds by numerical examples. In the first example, we consider the case when the exact value of $\left[\begin{array}{c|c} n & \omega \\ k & \end{array}\right]$ is known, and in the second one we take such values of k and ω for which the bounds for $\left[\begin{array}{c|c} n \\ k \end{array} \omega\right]$ are not known. The use of fast simulation algorithms on the multiprocessor complex SKIT-3 made it possible to obtain high-accuracy estimates for rather large values of n, k and ω. We consider the case when $n = 50$ and $q = 2^5 = 32$. Let us introduce the following notations:

$\hat{\theta}_\omega^{(low)}(n,k)$ and $\hat{\theta}_\omega^{(up)}(n,k)$ are the upper and lower estimates for $\left[\begin{array}{c|c} n \\ k \end{array} \omega\right]$ constructed with the relative error $\varepsilon = 1\%$ and the confidence level 0.99;

$\hat{\theta}_\omega^{(appr)}(n,k)$ is the approximation for $\left[\begin{array}{c|c} n \\ k \end{array} \omega\right]$ obtained due to the formula [2.46].

EXAMPLE 6.1.– We consider the following two cases when the exact value of $\left[\begin{array}{c|c} n \\ k \end{array} \omega\right]$ can be found: (a) $k = 2$ and ω is arbitrary (see equation [2.16]), and (b) $\omega = 2$ and k is arbitrary (see the recurrent formulae from Masol (1989)). The corresponding estimates are presented in Tables 6.1 and 6.2.

ω	$\left[\begin{array}{c\|c} n \\ k \end{array} \omega\right]$	$\hat{\theta}_\omega^{(low)}(n,k)$	$\hat{\theta}_\omega^{(up)}(n,k)$	$\hat{\theta}_\omega^{(appr)}(n,k)$
6	$8.30 \cdot 10^{86}$	$8.24 \cdot 10^{86}$	$8.32 \cdot 10^{86}$	$8.32 \cdot 10^{86}$
12	$5.63 \cdot 10^{99}$	$5.33 \cdot 10^{99}$	$5.61 \cdot 10^{99}$	$5.61 \cdot 10^{99}$
18	$7.42 \cdot 10^{110}$	$6.48 \cdot 10^{110}$	$7.33 \cdot 10^{110}$	$7.33 \cdot 10^{110}$
24	$4.44 \cdot 10^{120}$	$3.41 \cdot 10^{120}$	$4.41 \cdot 10^{120}$	$4.41 \cdot 10^{120}$
30	$1.53 \cdot 10^{129}$	$0.97 \cdot 10^{129}$	$1.52 \cdot 10^{129}$	$1.52 \cdot 10^{129}$
36	$2.70 \cdot 10^{136}$	$1.31 \cdot 10^{136}$	$2.70 \cdot 10^{136}$	$2.70 \cdot 10^{136}$
42	$1.37 \cdot 10^{142}$	$0.40 \cdot 10^{142}$	$1.36 \cdot 10^{142}$	$1.36 \cdot 10^{142}$
48	$5.26 \cdot 10^{138}$	$5.00 \cdot 10^{138}$	$8.27 \cdot 10^{143}$	$5.26 \cdot 10^{138}$

Table 6.1. $n = 50,\ k = 2,\ q = 2^5,\ \varepsilon = 1\%$

In both cases, a high accuracy of estimates is demonstrated by the approximation $\hat{\theta}_\omega^{(appr)}(n,k)$ for very different values of ω $(k = 2)$ and k $(\omega = 2)$. Both the lower and upper bounds ensure the same order of magnitude of estimates, which was not easy to reach for such a complicated problem (with the exception of the case $\omega = 48$ in Table 6.1 when the upper bound is overestimated).

EXAMPLE 6.2.– Let us consider the case when $k = 10$. If $\omega > 2$, then there are no analytical or approximate formulae for the evaluation of $\left[\begin{array}{c|c} n \\ k \end{array} \omega\right]$. The accuracy of estimates can be judged only by indirect data. In particular, according to the formula [2.2] the sum of values of $\left[\begin{array}{c|c} n \\ k \end{array} \omega\right]$ taken over ω is equal to $\left[\begin{array}{c} n \\ k \end{array}\right]$, which can be computed in explicit form by the formula [2.1]. We verify this statement for n and k being considered. In fact, we have $\begin{bmatrix} n \\ k \end{bmatrix} = 1.19{\cdot}10^{602}$. The results of computations for various ω are presented in Table 6.3. According to these data, $\sum_{\omega=1}^{41} \hat{\theta}_{\omega}^{(appr)}(50,10) \approx 1.22 \cdot 10^{602}$, i.e. the results are very close.

k	$\begin{bmatrix} n \\ k \end{bmatrix}_\omega$	$\hat{\theta}_{\omega}^{(low)}(n,k)$	$\hat{\theta}_{\omega}^{(up)}(n,k)$	$\hat{\theta}_{\omega}^{(appr)}(n,k)$
2	$6.93 \cdot 10^{76}$	$6.86 \cdot 10^{76}$	$6.96 \cdot 10^{76}$	$6.96 \cdot 10^{76}$
8	$1.28 \cdot 10^{447}$	$1.25 \cdot 10^{447}$	$1.28 \cdot 10^{447}$	$1.28 \cdot 10^{447}$
14	$1.01 \cdot 10^{709}$	$0.94 \cdot 10^{709}$	$1.01 \cdot 10^{709}$	$1.01 \cdot 10^{709}$
20	$3.38 \cdot 10^{862}$	$2.86 \cdot 10^{862}$	$3.36 \cdot 10^{862}$	$3.37 \cdot 10^{862}$
26	$4.83 \cdot 10^{907}$	$3.53 \cdot 10^{907}$	$4.81 \cdot 10^{907}$	$4.82 \cdot 10^{907}$
32	$2.93 \cdot 10^{844}$	$1.74 \cdot 10^{844}$	$2.92 \cdot 10^{844}$	$2.93 \cdot 10^{844}$
38	$7.59 \cdot 10^{672}$	$3.23 \cdot 10^{672}$	$7.61 \cdot 10^{672}$	$7.60 \cdot 10^{672}$
44	$8.36 \cdot 10^{392}$	$1.90 \cdot 10^{392}$	$8.33 \cdot 10^{392}$	$8.36 \cdot 10^{392}$
49	$1.19 \cdot 10^{73}$	$1.19 \cdot 10^{73}$	$1.19 \cdot 10^{73}$	$1.19 \cdot 10^{73}$

Table 6.2. $n = 50,\ \omega = 2,\ q = 2^5,\ \varepsilon = 1\%$

ω	$\hat{\theta}_{\omega}^{(low)}(n,k)$	$\hat{\theta}_{\omega}^{(up)}(n,k)$	$\hat{\theta}_{\omega}^{(appr)}(n,k)$
5	$1.07 \cdot 10^{554}$	$1.46 \cdot 10^{554}$	$1.46 \cdot 10^{554}$
10	$7.02 \cdot 10^{564}$	$2.02 \cdot 10^{565}$	$2.02 \cdot 10^{565}$
15	$1.53 \cdot 10^{574}$	$1.29 \cdot 10^{575}$	$1.29 \cdot 10^{575}$
20	$2.32 \cdot 10^{582}$	$7.56 \cdot 10^{583}$	$7.52 \cdot 10^{583}$
25	$3.07 \cdot 10^{589}$	$5.78 \cdot 10^{591}$	$5.79 \cdot 10^{591}$
30	$3.14 \cdot 10^{595}$	$6.16 \cdot 10^{598}$	$6.19 \cdot 10^{598}$
31	$3.55 \cdot 10^{596}$	$1.24 \cdot 10^{600}$	$1.23 \cdot 10^{600}$
32	$3.45 \cdot 10^{597}$	$2.27 \cdot 10^{601}$	$2.14 \cdot 10^{601}$
33	$2.19 \cdot 10^{598}$	$2.45 \cdot 10^{602}$	$9.92 \cdot 10^{601}$
34	$3.79 \cdot 10^{596}$	$5.07 \cdot 10^{602}$	$3.11 \cdot 10^{599}$
35	0	$7.02 \cdot 10^{602}$	0

Table 6.3. $n = 50,\ k = 10,\ q = 2^5,\ \varepsilon = 1\%$

The calculations for numerous values of n, k and q in a wide range of their change ($5 \le n \le 100,\ 3 \le k \le n,\ 2 \le q \le 2^{10}$) have shown that in all the cases the value $\sum_{\omega=1}^{n-k+1} \hat{\theta}_{\omega}^{(appr)}(n,k)$ is an upper bound for $\begin{bmatrix} n \\ k \end{bmatrix}$, and the excess is not more than 15%. Let us consider the data presented in Table 6.3. We see that for the majority of the values of ω, the estimates $\hat{\theta}_{\omega}^{(appr)}(n,k)$ and $\hat{\theta}_{\omega}^{(up)}(n,k)$ are practically identical; only when the values of ω are close to maximally admissible ones do the upper estimates begin to lose their accuracy abruptly. Unfortunately, the accuracy of the lower estimates decreases with the increasing ω and with the increasing k. Nevertheless, the upper and lower estimates determine the range of changes in $\left[\begin{smallmatrix} n \\ k \end{smallmatrix}\middle|\,\omega\right]$ for $\omega > 3$ and $k > 2$, which was not achieved earlier. In this case, there is a reason to believe that the estimate $\hat{\theta}_{\omega}^{(appr)}(n,k)$ gives a sufficiently accurate approximation for $\left[\begin{smallmatrix} n \\ k \end{smallmatrix}\middle|\,\omega\right]$.

6.3. Evaluation of the number of "good" permutations by fast simulation on the SCIT-4 multiprocessor computer complex

One of the most difficult classical problems in discrete mathematics, which is still important, is the problem of enumeration of all the complete mappings of the algebraic structure $(G, +)$ (Sachkov 1996). The mapping $f : G \to G$ is called *complete* if $f(\cdot)$ is a bijection and mapping $h(x) = x + f(x)$ is also a bijection. Hsiang *et al.* (2004) analyzed the complexity of finding the number of all the complete mappings for various structures $(G, +)$. For closed structures, this problem is proved to be NP-complete.

Let $\overline{s} = (s_0, s_1, \ldots, s_{N-1})$ be an arbitrary permutation of symbols $0, 1, \ldots, N-1$. Construct a new set $\overline{t} = (t_0, t_1, \ldots, t_{N-1})$ using the rule $t_i = i + s_i \pmod N), i = 0, 1, \ldots, N-1$. If the set $\overline{t}$ is a permutation, then the original permutation $\overline{s}$ is called *"good"* (Cooper *et al.* 1999). In Hsu and Keedwell (1985), such a mapping is called *a strong complete mapping*.

An exhaustive analysis of the fundamentals of the application of "good" permutations in cryptography is presented in Cooper *et al.* (1999). The principles of using such permutations in the rotor cryptographic systems are described by Konheim (1991).

Despite numerous attempts (Cooper *et al.* 1999; Cooper and Kovalenko 1995; Shieh 2001), an analytical approach to finding the number M_N of "good" permutations has not been found yet. It is only known that $M_N = 0$ for even N. The use of fast methods of direct enumeration and modern multiprocessor computer complexes made it possible to calculate M_N for all the odd $N \leq 25$. For example, the values of M_N for $N \leq 19$ are presented in Cooper *et al.* (1999), for $N \leq 23$ in Shieh (2001) and for $N \leq 25$ in The On-Line Encyclopedia of Integer Sequences. Further, an increase in N by two increases the computation time by two orders of magnitude, which makes finding the exact values of M_N for odd $N > 25$ an unrealistic task, even in case of fast development of computational capabilities. Therefore, the main efforts in the analysis of M_N are concentrated on approximate computing methods, in particular, asymptotic and statistical ones.

Since the total number of permutations is $N!$, the deterministic problem of finding M_N is equivalent to the probabilistic problem of estimating the probability P_N that a permutation constructed at random is "good", in this case $P_N = M_N/N!$.

In Kuznetsov (2007, 2008), a new approach enabling the estimation P_N for large values of N has been proposed. The fast simulation algorithm makes it possible to evaluate P_N for any $N \leq 205$ with the relative error not greater than 5%. Further improvement of the method (see next sections) enables the reduction from 10 to 100 times the time needed to evaluate the probability P_N with the given relative error and confidence level. The implementation of a fast simulation algorithm on the multiprocessor computer complex SCIT-4 at the V.M. Glushkov Institute of Cybernetics of the NAS of Ukraine makes it possible to calculate P_N for $N = 305$ with the confidence level 0.99 and the relative error of only 1%.

The probability P_N can be represented as

$$P_N = e^{\alpha_N - \beta_N N}. \tag{6.48}$$

It is hypothesized (Cooper *et al.* 1999) that

$$\beta_N \to 1 \text{ and } \alpha_N \to const \text{ as } N \to \infty. \tag{6.49}$$

As can be seen from the calculations presented above, the statistical data do not contradict the hypothesis ([3.1] and [3.2]).

6.3.1. *Modified fast simulation method*

The fast simulation algorithm (Kuznetsov 2007) is based on the idea of elaboration of some rules for the direct arrangement of symbols $0, 1, \ldots, N-1$ at one of N positions (for an improved version of the algorithm, see Kuznetsov (2008)). The main objective was to substantially increase the probability of constructing a "good" permutation. The unbiasedness of the estimate is attained due to the appropriate normalizing multiplier. The modified fast simulation algorithm described above increases this probability even more. The algorithm consists of three stages.

Let r and m be such integer parameters that $1 \leq r \leq m \leq N-1$. We suppose that these parameters are fixed (the considerations concerning their rational choice are stated at the end of this section). At the first stage, we arrange symbols at positions with numbers $m, \ldots, N-1$. Since the probability of obtaining a "bad" permutation is rather small, we use the simplest (and the fastest) algorithm. As the positions are filled in, the probability of obtaining a "bad" permutation essentially increases. Therefore, more careful analysis is needed at the second stage in order to determine symbols, which can be disposed at one position or another. At the same time, it is analyzed even if the partially filled permutation is deadlocked, and it is impossible to obtain any "good" permutation from it. At this stage, $m - r$ positions are filled in. At the third stage (deterministic algorithm), the last r positions are filled in so as to obtain all the possible "good" permutations. The estimate remains unbiased due to the appropriate normalizing factors.

The N-dimensional vector $\overline{s} = (s_0, s_1, \ldots, s_{N-1})$ is called the *current state* of the permutation. The notation $s_i = -1$ means that the position i has not been occupied yet by any of the symbols; if $s_i = k \geq 0$, then the symbol k is placed at the position i. The state $\overline{s}$ is called *"deadlocked"* if no arrangement of symbols allows obtaining a "good" permutation from this state. Denote

$$\nu_i(\overline{s}) = \begin{cases} 1 \text{ if } i = s_k \neq -1 \text{ for some } k, \\ 0 \text{ otherwise,} \end{cases} \qquad i = 0, 1, \ldots, N-1;$$

$$\mu_j(\overline{s}) = \begin{cases} 1 \text{ if } j = k + s_k (\text{mod } N), \ s_k \neq -1, \text{ for some } k, \\ 0 \text{ otherwise,} \end{cases} \quad j = 0, 1, \ldots, N-1.$$

The value of $\nu_i(\overline{s})$ indicates that for the state $\overline{s}$ the ith symbol is already placed at one of the positions of the vector $\overline{s}$, $\mu_j(\overline{s})$ indicates that for the state $\overline{s}$ the jth symbol is already placed at one of the positions of the vector $\overline{t} = (t_0, t_1, \ldots, t_{N-1})$, $t_i = i + s_i \ (\text{mod } N)$, $i = 0, 1, \ldots, N-1$. In the algorithm presented below, we use the notation $\hat{P}_{N1}$ for the estimate in one trial for P_N.

Stage 1 (a phone dial algorithm).

1) Let $\hat{P}_{N1} = 1$ be the initial value of the estimate. Without the loss of generality, as the initial state $\overline{s}$ choose: $\overline{s} = (0, -1, \ldots, -1)$, $\nu_i(\overline{s}) = 0$, $\mu_j(\overline{s}) = 0$, $i, j = 1, \ldots, N-1$, $\nu_0(\overline{s}) = 1$, and $\mu_0(\overline{s}) = 1$.

2) Assume that l (position number) sequentially takes the values $m + 1$, $m+2, \ldots, N-1$. Determine the set of symbols which can be disposed at the position l:

$$A_l(\overline{s}) = \{i : \nu_i(\overline{s}) = 0, \ \mu_k(\overline{s}) = 0, \ k = l + i \ (\text{mod } N)\}. \qquad [6.50]$$

Denote $|A_l(\overline{s})|$ the number of symbols in the set $A_l(\overline{s})$. If $|A_l(\overline{s})| = 0$, then both stage 1 and the algorithm are ended, and as the estimate we take $\hat{P}_{N1} = 0$ (in this trial, it is impossible to construct a "good" permutation). If $|A_l(\overline{s})| > 0$, then with the probability $\frac{1}{|A_l(\overline{s})|}$ choose one of the symbols from the set $A_l(\overline{s})$. If this is the symbol i, then we set

$$\hat{P}_{N1} := \hat{P}_{N1} \cdot \frac{|A_l(\overline{s})|}{N + m - l}, \quad s_l = i, \ \nu_i = 1, \ \mu_k = 1, \qquad [6.51]$$

where $k = l + i \pmod N$ (a symbol $:=$ means that the new value of $\hat{P}_{N1}$ is the product of the old value $\hat{P}_{N1}$ by the appropriate factor). Note that $N + m - l$ is the number of symbols that have not been allocated yet for the state $\overline{s}$. Incrementing l by one, repeat step 2 of the algorithm until the $(N-1)$th position is filled.

Stage 2 (the algorithm of directed enumeration).

Let $\overline{s}$ be some state in which L positions are not filled. This state is uniquely determined by the vector $\overline{W}$ of the following structure:

$$\overline{W} = (\theta_1, \kappa_1, \gamma_1^{(1)}, \ldots, \gamma_{\kappa_1}^{(1)}; \quad \ldots; \quad \theta_L, \kappa_L, \gamma_1^{(L)}, \ldots, \gamma_{\kappa_L}^{(L)}), \qquad [6.52]$$

where $\theta_1, \ldots, \theta_L$ are the numbers of non-filled positions, κ_l is the number of symbols that can be placed at the position θ_l, and $\gamma_1^{(l)}, \ldots, \gamma_{\kappa_l}^{(l)}$ are these symbols. Just the vector $\overline{W}$ will characterize the current state of permutation. The algorithm in Stage 2 is formulated as follows:

1) let $\overline{s}$ be the state resulted from the Stage 1 of the algorithm. Construct the vector $\overline{W}$ of the form [3.5], where $L = m, \quad (\theta_1, \ldots, \theta_L) = (1, \ldots, m)$;

2) let $\overline{W}$ be the current state. Calculate $l^* = \arg\min\{\kappa_1, \ldots, \kappa_L\}$ (if the minimum is attained on several values of $\{\kappa_j\}$, then choose the least value of j). If $\kappa_{l^*} = 0$, then none of the symbols can be placed at the position l^*. In this case, both Stage 2 and the algorithm are ended, and as the estimate we take $\hat{P}_{N1} = 0$ (in this trial it is impossible to construct a "good" permutation). If $\kappa_{l^*} > 0$, then with probability $\frac{1}{\kappa_{l^*}}$ choose one of the symbols $\gamma_1^{(l^*)}, \ldots, \gamma_{\kappa_{l^*}}^{(l^*)}$. If this is the symbol i, then place it at the position θ_{l^*} and construct a new state $\overline{W}$. Moreover, eliminate from the consideration all such $\gamma_j^{(l)}$ that $\gamma_j^{(l)} = i$ or $\gamma_j^{(l)} + \theta_l = i + \theta_{l^*} \pmod N$. Recalculate the normalizing factor:

$$\hat{P}_{N1} := \hat{P}_{N1} \cdot \frac{\kappa_{l^*}}{L}. \qquad [6.53]$$

Repeat $m - r$ times step 2 of Stage 2 of the algorithm. As the result, we obtain the state $\overline{W}$ of the form [3.5] in which there are exactly $L = r$ non-filled positions.

Stage 3 (the deterministic algorithm of the enumeration of all "good" permutations that can be obtained from the state $\overline{W}$).

Each state [3.5] can generate no more than l^* new states (some of them can be deadlocked). In turn, each of the newly obtained states generates a set of new states, etc. Repeat this recurrent algorithm $r-1$ times. If at one of the intermediate steps of the algorithm all the states appear deadlocked, then $\hat{P}_{N1} = 0$. If all $r - 1$ steps have been made, and $R > 0$ states were obtained (in this trial R "good" permutations have been constructed), then the algorithm is ended, and as an unbiased estimate constructed in one trial we take

$$\hat{P}_{N1} := \hat{P}_{N1} \cdot \frac{R}{r!}. \tag{6.54}$$

The algorithm described above depends on two parameters: r and m, $1 \leq r \leq m \leq N-1$. Actually, each pair (r, m) determines its own algorithm of the construction of the unbiased estimate. The time needed to construct an estimate with the given accuracy is known to be proportional to the product of average time of one trial by the variance of the estimate. The problem of optimal choice of r and m is to minimize this product. Since both r and m increase, the probability of the construction of at least one "good" permutation increases and it is logical to suppose that the variance of the estimate monotonically decreases with respect to r and m. At the same time, with the increase of r and m, the amount of the processed information increases, hence the average time of the trial increases. One more basic complexity of determining the optimal values of r and m is that the exact value of the variance is not known; therefore, sampling variance is used, which is not very stable. A practical recommendation is to carry out preliminary evaluations (with a rather small number of trials), which make it possible to find nearly optimal values of r and m. Then for several pairs (r, m) we construct estimates with the given accuracy and choose the pair for which the time spent is minimum. It is possible to determine a certain regularity of the choice of r and m depending on N. This approach was used to obtain the numerical data presented in the next section.

6.3.2. *Numerical results*

Let us present the estimates of the probability P_N, $N = 25, 35, \ldots, 305$, obtained by the method described above on the SCIT-4 multiprocessor computer complex (see Table 6.4). All the estimates are constructed with the use of only 20 processors.

We use the following notations: ε is the relative error of the estimate (in percentage); r and m are the parameters of the method for which the estimates are constructed; $\hat{P}_N$ and $\hat{M}_N$ are the estimates constructed for P_N and M_N with the confidence level 0.99 and the relative error ε; $\hat{C}_N$ is the estimate of the relative mean square error (the ratio of the root of the sampling variance to the estimate $\hat{P}_N$); $\hat{K}_N$ is the estimate of the "successful trials" frequency, i.e. it is the estimate of the proportion of the number of trials in which it was possible to construct "good" permutations (in percentage); $\hat{T}_N$ is the time spent by the SCIT-4 supercomputer to construct the estimate $\hat{P}_N$ (e.g. the notation $ww : xx : yy : zz$ means that ww days, xx hours, yy minutes and zz seconds were used).

The authors do not state that the above values of r and m are optimal. However, even for these values a high accuracy of estimates is attained for quite comprehensive time (the greatest time is 39 h for the estimate P_{125} with the relative error of 0.03%). The relative mean squared error $\hat{C}_N$ increases with the increase in N (there are exceptions due to the not optimal choice of r and m). It can be easily seen that the following rule was applied to choose r and m: the frequency of "successful" trials is approximately 20%.

N	ε, %	r	m	$\hat{P}_N$	$\hat{M}_N$	$\hat{C}_N$	$\hat{K}_N$, %	$\hat{T}_N$
25	0.01	9	12	$2.682 \cdot 10^{-9}$	$4.161 \cdot 10^{16}$	1.39	62.3	19:51
35	0.01	11	15	$2.008 \cdot 10^{-12}$	$2.075 \cdot 10^{27}$	1.89	50.9	01:08:48
45	0.01	12	20	$1.326 \cdot 10^{-17}$	$1.586 \cdot 10^{39}$	2.55	40.5	02:57:48
55	0.01	14	22	$8.117 \cdot 10^{-22}$	$1.031 \cdot 10^{52}$	2.97	39.1	06:54:23
65	0.01	15	26	$4.727 \cdot 10^{-26}$	$3.899 \cdot 10^{65}$	3.70	33.1	13:42:18
75	0.03	16	28	$2.660 \cdot 10^{-30}$	$6.600 \cdot 10^{79}$	4.81	25.1	02:59:05
85	0.03	17	33	$1.453 \cdot 10^{-34}$	$4.094 \cdot 10^{94}$	5.49	24.6	05:20:13
95	0.03	18	35	$7.800 \cdot 10^{-39}$	$8.057 \cdot 10^{109}$	6.63	20.4	09:12:43
105	0.03	19	39	$4.117 \cdot 10^{-43}$	$4.452 \cdot 10^{125}$	7.48	19.9	15:14:42
115	0.03	20	45	$2.137 \cdot 10^{-47}$	$6.250 \cdot 10^{141}$	7.54	22.0	21:51:40
125	0.03	21	50	$1.104 \cdot 10^{-51}$	$2.078 \cdot 10^{158}$	8.65	23.2	01:15:08:35
135	0.1	22	53	$5.604 \cdot 10^{-56}$	$1.508 \cdot 10^{175}$	9.66	22.9	05:41:57
145	0.1	23	53	$2.831 \cdot 10^{-60}$	$2.279 \cdot 10^{192}$	10.61	20.1	08:11:03
155	0.1	23	65	$1.414 \cdot 10^{-64}$	$6.773 \cdot 10^{209}$	13.06	18.0	12:32:45
165	0.1	24	69	$7.060 \cdot 10^{-69}$	$3.829 \cdot 10^{227}$	13.00	19.4	16:23:12
175	0.1	25	71	$3.519 \cdot 10^{-73}$	$3.957 \cdot 10^{245}$	13.74	19.9	23:43:02
185	0.1	26	72	$1.736 \cdot 10^{-77}$	$7.159 \cdot 10^{263}$	14.91	19.9	01:12:57:21
195	0.3	26	91	$8.484 \cdot 10^{-82}$	$2.199 \cdot 10^{282}$	16.22	19.8	05:17:21
205	0.3	27	88	$4.165 \cdot 10^{-86}$	$1.132 \cdot 10^{301}$	18.55	19.1	08:06:52
215	0.3	28	89	$2.030 \cdot 10^{-90}$	$9.416 \cdot 10^{319}$	18.89	20.2	11:35:26
225	0.3	28	111	$9.843 \cdot 10^{-95}$	$1.240 \cdot 10^{339}$	20.91	19.8	15:13:13
235	0.3	29	108	$4.796 \cdot 10^{-99}$	$2.555 \cdot 10^{358}$	22.72	20.4	22:25:01
245	0.3	30	103	$2.305 \cdot 10^{-103}$	$7.942 \cdot 10^{377}$	22.16	19.8	01:03:42:17
255	1	30	129	$1.115 \cdot 10^{-107}$	$3.736 \cdot 10^{397}$	26.73	20.2	03:52:18
265	1	31	122	$5.337 \cdot 10^{-112}$	$2.572 \cdot 10^{417}$	24.45	20.4	04:06:24
275	1	31	151	$2.573 \cdot 10^{-116}$	$2.600 \cdot 10^{437}$	31.16	19.7	07:17:39
285	1	32	143	$1.223 \cdot 10^{-120}$	$3.724 \cdot 10^{457}$	28.47	20.8	07:33:13
295	1	33	126	$5.868 \cdot 10^{-125}$	$7.642 \cdot 10^{477}$	31.33	18.3	11:04:14
305	1	34	128	$2.810 \cdot 10^{-129}$	$2.196 \cdot 10^{498}$	33.70	21.2	22:22:45

Table 6.4. *Estimates of the number of "good" permutations by the fast simulation method*

At the beginning of this section, the following hypothesis was advanced: if the probability P_N is represented as [3.1], the relation [3.2] holds true. Let us test whether the obtained statistical data does not contradict this hypothesis.

As the initial statistical data, we have 29 observations: $\hat{P}_{25}, \hat{P}_{35}, \ldots, \hat{P}_{305}$. In order to evaluate the asymptotic values of the parameters α_N and β_N, let us use a linear regression and construct the estimates of these parameters using the residual of the series $\hat{P}_s$, $\hat{P}_{s+10}$, $\hat{P}_{s+20}$, ..., $\hat{P}_{305}$ for different values of s. Denote the corresponding estimates by $\hat{\alpha}(s)$ and $\hat{\beta}(s)$. Using the least squares method, we obtain the estimates presented in Table 6.5.

s	25	75	115	155	195	235	275
$\hat{\alpha}$(s)	5.9422	6.5357	6.8086	7.0097	7.1731	7.3186	7.4194
$\hat{\beta}$(s)	0.9887	0.9914	0.9925	0.9933	0.9940	0.9945	0.9948

Table 6.5. *Estimates for the parameters $\alpha(s)$ and $\beta(s)$*

As s increases, we observe the monotone increase in $\alpha(s)$ and $\beta(s)$, and in both cases it is a growth with deceleration, which allows us to suppose that the limits for $\alpha(s)$ and $\beta(s)$ exist, and the limit for $\alpha(s)$ is finite, and for $\beta(s)$ is rather close to one. Thus, the results of the computations show that the obtained statistical data do not contradict the hypothesis [3.1] and [3.2].

The algorithm in the previous section includes three stages. The first two are based on a random choice of the symbol and position for it, and the third one is deterministic. It is possible to omit the first two stages and use the deterministic part only. In this case, we obtain exact values of the number of "good" permutations. The exact values of M_N calculated by SCIT-4 for $N = 9, 11, \ldots, 23$ are presented in Table 6.6. The estimates obtained by fast simulation on a personal computer (actually one processor) are also shown for comparison.

N	M_N	Number of Processors	Time (SCIT-4)	PC Estimate (relative error 0.1%)	Time (PC)
9	2 025	2	0.00	$2.0242 \cdot 10^3$	9
11	37 851	2	0.00	$3.7878 \cdot 10^4$	15
13	1 030 367	2	0.00	$1.0309 \cdot 10^6$	36
15	36 362 925	2	0.00	$3.6371 \cdot 10^7$	01:11
17	1 606 008 513	2	02:08	$1.6069 \cdot 10^9$	02:37
19	87 656 896 891	10	25:36	$8.7677 \cdot 10^{10}$	03:06
21	5 778 121 715 415	50	04:39:28	$5.7808 \cdot 10^{12}$	06:35
23	452 794 813 251 584	100	10:08:10:40	$4.5328 \cdot 10^{14}$	15:47

Table 6.6. *Comparison of exact values with statistical estimates*

The statistical estimates are close to the exact values of M_N. While for $N \leq 15$, the exact values are calculated almost instantly, as N increases, the computing time increases sharply. Thus, for $N = 23$ more than 10 days with 100 processors were required. At the same time, for the approximate estimate with the relative error of 0.1% less than 16 min on one processor was needed. As N increases by two, the computing time increases by two orders of magnitude. This can explain why the exact value of the number of “good” permutations for $N = 27$ has not been calculated yet.

Thus, the proposed fast simulation method is a real tool to estimate the number of “good” permutations for $N > 300$ in a reasonable time with rather a small relative error.

6.4. References

Andrews, G.E. (1998). *The Theory of Partitions*. Cambridge University Press, Cambridge.

Blanchet, J., Lam, H. (2011). Rare event simulation techniques. *Proc. 2011 Winter Simul. Conf.*, Las Vegas, 217–231.

Buslenko, N.P., Kovalenko, I.N., Kalashnikov, V.V. (1973). *Lectures on the Theory of Complex Systems*. Sov. Radio, Moscow (in Russian).

Calabi, E., Wilf, H. (1977). On the sequential and random selection of subspaces over a finite field. *J. Combin. Theory*, 22, 107–109.

Cooper, R., Kovalenko, I.N. (1995). An upper bound for the number of complete mappings. *Teoriya Veroyatn. i Mat. Statistika*, 53(5), 69–75 (in Russian).

Cooper, R., Gilchrist, R., Kovalenko, I.N., Novakovic, D. (1999). Estimation of the number of “good” permutations, with application to cryptography. *Cybernet. Syst. Anal.* 35(5), 688–693.

Fox, B.L., Glynn, P.W. (1990). Discrete-time conversion for simulating finite-horizon Markov processes. *SIAM J. Appl. Math.*, 50, 1457–1473.

Glasserman, P., Hedelberger, Ph., Shahabuddn, P., Zajc, T. (1999). Multilevel splitting for estimating rare event probabilities. *Oper. Res.*, 47, 585–600.

Glasserman, P. (2004). *Monte Carlo Methods in Financial Engineering*. Springer, New York.

Heidelberger, P. (1995). Fast simulation of rare events in queueing and reliability models. *ACM Trans. Modeling Comput. Simul.*, 5, 43–85.

Hsiang, J., Hsu, D.F., Shieh, Y.P. (2004). On the hardness of counting problems of complete mappings. *Discr. Math.*, 277(1–3), 87–100.

Hsu, D.F., Keedwell, A.D. (1985). Generalized complete mappings, neofields, sequenceable groups and block designs II. *Pacific J. Math.*, 117, 291–312.

Kleijnen, J.P. (1987). *Statistical Tools for Simulation Practitioners*. Marcel Dekker, New York.

Konheim, R. (1991). *Cryptography: A Primer*. Wiley, Chichester.

Kovalenko, I.N. (1975). *Investigations on the Analysis of Reliability of Complex Systems*. Naukova Dumka, Kiev (in Russian).

Kovalenko, I.N. (1980). *Analysis of Rare Events in Evaluating Systems Efficiency and Reliability*. Sov. Radio, Moscow (in Russian).

Kovalenko, I.N. (1980). Analysis of characteristics of highly reliable systems by the analytical-statistical method. *Electron. Model.* 2(4), 5–8 (in Russian).

Kovalenko, I.N. (1996). Upper bound on the number of complete maps. *Cybernet. Syst. Anal.*, 32(1), 65–68.

Kovalenko, I.N., Kuznetsov, N.Y., Pegg, P.A. (1997). *Mathematical Theory of Reliability of Time Dependent Systems with Practical Applications*. Wiley, Chichester.

Kovalenko, I.N., Kuznetsov, N.Y. (1988). *Methods for Evaluation of Highly Reliable Systems*. Radio i Svyaz, Moscow (in Russian).

Kuznetsov, N.Y. (1996). Computing the permanent by importance sampling method. *Cybernet. Syst. Anal.*, 32, 749–755.

Kuznetsov, N.Y. (1998). Using stratified sampling to solve the knapsack problem. *Cybernet. Syst. Anal.*, 34, 61–68.

Kuznetsov, N.Y. (2007). Applying fast simulation to find the number of "good" permutations. *Cybernet. Syst. Anal.* 43(6), 830–837.

Kuznetsov, N.Y. (2008). Estimating the number of good permutations by a modified fast simulation method. *Cybernet. Syst. Anal.* 44(4), 547–554.

Kuznetsov, N.Y. (2009). Estimating the number of Latin rectangles by the fast simulation method. *Cybernet. Syst. Anal.*, 45, 69–75.

Kuznetsov, N.Y. (2009). Fast simulation technique in reliability evaluation of Markovian and non-Markovian systems. In *Simulation and Optimization Methods in Risk and Reliability Theory*, Knopov, P.S., Pardalos, P.M. (eds). Nova Science Publishers, New York, 69–112.

Kuznetsov, I.N. (2010). Upper and lower estimates for the number of some k-dimensional subspaces of a given weight over a finite field. *Cybernet. Syst. Anal.*, 46, 900–912.

Kuznetsov, N.Y. (2014). Evaluation of the reliability of repairable $s - t$ networks by fast simulation method. In *Examining Robustness and Vulnerability of Networked Systems*, Butenko, S., Pasiliao, E.L., Shylo, V. (eds). IOS Press, Amsterdam, 120–140.

Kuznetsov, N.Y. (2016). Using the Monte Carlo method for fast simulation of the number of "good" permutations on the SCIT-4 multiprocessor computer complex. *Cybernet. Syst. Anal.*, 52, 52–57.

Kuznetsov, N.Y., Khomyak, O.N. (2017). Fast simulation of highly reliable networks with varying random external load. In *Optimization Methods and Applications*, Butenko, S., Pardalos, P., Shylo, V. (eds). Springer, New York, 253–272.

Kuznetsov, N.Y., Shumskaya, A.A. (2013). Evaluation of the hazard of failure of the repairable system by fast simulation method. *J. Autom. Inf. Sci.*, 45(5), 38–51.

Lagnoux, A. (2006). Rare event simulation. *Probab. Eng. Inf. Sci.*, 20, 45–66.

Li, J., Mosleh, A., Kang, R. (2011). Likelihood ratio gradient estimation for dynamic reliability applications. *Reliab. Eng. Syst. Safety*, 96, 1667–1679.

MacWilliams, F.J., Sloane, N.J.A. (1977). *The Theory of Error-Correcting Codes.* Elsevier, Amsterdam, 18.

Masol, V.I. (1989). Some applications of algorithms for constructing subspaces over a finite field. *Ukrainian Math. J.*, 41(8), 1146–1148.

Masol, V.I. (1996). Asymptotics behavior of the number of certain k-dimensional subspaces over a finite field. *Math. Notes*, 59(5), 525–530.

Masol, V.I. (2001). Investigation of linear codes possessing some extra properties. In *Cryptography and Coding.* Springer-Verlag, Berlin, 301–306.

Masol, V.I., Kuznetsov, I.N. (2010). Application of a fast simulation method to the estimation of the number of some k-dimensional subspaces over a finite field. *Cybernet. Syst. Anal.*, 46, 405–419.

Novakovich, D. (2000). Calculation of the number of complete mappings for permutations. *Cybernet. Syst. Anal.*, 36(2), 244–247.

Plaks, B.I. (1984). Calculation of the reliability of systems of complex structure by an accelarated Monte Carlo method. *Eng. Cybernetics*, 22, 53–58.

Rubinstein, R. (1981). *Simulation and Monte Carlo Method.* Wiley, New York.

Sachkov, V.N. (1996). *Combinatorial Methods in Discrete Mathematics.* Cambridge University Press, Cambridge.

Shieh, Y.P. (2001). Partition strategies for $\#P$-complete problem with applications to enumerative combinatories. PhD Thesis, National Taiwan University, 2001.

Shumskaya, A.A. (2003). Fast simulation of unreliability of a repairable system with a bounded relative error of estimate. *Cybernet. Syst. Anal.*, 39, 357–366.

Sloane, N.J.A. (2019). The On-Line Encyclopedia of Integer Sequences [Online]. Available: http://oeis.org/A003111.

Zavadskaya, L.A. (1981). Assessing the reliability of a system with checking and maintenance by the analytical-statistical method. *Cybernetics*, 17, 219–223.

7

Diffusion and Gaussian Limits for Multichannel Queueing Networks

Eugene LEBEDEV and Hanna LIVINSKA
National Taras Shevchenko University, Kiev, Ukraine

This chapter gives some results in diffusion and Gaussian approximation of multichannel queueing networks under heavy traffic conditions. Input flow of calls has a complicated structure. Moreover, its rate is allowed to be dependent on time. A local approach is used for the diffusion approximation of service process in multichannel networks with input flows of clearly defined structure. This approach enables us to prove limit theorems and establish convergence to a diffusion process with local characteristics, which can be specified explicitly via the network parameters. If there are no restrictions on the structure of the input flow, then the analysis procedure of a service process in multichannel networks changes. In this case, the limit process is described as a multidimensional Gaussian process. Proofs of the corresponding limit theorems are based on functional limit theorems for input flows as well as on the characteristic functions method. Often, the limit process turns out to be a Gaussian diffusion process. However, in this chapter, we mention conditions when the limit Gaussian process loses Markov property and it is not a diffusion process.

7.1. Introduction

In recent years, network research has acquired new practical importance as a primary tool for studying, designing and optimizing real-world systems with interconnected components. Queueing networks provide for them simple but extremely useful representation. The structure of such networks is determined by the probabilistic characteristics of the input flows and data processing algorithms. It is necessary to find their characteristics, optimize them and develop the corresponding control algorithms. Typically, the information processing in queueing networks is described by multidimensional vectors with interconnected entries, and the complex system of stochastic relations that define the process.

The main approach to the investigation of queueing networks is based upon the direct method of finding expressions for the network state probabilities. It allows us to find the exact solution for multiplicative or, as they are often called, locally balanced networks, with their stationary probabilities having a multiplicative form (see, for example, (Gnedenko and Kovalenko 2013, p. 84)). An alternative approach is to use the method of the asymptotic analysis for investigating complex systems and networks.

By asymptotic methods in queueing theory, one usually means the directions of investigations with their main objective to study complex multidimensional service processes using their approximation by processes with simpler structure. There are different versions of the limit conditions under consideration: extremely large interval of observations, small rate of service, increasing rate of input flows of calls, etc.

One of the first works in this direction is of Prokhorov (1956), where a Wiener process appears as an approximative process.

Diffusion approximation (DA) is a technique where a complicated and analytically intractable stochastic process is replaced by an appropriate diffusion process. A diffusion process is a Markov process having continuous sample paths. Diffusion processes considerably have an analytical structure and are, therefore, more mathematically tractable than the original process with which the user starts. Different approaches to DA of stochastic processes are elaborated. The traditional procedure for proving convergence of the sequence of stochastic processes to a limit process is to prove convergence of characteristic functions of finite-dimensional distributions of the sequence to the corresponding characteristic functions of the limit process. An essential fault of this procedure is that deriving characteristic functions of finite-dimensional distributions involves considerable technical difficulties and yields cumbersome expressions. Moreover, attempts in finding explicit form of finite-dimensional distributions of the limit process are not necessarily successful. In case of proving convergence to a diffusion process, these difficulties can be overcome by using the "local approach" to prove limit theorems (see, for example, (Borovkov 1980; Gikhman 1969)).

Convergence to diffusion processes using the local approach was first used by Bernstein (1934) and Khinchin (1936). Since general conditions for convergence to diffusion processes have been obtained in works of Gikhman, Skorokhod and Borovkov (Borovkov 1980; Gikhman 1969; Gikhman and Skorokhod 1982), the local approach became one of the main tools for strict mathematical justification of DA.

A large number of papers is devoted to the analysis of queueing models in heavy traffic conditions. There are several directions oriented to different classes of queueing models. Many authors deal with single-server nodes networks, renewal input process, independent service times and the routing processes that are independent of the current size of a queue or a workload process. For this case, the convergence of a normalized queue length or a workload process to a solution of a differential equation (fluid limits) or to a reflecting Brownian motion in a corresponding domain (Brownian approximation) is proved for a single-class network (Reiman 1984) and for various classes of multiclass networks (see surveys in (Williams 1996, 2016)).

Another direction is related to the analysis of Markov state-dependent queueing models. The method of analysis here is mainly based on a martingale technique and uses the continuous mapping theorems as well. This technique is used for the DA of some state-dependent networks in Basharin *et al.* (1989); Mandelbaum and Pats (1995) and Mandelbaum *et al.* (1998).

One more approach to study the fluid limits and the DA for some classes of state-dependent Markov queueing systems and networks is based on using the averaging principle (AP) and DA type results for so-called switching processes and for a special subclass of switching processes, recurrent processes of a semi-Markov type, introduced and developed in Anisimov (1995). Note that a switching process has the property that the character of its operation varies spontaneously (switches) at some epochs of time, which can be random functionals of the previous trajectory or possibly jumps of some random environment. Thus, this class serves as a convenient tool to describe wide classes of queueing models with state-dependent structure and operating in a random environment. The results of Anisimov (1995) are used and extended further to study AP and DA for rather general classes of queueing models of a switching structure (Anisimov 2002, 2008, 2010), etc.

Queueing Markov models whose evolution develops under the influence of random factors are studied in averaging and DA scheme in Korolyuk (1999). Such models (called Markov random evolution) are determined by two processes: a switched process describing the evolution of the system and a switching process describing changes of the random environment. Techniques for the analysis of random evolutions are based on the theory of singular perturbed reducibly invertible operators and on the martingale characterization of Markov stochastic processes. In Korolyuk *et al.* (2009), asymptotic average and DA schemes for semi-Markov queueing systems are studied by a random evolution approach and by using the compensating operator of the corresponding extended Markov renewal process.

We focus in this chapter on the analysis of the class of open queueing networks with an unlimited number of servers in each node (multichannel queueing networks). Since a direct analytical study of such models is not possible in most cases, we use methods of asymptotic analysis under heavy traffic conditions.

Queueing models with an infinite number of servers play an important role in the simulation of real systems. For example, similar models are considered: for the analytical modeling of information networks (Nazarov and Tuenbaeva 2009) and cellular communication networks (Massey and Whitt 1993, 1994); in the study of the parameters of the ionization track in tracking chambers (Dvurečenskij *et al.* 1981); in pharmacokinetics (Matis and Wehrly 1990); in economic models such as models in logistics, insurance, demography, etc.

Models with an infinite number of servers are widely covered in the scientific literature (see, for example, (Brian *et al.* 2009; Decreusefond and Moyal 2008; Moiseev and Nazarov 2015; Puhalskii and Reed 2008; Whitt 1982, 2016)). Asymptotic analysis of some multichannel queueing networks under condition of high-intensity input flow is considered in the papers by Moiseev and Nazarov (2016), of Nazarov and Moiseev (2014) and in the book by Moiseev and Nazarov (2015). The authors used their original methods: method of the first jump selecting and the method of multidimensional dynamic sifting.

In this chapter, we pay attention to studies of multichannel queueing networks under heavy traffic conditions. Input flow of calls can have a complicated enough structure, moreover, its rate is allowed to be dependent on time. Despite a great deal of recent works in the area of queueing networks, this class of models has not yielded to an exact analysis.

In the case when input flow has a clearly defined structure, it is efficient to use the local approach for DA of a service process in multichannel networks. This approach enables us to prove limit theorems and establish convergence to a diffusion process with its local characteristics, which can be specified explicitly via the network parameters. Throughout the chapter, convergence of processes implies convergence in the uniform topology. This is convenient because it yields convergence of the main functionals of stochastic processes paths. This feature allows us to formulate and solve problems connected with optimal choice of the network parameters.

If there are no restrictions on the structure of the input flow, then one should alter the analysis procedure of a service process in multichannel networks. In this case, it is suitable to describe the limit process as a multidimensional Gaussian process. Proofs of the corresponding limit theorems are based on functional limit theorems for input flows as well as on the characteristic functions method. Eventually, often the limit process turns out to be a Gaussian diffusion process. However, in the chapter we point out conditions when the limit Gaussian process loses Markov property and therefore is not a diffusion process.

This chapter is written based on authors' publications in the scientific journals (Lebedev 2001, 2002b, 2003, 2017; Lebedev and Livinska 2017, 2018, 2019; Livinska and Lebedev 2016, 2018; Livinska 2016) and the conference papers (Lebedev 2002a; Livinska and Lebedev 2015). It is organized as follows:

In section 7.2, the main mathematical model is described in details.

Section 7.3 focuses on the local approach for proving limit theorems, and the main idea of the approach is described. In section 7.3.1, this technique is used for the DA of a service process of a multichannel network with recurrent input flow and exponential service.

In section 7.4, networks with a more complicated structure of input flow are studied. In section 7.4.1, for a $[SM|M|\infty]^r$-network with input flow controlled by a semi-Markov process, DA is provided. Analysis of stationary regime for a network with generally distributed service times is given in section 7.4.2. In terms of these results, in section 7.4.3, it is established that the service process of the network operating in stationary regime converges to an Ornstein–Uhlenbeck process.

Section 7.5 is devoted to Gaussian approximation of multichannel networks of $[G|M|\infty]^r$-type. In section 7.5.1, for such networks operating in a heavy traffic regime, an approximative Gaussian process is built, and its characteristics are specified via the network parameters. In section 7.5.2, the criterium of Markovian behavior is derived for multidimensional Gaussian processes. Moreover, it is found that the approximative Gaussian process from the previous section is a diffusion process. In section 7.5.3, for networks of $[G|GI|\infty]^r$-type, convergence of the service process to a multidimensional Gaussian process is proved. In this case, under general assumptions, the limit process is a non-Markov one.

Section 7.6 provides Gaussian approximation for networks with a time-dependent input flow operating in a heavy traffic regime. In this, two cases with different starting loads are considered. For both cases, the Gaussian limit process is constructed. In the case of asymptotically large starting load considered in section 7.6.2, it is proved that the limit process is an Ornstein–Uhlenbeck process.

Section 7.7 concludes the chapter.

7.2. Model description and notation

Throughout the chapter, we will study similar networks. Therefore, in this section, our model for an open multichannel queueing network is described. More precisely, in the chapter, a family of networks is considered, each member of which may have a different collection of primitive processes. So, here we define: the structure of the network and routing algorithm, which are the same over all sections; the process of calls arrived from the outside, service time and initial conditions, which may be different in different sections.

Queueing networks under consideration consist of a fixed finite set r of service nodes (stations) through which calls (customers, jobs) are processed. A call visits nodes according to the routing algorithm below and receives service at each node visited. Each of the r nodes is a multichannel queueing system for serving calls. It means that each node has an infinite number of servers, thus, if a call arrives in such a system, then its service begins immediately and there is no queue of calls waiting for their service.

Routing algorithms through the network can be described as follows. Once the service of a call is completed in the ith node, the call is directed to the jth node with probability p_{ij}, or it leaves the network with probability $p_{ir+1} = 1 - \sum_{j=1}^{r} p_{ij}$. The $r \times r$ matrix $P = \|p_{ij}\|_1^r$ with entries p_{ij} is called the switching (routing) matrix of the network. An additional network node numbered by $r+1$ is interpreted as an "exit" from the network. To satisfy our open queueing network assumption, the matrix P is assumed to have spectral radius strictly less than one.

We assume that there is an r-dimensional external arrival process

$$\nu(t) = \left(\nu^{(1)}(t), ..., \nu^{(r)}(t)\right)', \quad t \geq 0,$$

such that for each $i = 1, 2, ..., r$ $\nu^{(i)}(t)$ is the number of calls arrived to the ith node from the outside of the network that have occurred by time t. Here and throughout the chapter, the superscript $'$ is used to denote the operation of taking the transpose of a vector or matrix. For $\nu(t)$, we consider some options.

– A recurrent flow of calls arrives to the ith node. Interarrival times are independent and identically distributed (i.i.d.) random variables with a distribution function $F_i(t)$. We use notation $[GI|\cdot|\infty]^r$ for networks with this type of input flow.

– A common input flow controlled by a semi-Markov process arrives from the outside to the network denoted by $[SM|\cdot|\infty]^r$. The structure of such a flow is more complicated. A detailed description of this case is given in a relevant section.

– A case without any restrictions imposed on the structure of the input flow $\nu(t) = (\nu^{(1)}(t), ..., \nu^{(r)}(t))'$. We denote this type of multichannel queueing network by $[G| \cdot |\infty]^r$.

– Calls arrive to the ith network according to the Poisson process with its rate dependent on time. The correspondent leading function $\Lambda_i(t), \quad i = 1, 2, ..., r, t \geq 0$, of the process $\nu^{(i)}(t)$ is a positive non-decreasing right-continuous function. In this case, the network is denoted by $\left[\overrightarrow{M}_t| \cdot |\infty\right]^r$.

As for the calls' service times at the ith node, they are i.i.d. random variables. In the chapter, two types of service times distribution are studied:

– service time at the ith node is exponentially distributed with the rate μ_i, $i = 1, 2, ..., r$, and the network is denoted by $[\,\cdot\,|M|\infty]^r$;

– service time at the ith node is distributed in accordance with the distribution function of general type $G_i(t)$, $i = 1, 2, ..., r$. Such a network is denoted by $[\,\cdot\,|GI|\infty]^r$.

The following descriptive process $Q(t)$ is used to measure the performance of our queueing networks. For each node $i = 1, 2, ..., r$, let $Q_i(t)$ define the number of calls that are being served at the ith node at the instant t. The r-dimensional process $Q(t) = (Q_1(t), ..., Q_r(t))'$, $t \geq 0$, is called the service process of calls for the queueing network.

For most results in the chapter, it is supposed that at the initial instant a network is empty, and there are no calls being served in the network nodes. This means that the network operates in a transient regime. Actually, it is not a crucial assumption, i.e. formulated results are valid not only for an absolutely empty network, but also for the network that is "moderately" loaded from the start. This implies that the number of calls at the initial instant is equal to the certain constant value that does not depend on a series number and, therefore, cannot affect the limit service process. A totally different case is when the starting load of the network is asymptotically increasing in series scheme. In such a situation, the loading process can yield to substantial changes in the representation of the limit process. For this case, we provide conditions under which a multidimensional Ornstein–Uhlenbeck process can be obtained as the limit of the service process.

We now consider a sequence of open multichannel queueing networks indexed by n (series number), where n tends to infinity. Each of these networks is to have the same basic structure as described above. The number of networks nodes remains fixed for all n. On the other hand, the basic stochastic input process $\nu(t)$, service times distribution and network initial load are allowed to change with n for different cases. This dependence is indicated by upper index $\cdot^n$ near the correspondent letter defining input flow or service time distribution.

So, in this section, a class of multichannel networks under consideration is defined. Often for a set of these models, convergence of the service process to a diffusion process can be proved using the local approach, like it is done in the following section.

7.3. Local approach to prove limit theorems

Recall that the diffusion process $\xi'(t) = (\xi_1(t), ..., \xi_r(t)) \in \mathbb{R}^r$ is defined as a Markov process exhibiting the following local properties for all $\varepsilon > 0$:

$$\mathbf{E}\left\{\delta_\varepsilon(\Delta\xi)\Delta\xi \mid \xi(s) = x\right\} = a(s,x)\cdot(t-s) + o(t-s), \qquad [7.1]$$

$$\mathbf{E}\left\{\delta_\varepsilon(\Delta\xi)\left(z'\Delta\xi\right)^2 \mid \xi(s) = x\right\} = z'\cdot b(s,x)\cdot z(t-s) + o(t-s), \qquad [7.2]$$

$$\mathbf{E}\left\{(1-\delta_\varepsilon(\Delta\xi)) \mid \xi(s) = x\right\} = o(t-s), \qquad [7.3]$$

$$\delta_\varepsilon(z) = \begin{cases} 1, & \text{if } |z| \le \varepsilon, \\ 0, & \text{if } |z| > \varepsilon, \end{cases}$$

where $a(s,t)$ is a drift vector, $b(s,t)$ is a diffusion matrix, $s, t \in [0,\infty)$, $s < t$, $x, z \in \mathbb{R}^r$, $\Delta\xi = \xi(t) - \xi(s)$.

Taking into account the local properties, it would be reasonable to seek convergence conditions for a sequence of stochastic processes

$$\xi^n(t) = (\xi_1^n(t), ..., \xi_r^n(t))' \in \mathbb{R}^r$$

to a diffusion process $\xi(t)$ in terms of increments of $\xi^n(t)$.

In this point, we formulate Gihman's variant of sufficient conditions of convergence to a diffusion (Gikhman 1969).

Let $\xi^n(t)$, $t \in [0,T]$, $n \ge 1$, be a sequence of stochastic processes and let $\xi(t)$, $t \in [0,T]$, be a diffusion with the drift vector $a(s,t)$ and the diffusion matrix $b(s,t)$. In order that $\xi^n(t)$, $n \ge 1$, converges to $\xi(t)$ in distribution, it is sufficient that $\xi^n(0)$ converges to a certain random vector ξ^0 in distribution, and there exists a sequence of partitions of the interval $[0,T]$ by points

$$\{t_{n_k},\ k = 0,1,...,m_n\},\quad n = 1,2,...,\quad 0 = t_{n_0} < t_{n_1} < ... < t_{n_{m_n}} = T,$$

$$\max_k \Delta t_{n_k} \to_{n\to\infty} 0, \quad \Delta t_{n_k} = t_{n_{k+1}} - t_{n_k},$$

and a sequence of families of monotonically increasing $\sigma-$algebras $\{\mathfrak{F}_n(t), t \in [0,T]\}$ ($\mathfrak{F}_n(t_1) \subset \mathfrak{F}_n(t_2)$, if $t_1 < t_2$) such that the following conditions hold:

– process $\xi^n(t)$ is $\mathfrak{F}_n(t)$- measurable for $t \in [0,T]$, and $n = 1, 2, \ldots$;

– for any $\varepsilon > 0$, $z \in \mathbb{R}^r$ and $n \to \infty$

$$\sum_{k=0}^{m_n-1} \mathbf{E}\left|\mathbf{E}\left\{\Delta\xi_{n_k} \mid \mathfrak{F}_{n_k}\right\} - a(t_{n_{k+1}}, \xi_{n_k}) \cdot \Delta t_{n_k}\right| \to 0, \qquad [7.4]$$

$$\sum_{k=0}^{m_n-1} \mathbf{E}\left|\mathbf{E}\left\{(z'\Delta\xi_{n_k})^2 \mid \mathfrak{F}_{n_k}\right\} - z'b(t_{n_{k+1}}, \xi_{n_k}) z \Delta t_{n_k}\right| \to 0, \qquad [7.5]$$

$$\sum_{k=0}^{m_n-1} \mathbf{P}\left\{|\Delta\xi_{n_k}| \geq \varepsilon\right\} \to 0, \qquad [7.6]$$

where $\xi_{n_k} = \xi^n(t_{n_k}), \quad \Delta\xi_{n_k} = \xi_{n_{k+1}} - \xi_{n_k}, \quad \mathfrak{F}_{n_k} = \mathfrak{F}_n(t_{n_k})$.

Conditions [7.4]–[7.6] correspond to the local properties [7.1]–[7.3] and mean proximity of the local characteristics (in terms of increments) of the prelimit and limit processes.

Note that in order to provide such approximation, in addition to conditions [7.4]–[7.6], it is necessary for $a(s,x)$ and $b(s,x)$ to be "fairly smooth" functions. The rigorous statement of these conditions can be found in Gikhman (1969). In the current section, all coefficients $a(s,x)$ and $b(s,x)$ of all diffusion processes approximating service processes of queueing networks satisfy the smooth conditions. Therefore, later we always imply that smooth conditions are fulfilled without emphasizing on it.

With that said, checking conditions [7.4] and [7.5] for the service processes of queueing systems and networks reduces to computation and analysis of the first and second conditional moments of increments of the prelimit process. Proof of [7.6] is usually equivalent to testing the Lindeberg condition for some sequence of random variables constructed on the service process.

7.3.1. *Network of the $[GI|M|\infty]^r$-type in heavy traffic*

The local approach is a suitable tool to prove limit theorems provided that the input flow of the network can be defined in a constructive way. Let us consider a multichannel network of the $[GI|M|\infty]^r$-type.

To obtain a heavy traffic limit theorem for our sequence of service processes, we need to impose some additional conditions of overloading on the network parameters as $n \to \infty$. So, service rates μ_i, $i = 1, 2, ..., r$, are assumed to be dependent on the series number n so that the following condition holds.

CONDITION 7.1.– $\lim_{n\to\infty} n\mu_i^{(n)} = \mu_i, \quad i = 1, 2, ..., r.$

In this section, we assume that other parameters of the $[GI|M|\infty]^r$-network do not depend on n.

For recurrent input flows $\nu^{(i)}(t)$, $i = 1, 2, ..., r$, we require boundedness of the first and the second moments of interarrival times:

CONDITION 7.2.– For all $i = 1, 2, ..., r$

$$\lambda_i^{-1} = \int_0^\infty t dF_i(t) < \infty, \quad d_i = \int_0^\infty t^2 dF_i(t) < \infty.$$

Here, we suppose the network to be empty at the initial instant:

CONDITION 7.3.– $Q_i(0) = 0$ for all $i = 1, 2, ..., r$.

Under conditions 7.1–7.3 for the open $[GI|M|\infty]^r$-network, we consider the sequence of stochastic processes

$$\xi^n(t) = n^{-1/2}(Q^n(nt) - nq(t)), \quad t \geq 0, \qquad [7.7]$$

where

$$q'(t) = (q_1(t), ..., q_r(t)) = (\theta/\mu)'(I - P(t)), \quad (\theta/\mu)' = (\theta_1/\mu_1, ..., \theta_r/\mu_r),$$

$\theta' = (\theta_1, ..., \theta_r) = \lambda'(I - P)^{-1}$ is a solution of the balance equation for the $[GI|M|\infty]^r$-network, $P(t) = \exp\{\Delta(\mu)(P - I)t\}$, $\Delta(x) = \|\delta_{ij}x_i\|_1^r$ is a diagonal matrix for any vector $x' = (x_1, ..., x_r)$, $\mu' = (\mu_1, ..., \mu_r)$, $I = \|\delta_{ij}\|_1^r$ is the identity matrix.

For $\xi^n(t)$, the following result holds true.

THEOREM 7.1.– Let for a queueing network of the $[GI|M|\infty]^r$-type conditions 7.1–7.3 be satisfied, $F_i(t)$, $i = 1, 2, ..., r$, be a non-lattice function and the spectral radius of the switching matrix P be strictly less than one. Then on any finite interval $[0, T]$, the sequence of stochastic processes $\xi^n(t)$, $n \geq 1$, converges weakly in the uniform topology to a diffusion process $\xi(t)$ ($\xi(0) = 0$) with drift vector $A(x) = A'x$ and diffusion matrix

$$B(t) = \Delta[q'(t)A] - A'\Delta[q(t)] - \Delta[q(t)]A + \Delta(\sigma^2),$$

where $A = \Delta(\mu)(P - I)$, $\sigma^2 = (\sigma_1^2, ..., \sigma_r^2)'$, $\sigma_i^2 = \lambda_i^3(d_i - \lambda_i^{-2})$, $i = 1, 2, ..., r$.

SKETCH OF THE PROOF.– Let us check conditions [7.4]–[7.6] successively.

As a sequence of monotonically increasing σ-algebras $\mathfrak{F}_n(t)$, we can take the σ-algebras generated by the random variables

$$\xi^n(u), \;\; u \le t, \;\; \tau^{(i)}_{\nu^{(i)}(nt)+1}, \;\; i = 1, 2, ..., r,$$

where $\tau_k^{(i)}$ is the arrival moment of the kth call in the ith node.

As a sequence $\{t_{n_k}\}_{k=0}^{m_n}$, $n = 1, 2, ...$, we take the sequence of the uniform partitions

$$\Delta t_{n_k} = T/[n^\alpha],$$

for some $\alpha \in (0; 1/2)$.

Now, for our case, condition [7.4] takes the form:

$$\sum_{k=0}^{m_n - 1} \mathbf{E} \left| \mathbf{E}\left\{\Delta\xi_{n_k} \mid \mathfrak{F}_{n_k}\right\} - A'\xi_{n_k}\Delta t_{n_k} \right| \to_{n\to\infty} 0.$$

The path of the call arrived at the $[GI|M|\infty]^r$-network through the kth node can be described by a Markov chain $y^{(k)}(t) = \{1, 2, ..., r.r + 1\}$, $t \ge 0$, $y^{(k)}(0) = k$, which is defined by the infinitesimal matrix $\|a_{ij}\|_1^{r+1}$ with the following entries:

$$a_{ij} = \begin{cases} -\mu_i (1 - p_{ii}), & \text{if } \; i = j = 1, ..., r; \\ \mu_i p_{ij}, & \text{if } \; i \ne j, \; i = 1, ..., r, \; j = 1, ..., r, r + 1; \\ 0, & \text{if } \; i = r + 1, \; j = 1, ..., r, r + 1; \end{cases}$$

The state $r+1$ for the chain $y^{(k)}(t)$ is an absorbing one. Absorption in the state $r+1$ means that the call leaves the network. We denote by $p_j^i(t)$ the transient probabilities of the chain, $P(t) = \left\|p_j^i(t)\right\|_1^r = \exp{(At)}$.

Let us connect an r-dimensional process of indicator type

$$\chi^{(k)}(t) = \left(\chi_1^{(k)}(t), ..., \chi_r^{(k)}(t)\right)', \quad t \geq 0,$$

with the chain $y^{(k)}(t)$ as follows:

$$\chi^{(k)}(t) = \begin{cases} e_j, & \text{if } \; y^{(k)}(t) = j, \; j = 1, ..., r; \\ e_0, & \text{if } \; y^{(k)}(t) = r + 1; \end{cases}$$

where e_0 is an r-dimensional zero vector, e_j is an r-dimensional vector with its jth entry equal to one while the rest of the entries are zero.

If, at the initial instant, a call arrives to the network for its service through the ith node, then the process $\chi^{(i)}(t)$ points out the node where the call is staying at the instant t. If $\chi^{(i)}(t) = e_0$, it means that the call has already been served before the instant t, and it leaves the network through the $(r+1)$th node.

Denote by $\chi^{(i,j)}(t)$, $\widetilde{\chi}^{(i,j)}(t)$, $i = 1, ..., r$, $j = 1, 2, ...$, independent stochastic processes with the same finite-dimensional distributions as ones of $\chi^{(i)}(t)$. Then we have the following:

$$\mathbf{E}\left\{\Delta\xi_{n_k} \mid \mathfrak{F}_{n_k}\right\} = \frac{1}{\sqrt{n}}\mathbf{E}\left\{ \sum_{j=1}^{\sqrt{n}\xi_i^n(nt_{n_k})+nq_i(t_{n_k})} \left(\chi^{(i,j)}(\Delta t_{n_k}) - e_i\right) + \right.$$

$$\left. + \sum_{i=1}^{r} \sum_{j=\nu^{(i)}(nt_{n_k})+1}^{\nu^{(i)}(nt_{n_{k+1}})} \widetilde{\chi}^{(i,j)}\left(t_{n_k} - \frac{\tau_j^{(i)}}{n}\right) - n\left[q(t_{n_{k+1}}) - q(t_{n_k})\right] \mid \mathfrak{F}_{n_k} \right\} =$$

$$= (P'(\Delta t_{n_k}) - I)\xi^n(t_{n_k}) + \frac{1}{\sqrt{n}}\mathbf{E}\left\{ \int_{t_{n_k}}^{t_{n_{k+1}}} P'(t_{n_{k+1}} - \tau) d[\nu(n\tau) - \lambda n\tau] \mid \mathfrak{F}_{n_k} \right\}, \qquad [7.8]$$

$$\nu'(t) = (\nu^{(1)}(t), ..., \nu^{(r)}(t)).$$

Substituting $\mathbf{E}\{\Delta\xi_{n_k}/\mathfrak{F}_{n_k}\}$ in the form [7.8] into [7.4], we obtain:

$$\sum_{k=0}^{m_n-1} \mathbf{E}\left|\mathbf{E}\{\Delta\xi_{n_k} \mid \mathfrak{F}_{n_k}\} - A'\xi_{n_k}\Delta t_{n_k}\right| \leq$$

$$\leq \sum_{k=0}^{m_n-1} \mathbf{E}\left|(P'(\Delta t_{n_k}) - I)\xi_{n_k} - A'\xi_{n_k}\Delta t_{n_k}\right| +$$

$$+\frac{1}{\sqrt{n}} \sum_{k=0}^{m_n-1} \mathbf{E}\left|(\mathbf{E}\{\Delta\nu(nt_{n_k}) \mid \mathfrak{F}_{n_k}\} - \lambda n\Delta t_{n_k}) + (I - P'(\Delta t_{n_k}))\right.$$

$$\left.\times(\nu(nt_{n_k}) - \lambda nt_{n_k}) + \int_{t_{n_k}}^{t_{n_{k+1}}} P'(t_{n_k} - \tau)A'(\mathbf{E}\{\nu(n\tau) \mid \mathfrak{F}_{n_k}\} - \lambda n\tau)d\tau\right|. \quad [7.9]$$

Estimation [7.9] proves validity of [7.4]. Substantially, $(P'(\Delta t_{n_k}) - I)\xi_{n_k} - A'\xi_{n_k}\Delta t_{n_k}$ has order $(\Delta t_{n_k})^2$, therefore

$$\sum_{k=0}^{m_n-1} \mathbf{E}\left|\mathbf{E}\{\Delta\xi_{n_k} \mid \mathfrak{F}_{n_k}\} - A'\xi_{n_k}\Delta t_{n_k}\right| \to 0, \text{ as } n \to \infty.$$

By virtue of the above hypothesis on the sequence of partitions, the second term on the right side of [7.9] has order $n^{-(1/2-\alpha)}$ and converges to zero as well. Similarly, one can check condition [7.5] relating to the diffusion matrix of the limit process.

Condition [7.6] is equivalent to Lindeberg's condition in the central limit theorem. Thus, the proof of [7.6] can be obtained as a consequence of the following auxiliary result.

LEMMA 7.1.– (See (Anisimov and Lebedev 1992, p. 149) Let $\left\{\zeta_k^{(n)}\right\}_{k=1}^{\infty}$ be a sequence of i.i.d. random variables with their distribution dependent on the parameter n (series number), and let $\mathbf{E}\zeta_1^{(n)} = 0$, $\mathbf{Var}\zeta_1^{(n)} = 1$. Then for any $A > 0$ and any sequence $m_n \to \infty$ under $n \to \infty$, we have:

$$P\left\{\sup_{x\in[0,A]}\left|\frac{1}{\sqrt{n}}\sum_{k=1}^{nx}\zeta_k^{(n)}\right| \geq m_n\right\} \leq 2P\left\{\left|\frac{1}{\sqrt{n}}\sum_{k=1}^{[nA]}\zeta_k^{(n)}\right| \geq m_n - \sqrt{2A}\right\}.$$

An effective tool to prove the convergence of functionals given for the service process of a network is Billingsley's test.

LEMMA 7.2.– (See (Billingsley 1999, p.142) Let the sample functions of $\xi(t)$ and $\xi^n(t)$, $n \geq 1$, $t \in [0,T]$, belong to the Skorokhod space $D[0,T]$ with probability 1 and the finite-dimensional distributions of $\xi^n(t)$, $n \geq 1$, $t \in [0,T]$, are assumed to converge to the finite-dimensional distributions of $\xi(t)$, $t \in [0,T]$. If for any $\varepsilon > 0$

$$P\{|\xi^n(t) - \xi^n(t_1)| \geq \varepsilon, |\xi^n(t_2) - \xi^n(t)| \geq \varepsilon\} \leq \frac{1}{\varepsilon^{2\gamma}}[F(t_2) - F(t_1)]^{2\alpha}, \qquad [7.10]$$

$$t_1 \leq t \leq t_2, \ \ t_1, t, t_2 \in [0,T], \ \ n \geq 1, \ \ \gamma \geq 0, \ \ \alpha > 1/2,$$

where $F(\cdot)$ is a non-decreasing continuous function on $[0,T]$, then the sequence of stochastic processes $\xi^n(t)$, $n \geq 1$, converges to $\xi(t)$ in the J-topology (Skorokhod's topology).

In the framework of theorem 7.1, the convergence of $\xi^n(t)$ to the diffusion $\xi(t)$ can be derived by virtue of [7.10] and Chebyshov's inequality. Since the limit process $\xi(t)$ has a continuous choice function with probability one, the convergence is in the uniform topology.

This completes the proof of theorem 7.1.

7.4. Limit theorems for networks with controlled input flow

7.4.1. *Diffusion approximation of* $[SM|M|\infty]^r$*-networks*

Now, let us consider a network of the $[SM|M|\infty]^r$-type. In this model, from the outside, a common input flow arrives at the network. The flow is controlled by a semi-Markov process $x(t), t \geq 0$, with the state space $\{1, 2, ..., N\}$. It means that the calls' arrival moments coincide with the jumps time τ_n, $n = 1, 2, ...,$ of $x(t)$. If at τ_n the process $x(t)$ jumps to the state i, then the arrived call numbered n is directed to the node j with probability h_{ij}, $\sum_{j=1}^{r} h_{ij} = 1$. The matrix $H = \|h_{ij}\|$ has dimension $N \times r$. By $F(t) = \|F_{ij}(t)\|_1^N$, we denote the semi-Markov matrix of the process $x(t)$, and by $F_i(t) = \sum_{j=1}^{N} F_{ij}(t)$, the distribution function of the sojourn time in the state i. The service of calls and the routing algorithm are the same as in the previous model.

Note that input flows $\nu^{(i)}(t)$, $i = 1, 2, ..., r$, into different nodes of the $[SM|M|\infty]^r$-network are related.

Instead of condition 7.2 from the previous section, we assume for semi-Markov process $x(t)$ the following:

CONDITION 7.4.– The matrix $F = \|F_{ij}(+\infty)\|_1^N = \|f_{ij}\|_1^N$ of transition probabilities of the embedded Markov chain for $x(t)$ is irreducible, and sojourn times at each node have the first- and second-order moments:

$$m_i = \int_0^\infty t dF_i(t) < \infty, \quad d_i = \int_0^\infty t^2 dF_i(t) < \infty, \quad i = 1, 2, ..., N.$$

Under condition 7.4 for F, a stationary distribution $(\pi_1, \pi_2, ..., \pi_N)$ exists, and Π denotes the matrix of the dimension $N \times N$ with N identical lines that are $(\pi_1, \pi_2, ..., \pi_N)$.

The rate λ_i, $i = 1, 2, ..., r$, of the input flow arriving to the ith node, which is used in the centering procedure of the service process, has the form:

$$\lambda_i = \sum_{j=1}^N a_j h_{ji}, \text{ where } \quad a_j = \pi_j \left\{ \sum_{i=1}^N \pi_i \int_0^\infty t dF_i(t) \right\}^{-1}.$$

In order to formulate a theorem on DA for a $[SM|M|\infty]^r$-network, the following limit correlation matrix is needed:

$$\sigma^2 = H'CH + \sum_{j=1}^N a_j [\Delta(h_j) - h_j h_j'], \qquad [7.11]$$

$$C = \|c_{\alpha\beta}\|_1^N,$$

$$c_{\alpha\beta} = a_\alpha \sum_{j=1}^N r_{\alpha j} f_{j\beta} + a_\beta \sum_{j=1}^N r_{\beta j} f_{j\alpha} + a_\alpha a_\beta \frac{d - 2m^{(2)}}{m} + a_\alpha \delta_{\alpha\beta},$$

$$R_1 = \|r_{ij}\|_1^N = (I - \Pi\Delta(a)) \, R_0 \, (I - \Delta(a)\Pi),$$

where $m = \sum_{i=1}^N m_i \pi_i$, $d = \sum_{i=1}^N d_i \pi_i$, $m^{(2)} = \sum_{i=1}^N m_i^2 \pi_i$, $a' = (a_1, ..., a_N)$, $h_j' = (h_{j1}, ..., h_{jr})$ is the jth line of the matrix H, $R_0 = (I - F + \Pi)^{-1} - \Pi$ is the potential of the embedded Markov chain.

We obtain the following result of asymptotic analysis of the stochastic process [7.7] for the $[SM|M|\infty]^r$-network in transient and heavy traffic regimes.

THEOREM 7.2.– Let conditions 7.1, 7.3, 7.4 be satisfied for a queueing $[SM|M|\infty]^r$-network, and let the spectral radius of the switching matrix P be strictly less than one. Then the sequence of stochastic processes $\xi^n(t)$, $n \geq 1$, converges weakly in the uniform topology on any finite interval $[0, T]$ to a diffusion process $\xi(t)$ ($\xi(0) = 0$) with drift vector $A(x) = A'x$ and diffusion matrix $B(t) = \Delta[q'(t)A] - A'\Delta[q(t)] - \Delta[q(t)]A + \sigma^2$, where the matrix σ^2 is determined via the network parameters by relations [7.11].

Theorem 7.2 can be proved using the local approach.

Note that models with similar parameters control (the $[M_{SM,Q}|M_{SM,Q}|1|\infty]^r$-networks) are investigated under the heavy traffic conditions in Anisimov (2008, pp.164–168), using another technique.

7.4.2. *Asymptotics of stationary distribution for* $[SM|GI|\infty]^r$*-networks*

In this section, we consider a multichannel $[SM|GI|\infty]^r$-network. It means that service times in the jth node are i.i.d. random variables with a distribution function $G_j(t)$, $j = 1, 2, ..., r$, $t \geq 0$.

At this point, we consider a service process $Q^i(t) = \left(Q^i_1(t), ..., Q^i_r(t)\right)'$, $t \geq 0$, $i = 1, 2, ..., r$, where $Q^i_j(t)$ is the number of calls in the jth node at the instant t, provided that the network is empty at the initial instant $t = 0$, $x(0) = i$, the arrival moment of the first call is the same as an exit moment τ_1 from the state i, and τ_1 is distributed in accordance with $F_i(t)$.

Denote multivariate generating functions of the vectors $Q^i(t)$, $i = 1, 2, ..., N$, by $\Phi^i(t, z)$, $t \geq 0$, $|z| \leq 1$.

Let L_T, $0 < T < \infty$, be a complete metric space of vector functions

$$\Phi(t, z) = \left(\Phi^1(t, z), \Phi^2(t, z), ..., \Phi^N(t, z)\right)', \quad z = (z_1, ..., z_r)',$$

with components $\Phi^i(t, z)$ determined for $t \geq 0$, $|z| \leq 1$, measurable over t which are probability generating functions of z. A distance between $\Phi, \Phi^* \in L_T$ is equal

$$\rho(\Phi, \Phi^*) = \max_{1 \leq i \leq N} \sup_{\substack{0 \leq t \leq T \\ |z| \leq 1}} \left|\Phi^i(t, z) - \Phi^{i*}(t, z)\right|.$$

It can be shown that for any $T > 0$ the generating functions $\Phi^i(t, z)$, $0 \le t \le T$, $|z| \le 1$, are the unique solution of the following system of integral equations:

$$\Phi^i(t, z) = 1 - F_i(t) + \sum_{j=1}^{N} \int_0^t dF_{ij}(u)\Phi^j(t-u, z)\times$$

$$\times \sum_{k=1}^{r} h_{jk}[p_{r+1}^k(t-u) + p_1^k(t-u)z_1 + ... + p_r^k(t-u)z_r], \quad i = 1, 2, ..., N.$$

[7.12]

Under $0 \le t \le T$ the solution belongs to L_T.

In the right part of [7.12], $p_j^i(t)$ are the transient probabilities of the semi-Markov process $y(t) \in (1, 2, ..., r, r+1)$, $t \ge 0$, given by the semi-Markov matrix $\|G_{ij}(t)\|_1^{r+1}$:

$$G_{ij}(t) = \begin{cases} p_{ij}G_i(t), & i = 1, 2, ..., r, \quad j = 1, 2, ..., r, r+1; \\ \delta_{r+1,j}G_{r+1}(t), & i = r+1, \qquad j = 1, 2, ..., r, r+1, \end{cases}$$

$$G_{r+1}(t) = \begin{cases} 0, & t < 1; \\ 1, & t \ge 1. \end{cases}$$

Now we reference one result.

THEOREM 7.3.– (See (Lebedev 2002b)) Let the matrix $F(t)$ be non-lattice, the matrix $F = F(+\infty)$ be irreducible and $\pi_1, \pi_2, ..., \pi_N$ be the stationary probabilities of F. If

$$\max_{1 \le i \le N} \int_0^\infty tdF_i(t) < \infty, \quad \max_{1 \le i \le r} \int_0^\infty tdG_i(t) < \infty,$$

and the spectral radius of the switching matrix P is strictly less than one, then there exists a stationary regime for the service process $Q(t)$, and

$$\Phi(z) = 1 - \sum_{j=1}^{N} a_j \int_0^\infty \Phi^j(t, z) \left\{ \sum_{k=1}^{r} h_{jk}[p_1^k(t)(1 - z_1) + ... + p_r^k(t)(1 - z_r) \right\} dt.$$

[7.13]

For non-Poisson input flow, the stationary distribution has a complex structure. The stationary probabilities do not have a multiplicative form which is valid for the wide class of Markov stochastic networks. Properties of stationary distribution for some special cases of the $[SM|GI|\infty]^r$-models were studied in Lebedev (2002a). It was shown that using binomial moments is an effective tool for investigating the stationary regime of such multichannel networks.

7.4.3. *Convergence to Ornstein–Uhlenbeck process*

In this section, we consider $[SM|M|\infty]^r$-models. In conditions for existence of a stationary regime, we denote by $Q = (Q_1, ..., Q_r)'$ a random vector corresponding to the stationary distribution. Let us proceed with the asymptotic analysis of the vector sequence $Q = Q^n$ in heavy traffic for an $[SM|M|\infty]^r$-network.

THEOREM 7.4.– Suppose conditions 7.1 and 7.4 hold for an $[SM|M|\infty]^r$-network, $F(t)$ is a non-lattice function, and the spectral radius of the switching matrix P is strictly less than one. Then

$$\zeta^n = n^{-1/2} \left(Q^n - n(\theta/\mu)\right) \overset{d}{\Rightarrow}_{n\to\infty} \zeta^0, \qquad [7.14]$$

where the random vector ζ^0 has Gaussian distribution with $\mathbf{E}\zeta^0 = 0$, and the following correlation matrix:

$$R = \mathbf{E}\zeta^0\zeta^{0'} = \int_0^\infty P'(t)[\sigma^2+\Delta(\theta)(I-P)+(I-P')\Delta(\theta)-\Delta(\lambda)]P(t)dt. [7.15]$$

The matrix σ^2 is given by [7.11].

A proof of theorem 7.4 is based on the functional limit theorem 7.2.

Assume that a $[SM|M|\infty]^r$-network operates in a stationary regime while conditions 7.1 and 7.4 of overloading hold. Then the sequence

$$\zeta^n(t) = n^{-1/2} \left(Q^n(nt) - n(\theta/\mu)\right), \quad n \geq 1,$$

satisfies the following result.

THEOREM 7.5.– If a queueing $[SM|M|\infty]^r$-network operates in a stationary regime and the assumptions of theorem 7.4 are satisfied, then the sequence of stochastic processes $\zeta^n(t)$, $n \geq 1$, converges weakly in the uniform topology on any finite interval $[0, T]$ to an r-dimensional stationary diffusion Ornstein–Uhlenbeck process $\zeta(t)$ ($\zeta(0) \stackrel{d}{=} \zeta^0$) with the drift vector $A(x) = A'x$ and the diffusion matrix

$$B = \Delta(\theta)(I - P) + (I - P')\Delta(\theta) - \Delta(\lambda) + \sigma^2.$$

PROOF.– The r-dimensional Ornstein–Uhlenbeck process with the drift $A(x)$ and the diffusion matrix B given by theorem 7.5 has stationary Gaussian distribution with zero mean and correlation matrix [7.15]. Therefore, theorem 7.4 provides weak convergence of an initial distribution for the prelimit and limit processes ($\zeta^n(0) \stackrel{d}{\Rightarrow} \zeta^0$). Validity of other conditions of local approach can be established similarly to that in theorem 7.1.

7.5. Gaussian approximation of networks with input flow of general structure

7.5.1. *Gaussian approximation of* $[G|M|\infty]^r$*-networks*

Now we consider a case without any restrictions imposed on the structure of the input flow $\nu(t) = (\nu^{(1)}(t), ..., \nu^{(r)}(t))'$.

Condition 7.1 is supposed to be fulfilled for service rates μ_i, $i = 1, 2, ..., r$, and for the input flow we now assume that the following property holds instead of conditions 7.2 and 7.4.

CONDITION 7.5.– There exist constants $\lambda_i \geq 0$, $i = 1, 2, ..., r$, $\lambda_1 + ... + \lambda_r \neq 0$, such that

$$n^{-1/2}\left(\nu^{(1)n}(nt) - \lambda_1 nt, ..., \nu^{(r)n}(nt) - \lambda_r nt\right) \stackrel{U}{\Rightarrow} W'(t) = (W_1(t), ..., W_r(t)),$$

as $n \to \infty$, where $W(t)$ is an r-dimensional Brownian motion with zero mean $\mathbf{E}W(1) = \overline{0} = (0, ..., 0)$ and the correlation matrix $\mathbf{E}W(1)W'(1) = \sigma^2 = \|\sigma_{ij}\|_1^r$, symbol $\stackrel{U}{\Rightarrow}$ denotes weak convergence in the uniform topology, n is a series number.

In order to construct a limit process for the sequence of stochastic processes

$$\xi^n(t) = n^{-1/2}(Q^n(nt) - nq(t)), \quad t \geq 0, \quad n \geq 1,$$

$$q'(t) = (q_1(t), ..., q_r(t)) = (\theta/\mu)'(I - P(t)), \quad (\theta/\mu)' = (\theta_1/\mu_1, ..., \theta_r/\mu_r),$$

where $\theta' = (\theta_1, ..., \theta_r) = \lambda'(I - P)^{-1}$ is a solution of the balance equation for the $[G|M|\infty]^r$-network, $\lambda' = (\lambda_1, ..., \lambda_r)$, we consider two independent Gaussian processes $\xi^{(1)}(t)$ and $\xi^{(2)}(t)$ with zero means and the following correlation matrices:

$$R^{(1)}(t) = \int_0^t P'(u)\sigma^2 P(u)du,$$

$$R^{(2)}(t) = \int_0^t [\Delta(\lambda' P(u)) - P'(u)\Delta(\lambda)P(u)]\, du,$$

$$R^{(i)}(s,t) = R^{(i)}(s)P(t-s), \quad s,t \geq 0, \quad s < t, \quad i = 1, 2.$$

The summarized process $\xi^{(1)}(t) + \xi^{(2)}(t)$ describes asymptotic behavior of the sequence of stochastic processes $\xi^n(t)$.

THEOREM 7.6.– (See (Lebedev 2003)) Let conditions 7.1, 7.3 and 7.5 be satisfied for a $[G|M|\infty]^r$-network, and assume the spectral radius of the switching matrix P to be strictly less than one. Then, the sequence of stochastic processes $\xi^n(t)$, $n \geq 1$, converges weakly in the uniform topology on any finite interval $[0, T]$ to $\xi^{(1)}(t) + \xi^{(2)}(t)$.

The proof of theorem 7.6 is based on some auxiliary results.

LEMMA 7.3.– The finite-dimensional distributions of $\int_0^t dW'(u)P(t-u)$ coincide with finite-dimensional distributions of the Gaussian process $\xi^{(1)}(t)$.

This result can be proved in virtue of properties of stochastic integrals over a Gaussian process with independent increments.

Service of a call in nodes of our network is independent of other calls. In order to define the service process in a constructive way, we use the process of indicator type $\chi^{(i)\prime}(t) = \left(\chi_1^{(i)}(t), ..., \chi_r^{(i)}(t)\right)$, $t \geq 0, i = 1, ..., r$, introduced in section 7.3.1.

For any natural number N and $z'(k) = (z_1(k), ..., z_r(k))$, $k = 1, 2, ..., N$, $|z(k)| \leq 1$, we denote the joint generating function of the vectors $\chi^{(i)}(t_1), ..., \chi^{(i)}(t_N)$, $0 < t_1 < ... < t_N$, by $\Phi^{(i)} = \Phi^{(i)}(z(1), ..., z(N), t_1, ..., t_N)$, $i = 1, 2, ..., r$. The following formula takes place for $\Phi = (\Phi^{(1)}, ..., \Phi^{(r)})'$.

LEMMA 7.4.– For any $N = 1, 2, ...$ and $0 < t_1 < ... < t_N$

$$\Phi = \bar{1} + \sum_{k=1}^{N} P(\Delta t_1)\Delta[z(1)]...P(\Delta t_{k-1})\Delta[z(k-1)]P(\Delta t_k)(z(k) - \bar{1}),$$

where $\bar{1}$ is an r-dimensional vector consisting of ones, $\Delta t_k = t_k - t_{k-1}$ $(t_0 = 0)$, $k = 1, 2, ..., N$.

The proof of lemma 7.4 can be derived by induction over N.

Returning to theorem 7.6, we note that its proof can be obtained from lemmas 7.3 and 7.4, and it is composed of two parts. First, convergence of finite-dimensional distributions is established. Second, convergence in the uniform topology is verified. Arguments of the proof use the method of characteristic functions and representation of the service process under fixed path of input flow as a sum of indicator processes with necessary calculations.

It should be noted that the first term of the limit process is connected with fluctuations of the input flow, and the second one with fluctuations of service time in the network nodes.

Let us prove that the limit Gaussian process in theorem 7.6 is a diffusion. With this aim in view, Markovian property of the limit process has to be proved.

7.5.2. *Criterion of Markovian behavior for r-dimensional Gaussian processes*

In a one-dimensional case, there is a criterion for Gaussian processes in terms of necessity and sufficiency to verify the Markovian property (see (Feller 1966, theorem 1, p. 94)). The principle condition of the criterion is written in terms of correlations and corresponds to the characteristic property of the exponential function. It is quite convenient to verify it in practice. In the multidimensional case, the situation becomes more complicated and there is no general criterion. We present a variant of the sufficient conditions for the Markovian property of r-dimensional Gaussian processes from Lebedev (2001) and apply this criterion to the limit process in theorem 7.6.

Let $\xi'(t) = (\xi_1(t), \xi_2(t), \ldots, \xi_r(t)) \in R^r$ be an r-dimensional Gaussian process with zero mean

$$\mathbf{E}\xi_i(t) = 0, \quad i = 1, 2, ..., r,$$

and correlation matrices

$$R(t) = \mathbf{E}\xi(t)\xi'(t) - \mathbf{E}\xi(t)\mathbf{E}\xi'(t),$$

$$R(s,t) = \mathbf{E}\xi(s)\xi'(t) - \mathbf{E}\xi(s)\mathbf{E}\xi'(t), \quad s < t.$$

THEOREM 7.7.– (See (Lebedev 2001; Livinska and Lebedev 2016)) Let for some matrix A and for all s, t $(0 \leq s < t)$ the functions $R(s)$ and $R(s,t)$ be related in the following way:

$$R(s,t) = R(s)P(t-s), \text{ where } P(t) = \exp(At).$$

Then the r-dimensional Gaussian process $\xi(t)$ is a Markov process, and the conditional distribution $P(\xi(t) \in B \mid \xi(s) = x)$, $B \in \mathfrak{B}_{\mathbb{R}^r}$ (σ-algebra of Borel subsets of $\mathbb{R}^r$), is Gaussian with mean vector $P'(t-s)x$ and correlation matrix $R(t) - P'(t-s)R(s)P(t-s)$.

Note that if G_A is the set of Gaussian processes for which the condition of theorem 7.7 is valid and the corresponding matrices are the same (equal A), then G_A satisfies the closure condition: a linear combination of two independent processes from G_A belongs G_A . Thus, we obtain the following interesting fact: the sum of two independent Markov G_A-processes is a Markov process. Note also that the multidimensional Ornstein–Uhlenbeck process satisfies the condition of theorem 7.7.

Let us apply the criterion of Markov behavior given by theorem 7.7 to the limit Gaussian process $\xi^{(1)}(t) + \xi^{(2)}(t)$ in theorem 7.6.

COROLLARY 7.1.– Let conditions of theorem 7.6 be fulfilled. Then the sequence of stochastic processes $\xi^n(t)$, $n \geq 1$, converges weakly in the uniform topology on any finite interval $[0, T]$ to diffusion $\xi(t)$ $(\xi(0) = 0)$ with the drift vector $A(x) = A'x$ and the diffusion matrix

$$B(t) = \Delta[q'(t)A] - A'\Delta[q(t)] - \Delta[q(t)]A + \sigma^2.$$

If we denote the corresponding correlation matrices of the process $\xi^{(1)}(t) + \xi^{(2)}(t)$ by $R(t)$, $R(s,t)$, $s < t$, then the conditions of theorem 7.7 are satisfied for them with $A = \Delta(\mu)(P - I)$. So, $\xi^{(1)}(t) + \xi^{(2)}(t)$ is a Markov diffusion process. Drift vector and diffusion matrix are determined by the form of conditional distribution $P(\xi(t) \in B \mid \xi(s) = x)$, $B \in \mathfrak{B}_{\mathbb{R}^r}$.

7.5.3. *Non-Markov Gaussian approximation of $[G|GI|\infty]^r$-networks*

Now, in contrast to sections 7.5.1 and 7.5.2, the distribution function $G_i(t)$, $i = 1, 2, ..., r$, of service times in the network nodes has general form and the memorylessness property can be not fulfilled.

The path of a call in such a network can be described by a semi-Markov process $y(t), t \geq 0$, with the set of states $\{1, 2, ..., r, r+1\}$ and with its semi-Markov matrix $\|G_{ij}(t)\|_1^{r+1}$ defined in section 7.4.2.

In accordance with the construction, the state $r+1$ is an absorbing one for the process $y(t)$. Recall that transient probabilities of the semi-Markov process $y(t)$ are $p_j^i(t) = P\left(y(t) = j \mid y(0) = i\right), P(t) = \left\|p_j^i(t)\right\|_1^r$.

In order to define the heavy traffic transient regime for the $[G|GI|\infty]^r$-network, we keep conditions 7.3 and 7.5 for initial distribution and for input flow, but instead of condition 7.1 we require the following.

CONDITION 7.6.– $G_i^{(n)}(nt) \stackrel{d}{\Rightarrow} G_i(t)$, as $n \to \infty$, $\quad i = 1, 2, \ldots, r$,

where $\stackrel{d}{\Rightarrow}$ means weak convergence of distribution functions.

The main objective of this section is to study limit behavior of the sequence of stochastic processes

$$\xi^{n'}(t) = n^{-1/2}\left(Q^{n'}(nt) - n\lambda' \int_0^t P^{(n)}(u)du\right), \quad n \geq 1, \quad t \geq 0,$$

where $P^{(n)}(u) = \left\|p_j^{i(n)}(u)\right\|_{i,j=1}^r = \left\|p_j^{i,n}(nu)\right\|_{i,j=1}^r$, $p_j^{i,n}(t)$ are the transient probabilities of the semi-Markov process $y^n(t)$ defined likewise $y(t)$, $t \geq 0$, with changing $G_i(t)$ to $G_i^{(n)}(t)$, $i = 1, 2, ..., r$.

In this regard, let us consider two mutually independent Gaussian processes $\xi^{(1)}(t)$ and $\xi^{(2)}(t)$, $t \geq 0$, with zero means and correlation matrices in the following form:

$$R^{(1)}(t) = \int_0^t P'(u)\sigma^2 P(u)du, \quad R^{(1)}(s,t) = \int_0^s P'(u)\sigma^2 P(u+t-s)du, \quad s < t,$$

$$R^{(2)}(t) = \int_0^t \left[\Delta(\lambda' P(u)) - P'(u)\Delta(\lambda)P(u)\right] du,$$

$$R^{(2)}(s,t) = \sum_{m=1}^r \lambda_m \int_0^s \left[\Delta(p_m(u)) - p_m(u)p_m'(u)\right] E^{(m)}(u, u+t-s)du, \quad s < t,$$

where $p'_m(t) = (p_1^m(t), ..., p_r^m(t))$ is the mth-line of the matrix $P(t)$, and for $s < t$, $E^{(m)}(s,t) = \left\| E_{ij}^{(m)}(s,t) \right\|_{i,j=1}^{m}$,

$$E_{ij}^{(m)}(s,t) = \begin{cases} p_{ij}^m(s,t)/p_i^m(s), & \text{if } p_i^m(s) \neq 0; \\ 0, & \text{else,} \end{cases}$$

$$p_{ij}^m(s,t) = P\left(y(t) = j, y(s) = i/y(0) = m\right).$$

If the $[G|GI|\infty]^r$-network operates under heavy traffic transient conditions, then the next result holds for the normalized service process $\xi^n(t)$, $n \geq 1$.

THEOREM 7.8.– Let a queueing $[G|GI|\infty]^r$-network satisfy conditions 7.3, 7.5 and 7.6, and let the spectral radius of the switching matrix P be strictly less than one. Then the sequence of stochastic processes $\xi^n(t)$, $n \geq 1$, converges in the uniform topology on any finite interval $[0, T]$ to $\xi^{(1)}(t) + \xi^{(2)}(t)$.

Substantially, theorem 7.8 is an analogue to theorem 7.6. The main difference is that the limit process is a non-Markov process if the distribution function of service times is not exponential at least in one network node.

The sketch of the proof is similar as well. The statement of lemma 7.3 for the $[G|GI|\infty]^r$-network will be the same if we consider the matrix of transient probabilities of semi-Markov process $y(t)$ as the matrix $P(t)$.

Going forward, we need the following result instead of lemma 7.4.

Let $0 \leq t_1 < t_2 < ... < t_k$, and

$$p_{i_1 i_2 \ldots i_k}^m(t_1, t_2, ..., t_k) = P\left(y(t_1) = i_1, ..., y(t_k) = i_k \mid y(0) = m\right), \quad m = 1, 2, ..., r,$$

where the distribution function of sojourn time of $y(t)$ in the initial state m is assumed to be the same as $G_m(t)$.

Functions $p_{i_1 i_2 \ldots i_k}^m(t_1, t_2, ..., t_k)$ are uniquely determined by the sequence of systems of Markov renewal integral equations

$$p_{i_1}^m(t_1) = \sum_{j=1}^{r} \int_0^{t_1} dG_{m_j}(u) p_{i_1}^m(t_1 - u) + \delta_{mi_1}[1 - G_m(t_1)], \quad k = 1,$$

$$p^m_{i_1 i_2 \ldots i_k}(t_1, t_2, \ldots, t_k) =$$

$$\sum_{j=1}^{r} \int_0^{t_1} dG_{m_j}(u) p^m_{i_1 i_2 \ldots i_k}(t_1 - u, t_2 - u, \ldots, t_k - u) +$$

$$\sum_{\alpha=2}^{k} \delta_{m i_1} \ldots \delta_{m i_{\alpha-1}} \sum_{j=1}^{r} \int_{t_{\alpha-1}}^{t_\alpha} dG_{m_j}(u) p^j_{i_\alpha i_{\alpha+1} \ldots i_k}(t_\alpha - u, t_{\alpha+1} - u, \ldots, t_k - u) +$$

$$+ \delta_{m i_1} \ldots \delta_{m i_k} [1 - G_m(t_k)], \quad k > 1.$$

LEMMA 7.5.– If the spectral radius of the matrix P is strictly less than one, then for any $N = 1, 2, \ldots$ and $0 < t_1 < \ldots < t_N$ we have

$$\Phi^{(m)}(t_1, \ldots, t_N, z(1), \ldots, z(N)) = 1+$$

$$+ \sum_{k=1}^{N} \sum_{i_1, \ldots, i_k = 1}^{r} p^m_{i_1 i_2 \ldots i_k}(t_1, t_2, \ldots, t_k) z_{i_1}(1) \ldots z_{i_{k-1}}(k-1)[z_{i_k}(k) - 1].$$

Like the statement of lemma 7.4, the last result can be obtained by the method of mathematical induction with respect to N.

The above models have one point in common: their parameters do not depend on time. Surely, it constricts the applications field. Especially, this point is sensitive when one simulates input flows with their real-life rates dependent on time. Further, within the framework of the simplest Markov models, we remove this restriction.

7.6. Limit processes for network with time-dependent input flow

7.6.1. *Gaussian approximation of* $\left[\vec{M}_t|M|\infty\right]^r$*-networks in heavy traffic*

Let us consider a network with a non-homogeneous Poisson input flow of calls $\nu^{(i)}(t)$ arriving at the ith node, $i = 1, 2, \ldots, r$. Suppose $\Lambda_i(t)$ is the leading function of $\nu^{(i)}(t)$. Service times in the ith node are independent exponentially distributed random variables with rate μ_i. As before, the r-dimensional service process of calls $Q(t) = (Q_1(t), \ldots, Q_r(t))'$, $t \geq 0$, is studied.

The heavy traffic regime for the $\left[\vec{M}_t|M|\infty\right]^r$-network is determined by condition 7.1 and the following condition on input flow parameters.

CONDITION 7.7.– Input flows depend on n in such a way that on any finite interval

$$n^{-1}\Lambda_i^{(n)}(nt) \overset{U}{\Rightarrow} \Lambda_i^{(0)}(t) \in C[0, T], \text{ as } n \to \infty, \quad i = 1, 2, \ldots, r,$$

where $C[0, T]$ is the set of continuous on $[0, T]$ functions, the symbol $\overset{U}{\Rightarrow}$ stands for convergence in the uniform topology.

Let us consider two cases that are important for applications when condition 7.7 is valid. We temporarily assume that the Poisson flows $\nu^{(i)}(t)$, $i = 1, 2, \ldots, r$, are regular: $\Lambda_i(t) = \int_0^t \lambda_i(u)du$, where $\lambda_i(u)$ is the instant value of the parameter.

7.7a If $\lambda_i(t)$ is a periodic function with period T_i, then condition 7.7 holds with $\Lambda_i^{(0)}(t) = \left(\int_0^{T_i} \lambda_i(u)du\right) t$.

7.7b If $\lim_{t\to\infty} \lambda_i(t) = \lambda_i > 0$, then condition 7.7 holds with $\Lambda_i^{(0)}(t) = \lambda_i t$.

Note, that networks with a periodic input flow are studied in more details in Livinska and Lebedev (2018).

In the context of conditions 7.1 and 7.7, we consider the sequence of stochastic processes:

$$\xi^n(t) = n^{-1/2}(Q^{(n)}(nt) - q^{(n)}(nt)), \quad t \geq 0,$$

where $q^{(n)\prime}(nt) = (q_1^{(n)}(nt), \ldots, q_r^{(n)}(nt))$, $q_j^{(n)}(nt) = \sum_{i=1}^{r} \int_0^{nt} d\Lambda_i^{(n)}(u)p_j^{i(n)}(nt - u)$, $j = 1, \ldots, r$; $p_j^{i(n)}(\tau)$ are entries of the matrix

$$P^{(n)}(\tau) = \|p_j^{i(n)}(\tau)\|_1^r = \exp\{\Delta(\mu^{(n)})(P - I)\tau\},$$

$$\mu^{(n)\prime} = (\mu_1^{(n)}, \ldots, \mu_r^{(n)}).$$

In order to describe the limit of the sequence of stochastic processes $\xi^n(t)$ as $n \to \infty$, we introduce two independent Gaussian processes

$$\xi^{(i)\prime}(t) = (\xi_1^{(i)}(t), \ldots, \xi_r^{(i)}(t)), \quad i = 1, 2.$$

The process $\xi^{(1)}(t)$ is determined by zero mean:

$$\mathbf{E}\xi^{(1)}(t) = 0$$

and by the correlation matrices:

$$R^{(1)}(t) = \mathbf{E}\xi^{(1)}(t)\xi^{(1)\prime}(t) - \mathbf{E}\xi^{(1)}(t)\mathbf{E}\xi^{(1)\prime}(t) = \int_0^t P'(t-\tau)\Delta[d\Lambda^{(0)}(\tau)]P(t-\tau),$$

$$R^{(1)}(s,t) = \mathbf{E}\xi^{(1)}(s)\xi^{(1)'}(t) - \mathbf{E}\xi^{(1)}(s)\mathbf{E}\xi^{(1)'}(t) = R^{(1)}(s)P(t-s), \quad s < t,$$

where $\Lambda^{(0)'}(t) = (\Lambda_1^{(0)}(t), ..., \Lambda_r^{(0)}(t)), \quad P(\tau) = \exp\{\Delta(\mu)(P-I)\tau\}$.

For the process $\xi^{(2)}(t)$:

$$\mathbf{E}\xi^{(2)}(t) = 0,$$

$$R^{(2)}(t) = \int_0^t \left[\Delta[(d\Lambda^{(0)}(\tau))'P(t-\tau)] - P'(t-\tau)\Delta[d\Lambda^{(0)}(\tau)]P(t-\tau)\right],$$

$$R^{(2)}(s,t) = R^{(2)}(s)P(t-s), \quad s < t.$$

The following result gives an approximation for the service process of an $\left[\overrightarrow{M}_t|M|\infty\right]^r$-network under heavy traffic conditions 7.1 and 7.7.

THEOREM 7.9.– Let conditions 7.1 and 7.7 take place for an $\left[\overrightarrow{M}_t|M|\infty\right]^r$-network, and let the network be empty at the initial instant $t = 0 : Q_i(0) = 0, \ i = 1, 2, \ldots, r$. Then, the sequence of stochastic processes $\xi^n(t)$ converges on any finite interval $[0, T]$ as $n \to \infty$ to the process $\xi^{(1)}(t) + \xi^{(2)}(t)$ in the uniform topology.

SKETCH OF THE PROOF OF THEOREM 7.9.– The proof follows from some auxiliary results and is similar to the proof of theorem 7.6.

In the case of non-homogeneous Poisson input flow, we have to prove the next property which is parallel to condition 7.5 supposed *a priori*.

LEMMA 7.6.– Let $\nu^n(t)$ be a non-homogeneous Poisson process with the leading function $\Lambda^{(n)}(t)$ such that condition 7.7 is valid. Then the sequence of stochastic processes $W^{(n)}(t) = n^{-1/2}(\nu^n(nt) - \Lambda^{(n)}(nt))$, $n \geq 1$, converges in the uniform topology on any finite interval $[0, T]$ to a Wiener process $W^{(0)}(t)$ with $\mathbf{E}W^{(0)}(t) = 0$ and $\mathbf{Var}W^{(0)}(t) = \Lambda^{(0)}(t)$.

Convergence of finite-dimensional distributions of the process $W^{(n)}(t)$ to $W^{(0)}(t)$ follows from the fact that for any natural number N and instants $0 < t_1 < \ldots < t_N$ the joint characteristic function of the distribution of $\nu^n(t_1), \ldots, \nu^n(t_N)$ is equal to

$$E\exp\left\{i\sum_{k=1}^{N} s(k)\nu^n(t_k)\right\} =$$

$$= \prod_{k=0}^{N-1} \exp\left\{ \left[\Lambda^{(n)}(t_{k+1}) - \Lambda^{(n)}(t_k)\right] \left[\exp\left(i \sum_{m=k+1}^{N} s(m)\right) - 1\right]\right\},$$

where $(s(1), \ldots, s(N)) \in \mathbb{R}^N$, $t_0 = 0$.

Now, in order to prove the convergence in the uniform topology, it is sufficient to check the following condition for any $\varepsilon > 0$:

$$\lim_{h\to 0} \overline{\lim_{n\to\infty}} \sup_{|t_1-t_2|\le h} P\left\{\left|W^{(n)}(t_2) - W^{(n)}(t_1)\right| > \varepsilon\right\} = 0. \quad [7.16]$$

Based on Chebyshov inequality, we have the upper estimate:

$$P\left\{\left|W^{(n)}(t_2) - W^{(n)}(t_1)\right| > \varepsilon\right\} =$$
$$P\left\{\left|\nu^n(nt_2) - \nu^n(nt_1) - \left(\Lambda^{(n)}(nt_2) - \Lambda^{(n)}(nt_1)\right)\right| > \varepsilon\sqrt{n}\right\} \le$$
$$\le \varepsilon^{-2}\left|n^{-1}\Lambda^{(n)}(nt_2) - n^{-1}\Lambda^{(n)}(nt_1)\right|. \quad [7.17]$$

On the other hand, condition 7.7 implies that

$$\lim_{h\to 0} \overline{\lim_{n\to\infty}} \sup_{|t_2-t_1|\le h} \left|n^{-1}\Lambda^{(n)}(nt_2) - n^{-1}\Lambda^{(n)}(nt_1)\right| = 0. \quad [7.18]$$

So, [7.16] follows from [7.17] and [7.18].

Lemma 7.6 is proved.

Throughout the section, $W_i^{(0)}(t)$, $i = 1, 2, ..., r$, denote independent Wiener processes with zero mean and $\mathbf{Var}W_i^{(0)}(t) = \Lambda_i^{(0)}(t)$. If condition 7.7 holds, they approximate the input flows $\nu^{(i)n}(t)$.

For $W^{(0)'}(t) = (W_1^{(0)}(t), \ldots, W_r^{(0)}(t))$, we need the result given by lemma 7.3. Lemma 7.4 have to be satisfied as well. So, like in the case for theorem 7.6, the proof of theorem 7.9 can be obtained from lemmas 7.3, 7.4 and 7.6.

It is easily observed that the limit of $\xi^n(t)$ is a Markov Gaussian process. However, it can be presented as a diffusion process due to theorem 7.7.

THEOREM 7.10.– Let an $[\overrightarrow{M}_t|M|\infty]^r$-network be empty at the initial instant $t = 0$, let conditions 7.1 and 7.7 hold and $\Lambda_i^{(0)}(t) = \int_0^t \lambda_i^{(0)}(u)du$, $\lambda_i^{(0)}(u) \in C[0,T]$, $i = 1,2,\dots,r$. Then, the sequence of stochastic processes $\xi^n(t)$, $n \geq 1$, weakly converges on any finite interval $[0,T]$ in the uniform topology to the r-dimentional diffusion process $\xi^{(0)}(t)$ with the drift vector $A(x) = A'x$ and the diffusion matrix

$$B(t) = \Delta[\lambda^{(0)'}(t) + q'(t)A] - A'\Delta[q(t)] - \Delta[q(t)]A,$$

where $A = \Delta(\mu)(P - I)$, $q'(t) = \int_0^t \lambda^{(0)'}(\tau)P(t - \tau)d\tau$, $\lambda^{(0)'}(\tau) = (\lambda_1^{(0)}(\tau),\dots,\lambda_r^{(0)}(\tau))$.

The form of the limit process as a multidimensional diffusion is attractive because a diffusion process is determined only by its local characteristics and one can use the developed tools of Markov diffusion processes for the analysis of its functionals. However, now the limit process does not reflect in detail the structure of the prelimit service process.

7.6.2. *Limit process in case of asymptotically large initial load*

Let us consider a case when the number of calls being served in the $[\overrightarrow{M}_t|M|\infty]^r$-network at the initial instant of time depends on n and increases as $n \to \infty$. In this case, the service process of calls that were initially located at the network can converge to a non-zero limit, and this term can enter into the overall limit process. So, fulfillment of the following condition is required here instead of $Q_i^{(n)}(0) = 0$, $i = 1, 2, ..., r$:

CONDITION 7.8.– $Q_i^{(n)}(0) = \left[n\theta_i/\mu_i + \sqrt{n}\eta_i^{(0)}\right]$, $i = 1, 2, ..., r$, where $\eta^{(0)'} = \left(\eta_1^{(0)}, ..., \eta_r^{(0)}\right)$ is a fixed vector in $\mathbb{R}^r$, $[\cdot]$ is the integer part of a number.

Here, we will consider an open $[\overrightarrow{M}_t|M|\infty]^r$-network with periodic input flow. As before, let $\theta' = (\theta_1, ..., \theta_r) = \lambda'(I - P)^{-1}$ be a solution of the balance equation, where $\lambda' = (\lambda_1, ..., \lambda_r)$, and in this case $\lambda_i = \int_0^{T_i} \lambda_i(u)du$, $i = 1, 2, ..., r$.

Next, we study the asymptotic behavior of the sequence of stochastic processes

$$\eta^n(t) = n^{-1/2}\left(Q^n(nt) - n\left(\theta/\mu\right)\right), \, t \geq 0, \, n \geq 1,$$

$$(\theta/\mu)' = (\theta_1/\mu_1, ..., \theta_r/\mu_r)\,.$$

With this in mind, we introduce in addition a Gaussian process $\xi^{(3)}(t)$ that does not depend on the previously introduced processes $\xi^{(1)}(t)$ and $\xi^{(2)}(t)$. This process $\xi^{(3)}(t)$ is determined by the mean values

$$E\xi^{(3)}(t) = P'(t)\eta^{(0)},$$

and by the correlation matrices

$$R^{(3)}(t) = E\xi^{(3)}(t)\xi^{(3)'}(t) - E\xi^{(3)}(t)E\xi^{(3)'}(t) = \\ = \Delta\left((\theta/\mu)' P(t)\right) - P'(t)\Delta\left(\theta/\mu\right) P(t),$$

$$R^{(3)}(s,t) = E\xi^{(3)}(s)\xi^{(3)'}(t) - E\xi^{(3)}(s)E\xi^{(3)'}(t) = R^{(3)}(s)P(t-s),\ s < t.$$

THEOREM 7.11.– (See (Lebedev and Livinska 2018)) Suppose that an $\left[\overrightarrow{M}_t|M|\infty\right]^r$-network with a periodical input flow has the spectral radius of its switching matrix strictly less than 1, and let conditions 7.1, 7.7a and 7.8 be satisfied. Then, the sequence of stochastic processes $\eta^n(t)$ converges as $n \to \infty$ to $\xi^{(1)}(t) + \xi^{(2)}(t) + \xi^{(3)}(t)$ in the uniform topology on any finite interval $[0, T]$.

In comparison with the statement of theorem 7.9, the additional summand $\xi^{(3)}(t)$ of the limit process is associated with fluctuations of service times of calls located in network nodes at the initial instant $t = 0$.

It is not difficult to check that for correlation matrices, we have:

$$R^{(1)}(t) + R^{(2)}(t) + R^{(3)}(t) = \Delta(\theta/\mu) - P'(t)\Delta(\theta/\mu)P(t).$$

So, theorem 7.11 implies the following result for $\eta^n(t)$.

COROLLARY 7.2.– Let for an open $\left[\overrightarrow{M}_t|M|\infty\right]^r$-network conditions 7.1, 7.7a and 7.8 be satisfied. Then the sequence of stochastic processes $\eta^n(t)$ converges as $n \to \infty$ in the uniform topology on any finite interval $[0, T]$ to an r-dimensional Ornstein–Uhlenbeck diffusion process $\eta^{(0)}(t)$ in transient regime ($\eta^{(0)}(0) = \eta^{(0)}$) with the following correlation characteristics:

$$E\eta^{(0)}(t) = P'(t)\eta^{(0)},$$

$$R^{(0)}(t) = E\eta^{(0)}(t)\eta^{(0)'}(t) - E\eta^{(0)}(t)E\eta^{(0)'}(t) = \Delta(\theta/\mu) - P'(t)\Delta(\theta/\mu)P(t),$$

$$R^{(0)}(s,t) = E\eta^{(0)}(s)\eta^{(0)'}(t) - E\eta^{(0)}(s)E\eta^{(0)'}(t) = R^{(0)}(s)P(t-s), \quad s < t.$$

7.7. Conclusion

In this chapter, we investigate the asymptotic behavior of the service process of calls for some stochastic networks with infinite number of servers in each node under heavy traffic assumptions. It is convenient to use the local approach for proving DA for this class of network, or to use constructive probabilistic methods for the Gaussian approximation of service process of the network.

We consider networks with input flows of different structure and with exponential and general service. Networks with time-dependent Poisson input flow are considered in two different cases of starting load: first, the number of calls in the initial instant is assumed to be equal to zero; second, this number is assumed to increase with a series number increasing. For models with exponential service, DA is proved either by the local approach or by proving a functional limit theorem on convergence to Gaussian process and checking Markovian property of the limit process using correspondent criterium presented in the chapter. Note that if a distribution function of the service time is not exponential at least in one node, then an accumulation of aftereffect takes place during servicing, and the limit process is not a Markov process. More networks with such service are studied, for example, in Lebedev (2002a, 2002b) and in Livinska (2016).

As for Gaussian approximation provided in the chapter, it is proved that under formulated heavy traffic conditions, the normalized service process has Gaussian process as a limit in the uniform topology. So, jump-wise service processes are approximated by Gaussian processes with continuous paths and simpler structure. It is significant that the limit process is decomposed into the sum of independent Gaussian processes. In the case of the fixed initial load, it has two summands $\xi^{(1)}(t)$ and $\xi^{(2)}(t)$. The part $\xi^{(1)}(t)$ of the limit process is associated with fluctuations of an input flow, and $\xi^{(2)}(t)$ is associated with fluctuations of service times. In the case of asymptotically large initial load, the third term $\xi^{(3)}(t)$ appears in the sum. It is connected with fluctuations of service times of calls located in the network nodes at the initial instant. Moreover, in the case of asymptotically large load, the limit process is a multidimensional Ornstein–Uhlenbeck diffusion process. For all cases, the approximating Gaussian processes are constructed with their characteristics written in explicit form via the network parameters.

Note also that convergence to Gaussian process is proved in the uniform topology. This type of convergence of stochastic processes gives us a possibility to calculate functionals related with the processes and to solve optimization problems (see, for example, Lebedev and Makushenko (2012)).

Some studies of similar multichannel stochastic networks with time-dependent input flow and different types of service can be found in Lebedev (2017), Lebedev and Livinska (2018), Livinska (2016), Livinska and Lebedev (2018). Networks with input flow controlled by a Markov process were investigated in Lebedev and Livinska (2017).

7.8. Acknowledgements

We would like to thank the editors and the anonymous referees for their valuable suggestions.

7.9. References

Anisimov, V.V. (1995). Switching processes: Averaging principle, diffusion approximation and applications. *Acta Applicandae Mathematicae*, Kluwer, The Netherlands, 40(2), 95–141.

Anisimov, V.V. (2002). Diffusion approximation in overloaded switching queueing models. *Queueing Syst.*, 40(2), 141–180.

Anisimov, V.V. (2008). *Switching Processes in Queueing Models*. ISTE Ltd, London, and Wiley, New York.

Anisimov, V.V. (2010). Switching queueing networks. In *Performance Evaluation and Applications*, D. Kouvatsos (ed.), Springer-Verlag. 5233, 258–283.

Anisimov, V.V. and Lebedev, E.A. (1992). *Stochastic Queueing Networks. Markov Models*. Lybid, Kiev.

Basharin, G., Bocharov, P. and Kogan, J. (1989). *Analysis of Queues in Computing Networks*. Nauka, Moscow (in Russian).

Bernstein, S.N. (1934). Principes de la théorie des équations différentielles stochastiques. *Proceedings of the Phy. Math. Inst. Named Steklov*, 5, 95–124.

Billingsley, P. (1999). *Convergence of Probability Measures*. John Wiley & Sons, Inc., NewYork.

Borovkov, A.A. (1980). *Asymptotic Methods in Queueing Theory*. Nauka, Moscow (in Russian).

Decreusefond, L. and Moyal, P. (2008). A functional central limit theorem for the $M/GI/\infty$ queue. *Ann. Appl. Probab.*, 186, 156–178.

Dvurečenskij, A., Kulyukina, L.A. and Ososkov, G.A. (1981). Estimations of track ionization chambers. *Transactions of United Nuclear Research Institute*, Dubna.

Feller, W. (1966). *An Introduction to Probability Theory and Its Applications*, vol. 2. John Wiley and Sons, Inc., New York.

Fralix, B.H. and Adan, I.J.B.F. (2009). An infinite-server queue influenced by a semi-Markovian environment. *Queueing Syst.*, 61, 65–84.

Gikhman, I.I. (1969). On convergence to Markov processes. *Ukr. Math. J.*, 21(3), 312–320.

Gikhman, I.I. and Skorokhod, A.V. (1982). *Stochastic Differential Equations*. Naukova Dumka, Kiev (in Russian).

Gnedenko, B.V. and Kovalenko, I.N. (2013). *Introduction to Queueing Theory*. URSS, Moscow (in Russian).

Khinchin, A.Y. (1936). *Asymptotic Principles of Probability Theory*. ONTI NKTP USSR, Moscow (in Russian).

Korolyuk, V.S. and Korolyuk, V.V. (1999). *Stochastic Models of Systems*. Kluwer, Dordrecht.

Korolyuk, V.S. and Limnios N. (2004). Average and diffusion approximation for evolutionary systems in an asymptotic split phase state. *Ann. Appl. Prob.*, 14(1), 489–516.

Korolyuk, V.S. and Limnios, N. (2005). *Stochastic Systems in Merging Phase Space*. World Scientific, Singapore.

Korolyuk, V.S., Korolyuk, V.V. and Limnios N. (2009). Queuing systems with semi-Markov flow in average and diffusion approximation schemes. *Metodol. Comput. Appl. Probab.*, 11, 201–209.

Lebedev, E.A. (2001). On the Markov property of multivariate Gaussian processes. *Bulletin of Kiev Univ. Series: Phys.-Math.*, 4, 287–291.

Lebedev, E.A. (2002a). Networks of infinite server queues. *ESM'2001 Conference Proceedings*, Darmstadt, Germany.

Lebedev, E.A. (2002b) Stationary regime and binomial moments for networks of the type $[SM|GI|\infty]^r$. *Ukr. Math. J.*, 54(10), 1371–1380.

Lebedev, E.A. (2003). A limit theorem for stochastic networks and its applications. *Theor. Probab. and Math. Statist.*, 68, 81–92.

Lebedev, E.A. and Livinska, H. (2017). On stationary regime for queueing networks with controlled input. *WSEAS Trans. on Systems and Control*, 12, 9–13.

Lebedev, E.A. and Livinska, H. (2018). On Gaussian approximation of queueing networks with different starting load. *Commun. Comp. Inf. Sci.*, 912, 27–38.

Lebedev, E.A. and Livinska, H. (2019). Multi-channel queueing networks with input flow controlled by semi-Markov process, *Math. Comput. Sci.* [Online]. Available: https://doi.org/10.1007/s11786-019-00398-4.

Lebedev, E.A. and Makushenko, I.A. (2012). *Optimal Partitioning of External Load in Multi-channel Queuing Networks*. V.I. Vernadsykiy Library, Kyiv.

Lebedev, E.A., Chechelnitsky, A.A. and Livinska, H.V. (2017). Multi-channel networks with interdependent input flows in heavy traffic. *Theor. Probab. and Math. Statist.*, 97, 109–119.

Livinska H. (2016). A limit theorem for non-Markovian multi-channel networks under heavy traffic conditions. *Theor. Probab. and Math. Statist.*, 93, 113–122.

Livinska, H. and Lebedev E.O. (2015). Conditions of Gaussian non-Markov approximation for multi-channel networks. *Proceedings of the 29-th Euro. Conf. on Modelling and Simul. ECMS-2015*. Albena, Bulgaria, 642–649.

Livinska, H. and Lebedev, E. (2016). Gaussian and diffusion limits for multi-channel stochastic networks. *J. Math. Sci.*, 218(3), 328–334.

Livinska, H. and Lebedev, E. (2018). On transient and stationary regimes for multi-channel networks with periodic inputs. *Appl. Math. and Comput.*, 319, 13–23.

Mandelbaum, A. and Pats, G. (1995). State-dependent queues: Approximations and applications. In *Proc. of the IMA. Stochastic networks*, F.P. Kelly and R.J. Williams, (eds), Springer, New York, 239–282.

Mandelbaum, A. and Pats, G. (1998). State-dependent stochastic networks. Part I: Approximations and applications with continuous diffusion limits. *Ann. Appl. Prob.*, 8(2), 569–646.

Mandelbaum A., Masey W. and Reiman M. (1998). Strong approximation for Markovian service networks. *Queueing Syst.*, 30(1,2), 149–202.

Massey, W.A. and Whitt, W. (1993). Networks of infinite-server queues with nonstationary Poisson input. *Queueing Syst.*, 13, 183–250.

Massey, W.A. and Whitt, W. (1994). A stochastic model to capture space and time dynamics in wireless communication systems. *Prob. Eng. Inf. Sci.*, 8, 541–569.

Matis, J.H. and Wehrly, T.E. (1990). Generalized stochastic compartmental models with Erlang transit times. *J. Pharmac. Bioph.*, 18, 589–607.

Moiseev, A. and Nazarov, A. (2015). *Infinite-server Queueing Systems and Networks. NTL Publishing*, Tomsk (in Russian).

Moiseev, A. and Nazarov, A. (2016). Queueing network $MAP - (GI \mid \infty)^K$ with high-rate arrivals. *Eur. J. Oper. Res.*, 254(1), 161–168.

Nazarov, A. and Moiseev, A. (2014). Investigation of the queueing network $GI - (GI|\infty)^K$ by means of the first jump equation and asymptotic analysis. In *DCCN 2013, CCIS 279*, V. Vishnevsky *et al.* (ed.). Springer, 229–240.

Nazarov, A. and Tuenbaeva, A.N. (2009). Investigation of mathematical model of communication network with unsteady flow of requests. *Transp. and Telecom.*, 4, 28–34.

Prokhorov, Y.V. (1956). Convergence of stochastic processes and limit theorems of probability theory. *Theor. Probab. and Appl.*, 1(2), 177–238.

Puhalskii, A.A. and Reed, J.E. (2008). On many-server queues in heavy traffic. *Ann. Appl. Probab.*, 20, 129–195.

Reiman, M. (1984). Open queueing networks in heavy traffic. *Math. Oper. Res.*, 9(3), 441–458.

Skorokhod, A.V. (1956). Limit theorems for stochastic processes. *Theor. Probab. and Math. Statist.*, 1(3), 289–319.

Whitt, W. (1982). On the heavy-traffic limit theorem for $GI/G/\infty$ queues. *Adv. Appl. Probab.*, 14, 171–190.

Whitt, W. (2016). Heavy-traffic fluid limits for periodic infinite-server queues. *Queueing Syst.*, 84, 111–143.

Whitt, W. (2018). Time-varying queues. *Queueing Mod. Serv. Managm.*, 1(2), 79–164.

Williams, R.J. (1996). On the approximation of queueing networks in heavy traffic. In *Stochastic Networks: Theory and Applications, Proceedings of Royal Society Workshop, Edinburgh, 1995*, Kelly, F.P., Zachary, S., Ziedins, I. (eds). Oxford University Press, Oxford, 35–56.

Williams, R.J. (2016). Stochastic processing networks. *Ann. Rev. Stat. Appl.*, 3, 323–346.

8

Recent Results in Finite-source Retrial Queues with Collisions

Anatoly NAZAROV[1], **János SZTRIK**[2] **and Anna KVACH**[1]

[1] *National Research Tomsk State University, Russia*

[2] *University of Debrecen, Hungary*

The aim of this chapter is to give a review of recent results on single-server finite-source retrial queueing systems with collision of the customers. These include investigations of when the server is reliable and models of when the server is subject to random breakdowns and repairs depending on whether it is idle or busy. Tool-supported, numerical, simulation and asymptotic methods are considered under the condition of unlimited growing number of sources. Several cases and examples are treated and the results of different approaches are compared to each other showing the advantages and disadvantages of the given method. In general, we prove that the steady-state distribution of the number of customers in the service facility can be approximated by a normal distribution with given mean and variance. Using asymptotic methods under certain conditions in steady state, the distribution of the sojourn time in the orbit and in the system can be approximated by a generalized exponential one. Furthermore, it is proved that the distribution of the number of retrials until the successful service in the limit is geometrically distributed. With the help of stochastic simulation, several systems are analyzed showing directions for further analytic investigations. Tables and figures are presented to illustrate some special features of these systems.

8.1. Introduction

Finite-source retrial queues are very useful and effective stochastic systems to model several problems arising in telephone switching systems and

Queueing Theory 1,
coordinated by Vladimir ANISIMOV, Nikolaos LIMNIOS.

telecommunication networks including cellular networks, CSMA (carrier sense multiple access) local area computer networks, call centers and CSMA-based wireless mesh networks at the frame level. To see their importance, the interested reader is referred to the following work and references cited in them: Artalejo and Corral (2008); Falin and Artalejo (1998); Fiems and Phung-Duc (2017); Gómez-Corral and Phung-Duc (2016); Kim and Kim (2016).

Searching the scientific databases, we have noticed that just a relatively small number of papers have been devoted to queueing systems when the arriving calls (primary or secondary) cause collisions to the request under service and both go to the orbit (see, for example, Ali and Wei (2015); Choi *et al.* (1992); Kim (2010); Kumar *et al.* (2010); Lakaour *et al.* (2018); Peng *et al.* (2014); Takeda and Yoshihiro (2017)).

In real CSMA systems, collisions are unavoidable and they decrease the effectiveness of the system performance and that is why new protocols should be developed to avoid them (see Cao *et al.* (2018); Jinsoo *et al.* (2018); Kwak *et al.* (2018); Wentink (2017); Yeo *et al.* (2017)). Real situations where collisions may occur were modeled by Reith (2017) and Takeda and Yoshihiro (2017).

Stochastic modeling of systems with collisions is required not only from a technical point of view but it is a mathematical challenge since it needs more sophisticated approaches.

Nazarov and his research group developed a very effective asymptotic method (Nazarov and Moiseeva 2006) with the help of which various systems have been investigated. Concerning finite-source retrial systems with collision, we should mention the following papers: Kvach and Nazarov (2015b); Kvach (2014); Kvach and Nazarov (2015a,c); Nazarov *et al.* (2014).

Sztrik and his research group have been dealing with systems with unreliable server/s as can be seen in Almási *et al.* (2005); Sztrik (2005); Sztrik *et al.* (2006); Wüchner *et al.* (2010) and that is why it was understandable that the two research groups started cooperation in 2017.

Our investigations have been based on analytical, numerical, simulation and asymptotic approaches as treated in Anisimov and Sztrik (1989); Anisimov and Artalejo (2001); Anisimov (1999); Artalejo and Corral (2008); Bhat (2015); Bossel (2013); Falin and Templeton (1997); Harchol-Balter (2013); Kobayashi and Mark (2009); Kulkarni (2016); Lakatos *et al.* (2013); Law and Kelton (1991); Nazarov and Terpugov (2004); Nazarov and Moiseeva (2006); Rubinstein and Kroese (2016); Stewart (2009); Yao (2016).

The primary aim of this chapter is to give a survey on the results obtained in this field in the recent past by means of different methods. Doing so we have tried to unify the notation which has appeared in different publications and to use the standard notation of Western-style papers, which differs from the Russian-style papers. We are confident that our models with collision can be used in real situations to describe, for example, the operation of random access systems treated in Fiems and Phung-Duc (2017) without collision; CSMA-based wireless mesh networks at frame and packet level investigated in Takeda and Yoshihiro (2017); and systems with transmission errors analyzed in Lakaour *et al.* (2018). When a collision occurs, signals are superimposed, packets are distorted and therefore packets need to be retransmitted, for which they both should be sent to the orbit.

We agree that the light-traffic approach applied in Fiems and Phung-Duc (2017) to get the distribution of the number of customers in the service facility for systems without collisions could be integrated to investigate systems with collisions. However, we must emphasize that beside these we are able to obtain approximation for the distribution of a number of retrials and for the distribution of the response/waiting time of a customer. These measures were not treated in the above-mentioned paper. In addition, in our different cases the service time is generally distributed and the server is not reliable.

From the theoretical point of view, one of the main difficulties compared to the systems without collision is that the service process is interrupted many times by the collisions until it is successfully completed. Unreliable server failures also interrupt the service, resulting in a quite complicated total service time structure. In addition, in case of a collision the number of customers in the orbit increases by two. Using an algorithmic approach means that the nth iteration depends on not only the previous iteration but on two previous ones. Our most important contribution is that, besides obtaining the asymptotic distribution of the number of customers in the orbit, the asymptotic distribution of the number of retrials and the distribution of the response/waiting time of a call are determined under certain conditions.

The new numerical insights for models with collisions and unreliable server are obtained by the help of algorithmic, numeric (MOSEL), simulation approaches. The combination of these methods allows us to validate the accuracy of the asymptotic method. We show that under a certain parameter setup, the characteristic property of the finite-source retrial queues having a maximum of the mean waiting time for an increasing arrival rate remains valid even in case of collision. This means that the maximum is due to the finite source and not to the collision. This surprising feature was noticed by several authors and it was explained in several ways. However, in systems with collisions we have to deal with the total service time, which is many times interrupted by collisions and failures. This notion is due to the collision and the successful service time is the last phase of this measures of interest. According to our experiments, the mean of the total and successful service time heavily depends on the distribution of the service time and the minimum can be obtained only under very special parameter setup.

This chapter is an enlarged and modified version of our recent paper (Nazarov *et al.* 2018).

The rest of the paper is organized as follows. In section 8.2, a description of the model is given, and the corresponding multi-dimensional non-Markov process is defined. In sections 8.3 and 8.4, systems with a reliable and an unreliable server are treated, respectively. In these sections, models with exponentially and generally distributed service times are investigated, and then analyzed by means of tool supported, algorithmic, simulation and asymptotic methods, respectively. The main results of the papers are collected and several figures illustrate the most interesting features of the given system. Finally, this chapter ends with a conclusion and some future plans are highlighted.

8.2. Model description and notations

In the following, we introduce the model in the most general form as it was treated with the help of numerical and asymptotic methods.

Let us consider a retrial queuing system of type M/GI/1//N with collision of the customers and an unreliable server (Figure 8.1). The number of sources is N and each of them can generate a primary request during an exponentially distributed time with rate λ/N. A source cannot generate a new call until the end of the successful service of this customer.

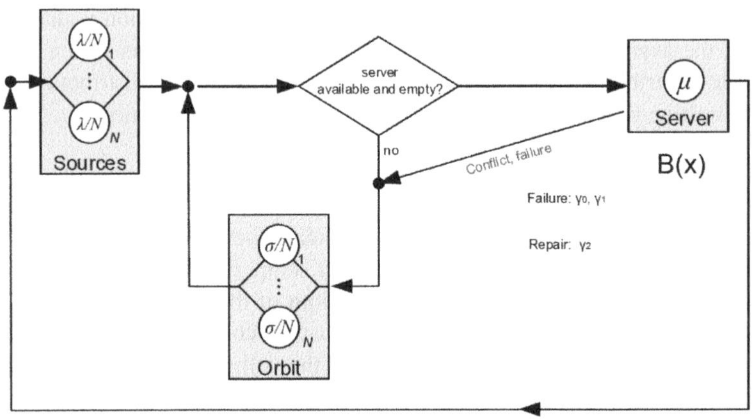

Figure 8.1. *Retrial queuing system of type M/GI/1//N with collisions of the customers and unreliable server*

If a primary request finds the server idle, he enters into service immediately, in which the required service time has a probability distribution function $B(x)$. Let us denote its service rate function by $\mu(y) = B^{'}(y)(1 - B(y))^{-1}$ and its Laplace -Stieltjes transform by $B^{*}(y)$, respectively. If the server is busy, an arriving (primary or repeated) customer goes into collision with a customer under service and they both move into the orbit. The interretrial times of customers are supposed to be exponentially distributed with rate σ/N, that is, it does not depend on how many times it has been blocked. We assume that the server is unreliable, that is its lifetime is supposed to be exponentially distributed with failure rate γ_0 if the server is idle and with rate γ_1 if it is busy. When the server breaks down, it is immediately sent for repair and the repair time is assumed to be exponentially distributed with rate γ_2. We deal with the case when the server is down, all sources continue generation of customers and send them to the orbit; similarly, customers may retry from the orbit to the server but all arriving customers immediately go into the orbit. Furthermore, in this unreliable model, we suppose that the interrupted request goes to the orbit immediately and its next service is independent of the interrupted one. Of course, in the case of a reliable server $\gamma_0 = \gamma_1 = 0$. All random variables involved in the model construction are assumed to be independent of each other.

Let $J(t)$ be the number of customers in the system at time t, that is, the total number of customers in the orbit and in service. Similarly, let $K(t)$ be the server state at time t, that is

$$K(t) = \begin{cases} 0, & \text{if the server is idle,} \\ 1, & \text{if the server is busy,} \\ 2, & \text{if the server is down (under repair).} \end{cases}$$

Thus, we will investigate the process $\{K(t), J(t)\}$, which is not a Markov process unless the service time is exponentially distributed. To be a Markov process we must use the method of supplementary variables, namely, we will consider two variants: the residual service time method and the elapsed service time method depending on what the aim of the investigation is.

Let us denote by $Y(t)$ the supplementary random process equals the elapsed service time of the customer till the moment t and by $Z(t)$ the residual service time, that is, the time interval from the moment t until the end of successful service of the customer, respectively. It is obvious that $\{K(t), J(t), Y(t)\}$ and $\{K(t), J(t), Z(t)\}$ are Markov processes. Let us note that $Y(t)$ and $Z(t)$ are defined only in those moments when the server is busy, that is, when $K(t) = 1$.

Let us define the stationary probabilities as follows:

$$\begin{aligned} P_0(j) &= P\{K=0, J=j\}, \\ P_1(j,y) &= P\{K=1, J=j, Y<y\}, \\ P_1(j,z) &= P\{K=1, J=j, Z<z\}, \\ P_2(j) &= P\{K=2, J=j\}. \end{aligned}$$

Of course, in the case of exponentially distributed service time the steady-state probabilities are denoted as follows:

$$P_k(j) = P\{K=k, J=j\}, \quad k=0,1,2, \quad j=0,...,N.$$

The steady-state distribution of the server's state is denoted by

$$R_k = P(K=k), k=0,1,2$$

and the distribution of the number of customers in the system is designated by

$$P(j) = P(J=j), j=0,...,N.$$

It is clear that in the case of a reliable server, all the probabilities where $K = 2$ are 0.

The main aim of the investigations is to get these distributions and other performance measures of the systems, such as the distribution of the sojourn time in the system, distribution of the total service time and distribution of the number of retrials. These are very complicated problems and to the best knowledge of the authors there are no exact analytical formulas to the solutions. That is the reason we have tried to obtain the characteristics of different systems by the help of tool supported, algorithmic, stochastic simulation and asymptotic methods.

Since in the following, we will use the gamma distribution to show special features of some systems we introduce it here to be sure of the form because it is defined in several ways.

A random variable X is said to have a **gamma distribution** with parameters (α, β) if its density function is given by

$$f(x) = \begin{cases} 0 & \text{, if } x < 0 \\ \frac{\beta(\beta x)^{\alpha-1} e^{-\beta x}}{\Gamma(\alpha)} & \text{, if } x \geq 0, \end{cases}$$

where $\alpha > 0$, $\beta > 0$, and

$$\Gamma(\alpha) = \int_0^\infty t^{\alpha-1}e^{-t}dt$$

is the so-called **complete gamma function**.

Its distribution function cannot be obtained in an explicit form except $\alpha = n$. In this case, it reduces to the Erlang distribution.

It should be noted that depending on the parameters it can take small values with high probability, that is when the shape parameter is low, in this case the squared coefficient of variation C_X^2 is high. Its density function has the form

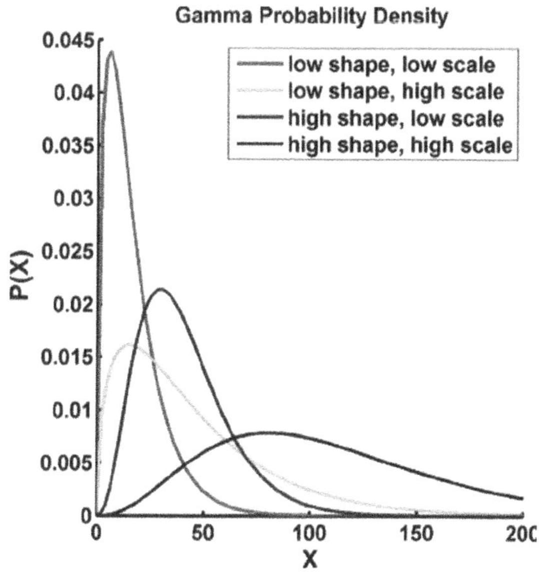

Figure 8.2. *Density function of the gamma distribution. For a color version of this figure, see www.iste.co.uk/anisimov/queueing1.zip*

It can be shown that

$$\mathsf{E}(X) = \frac{\alpha}{\beta}, \quad Var(X) = \frac{\alpha}{\beta^2}, \quad C_X^2 = \frac{1}{\alpha}.$$

α and β are called the *shape parameter* and *scale parameter*, respectively, and they can be expressed by the mean and variance in the following way if we would like to estimate the parameters from the sample

$$\alpha = \frac{(\mathsf{E}(X))^2}{Var(X)}, \quad \beta = \frac{\mathsf{E}(X)}{Var(X)}.$$

It can be shown that its Laplace transform has the form

$$L_X(s) = \left(\frac{\beta}{\beta + s}\right)^{\alpha} = \left(1 + \frac{s}{\beta}\right)^{-\alpha}.$$

8.3. Systems with a reliable server

8.3.1. *M/M/1 systems*

8.3.1.1. *Algorithmic approach*

In previous papers (Kvach 2014; Nazarov *et al.* 2014), the steady-state Kolmogorov equations were derived and the distribution of the system's state was obtained by an algorithmic approach. Then the distribution of the number of customers in the system were calculated and used to validate the asymptotic results.

8.3.1.2. *Asymptotic approach*

The main contribution of the paper by Nazarov *et al.* (2014) is that the steady-state characteristic function of prelimit distribution of the number of customers in the system can be approximated by the following third-order asymptotic characteristic function, that is

$$\mathsf{E}\exp\{iuJ\} \approx \exp\left\{iu\kappa_1 N + \frac{(iu)^2}{2}\kappa_2 N + \frac{(iu)^3}{3!}\kappa_3 N\right\}$$

where the distribution of the server's state is explicitly given in the following form:

$$R_1 = \frac{\sigma(2\lambda + \mu) - \sqrt{\sigma^2(2\lambda - \mu)^2 + 8\sigma\mu\lambda^2}}{4\mu(\sigma - \lambda)},$$

$$R_0 = 1 - \frac{\sigma(2\lambda + \mu) - \sqrt{\sigma^2(2\lambda - \mu)^2 + 8\sigma\mu\lambda^2}}{4\mu(\sigma - \lambda)},$$

and the cumulants (semi-invariants) (Nazarov and Sudyko 2010) are

$$\kappa_1 = \frac{2\mu R_1^2}{\sigma(1-2R_1)}, \qquad \kappa_2 = \mu R_1 \frac{1+(R_1-R_0)R_0}{\lambda-(\lambda-\sigma)(R_1-R_0)^2}.$$

$$a = R_0 R_1 (R_1 - R_0),$$

$$\kappa_3 = 2 \cdot \left\{ \frac{\left[(\lambda-\sigma)(R_0-R_1)^2\kappa_2 + \mu a\right] \cdot \left[\frac{1}{2} - ((R_0-R_1)(\sigma-\lambda)+\lambda)\frac{\kappa_2}{\mu R_1}\right]}{\lambda+(\sigma-\lambda)(R_0-R_1)^2} + \right.$$
$$\left. + \frac{\lambda\kappa_2 \cdot \left[\frac{a}{R_0}+\frac{1}{2}\right] + \mu\left[\frac{1}{R_0}-\frac{a}{2}\right]}{\lambda+(\sigma-\lambda)(R_0-R_1)^2)} \right\}.$$

Actually, κ_1 is the positive solution of the equation

$$\lambda(1-\kappa_1) = \mu R_1(\kappa_1),$$

and numerically it is easy to verify that

$$R_0(\kappa_1) = \frac{\delta(\kappa_1)+\mu}{2\delta(\kappa_1)+\mu},$$

$$R_1(\kappa_1) = \frac{\delta(\kappa_1)}{2\delta(\kappa_1)+\mu},$$

where $\delta(\kappa_1) = \lambda(1-\kappa_1) + \sigma\kappa_1$.

It is easy to see that the second-order approximation results in a normal distribution. In this case, let us denote by $G(x)$ the distribution function of the Gaussian distribution with mean $N\kappa_1$ and variance $N\kappa_2$.

Furthermore, let us denote by $P_{as}(j)$ the asymptotic discrete distribution obtained by the help of Gaussian approximation, that is

$$P_{as}(j) = \{G(j+0.5) - G(j-0.5)\}\left[G(N+0.5) - G(-0.5)\right]^{-1},$$
$$j = 0, ..., N \qquad [8.1]$$

which we will call the Gaussian approximation of the prelimit discrete distribution $P(j)$.

In the following, we show how we can get the distribution with the help of the inverse characteristic function (see Nazarov and Sudyko (2010)).

Let $D_m(j)$ be a probability distribution with $j = 0, 1, 2, ..., N$ and $h_m(u)$ its characteristic function, that is

$$h_m(u) = \sum_{j=0}^{N} e^{iuj} D_m(j)$$

where $i = \sqrt{-1}$ is the imaginary unit and $h_m(u)$ is the mth-order asymptotic characteristic function. Then using the inverse transformation $D_m(j)$ is given by the formula

$$D_m(j) = \frac{1}{2\pi} \int_{-\pi}^{\pi} e^{-iuj} h_m(u)\, du.$$

It is obvious that

$$\sum_{j=0}^{N} D_m(j) = h_m(0) = 1.$$

If the values of $D_m(j)$ are real and non-negative for all $j = 0, ..., N$, then

$$D_m(j) = P_m(j),$$

where $P_m(j)$ is called the mth-order approximation of the prelimit distribution $P(j)$.

Since $D_m(j)$ is calculated by numeric integration, among the numbers $D_m(j)$ there can appear complex numbers with small enough imaginary parts or real negative numbers with small enough absolute values.

In this case, we define real non-negative numbers $g_m(j)$ by the formula

$$g_m(j) = \frac{1}{4}\{D_m(j) + \overline{D_m(j)} + \left|D_m(j) + \overline{D_m(j)}\right|\},$$

where the over bar denotes complex conjugation. Then let us define the approximation of the prelimit distribution by the equality

$$P_m(j) = g_m(j) / \sum_{l=0}^{N} g_m(l).$$

To compare the probability distributions $P(j)$ and $P_m(j)$, we will use the Kolmogorov distance defined as

$$\Delta_m = \max_{0 \le n \le N} \left| \sum_{j=0}^{n} P(j) - \sum_{j=0}^{n} P_m(j) \right|.$$

By this distance, we can estimate the error of the mth-order approximation of the prelimit distribution.

In a previous paper (Nazarov *et al.* 2014), second- and third-order approximations of the prelimit distribution were compared to the exact distribution obtained by the algorithmic method. With a different parameter setup and for different N, the applicability of the asymptotic method was validated and some conclusions were drawn.

A more complicated problem, namely the distribution of the sojourn time in the service facility, was investigated by Kvach and Nazarov (2015b) with the help of asymptotic methods as N tends to infinity. It was proved that the characteristic function of the sojourn time T that a customer spends in the service facility can be approximated by

$$\mathsf{E} \exp\{iuT\} \approx q + (1-q)\frac{\sigma q/N}{\sigma q/N - iu},$$

where $q = \frac{\mu R_0}{\delta + \mu}$.

Then it is easy to show that the approximation of the prelimit distribution function of the sojourn time can be written as

$$A(x) \approx 1 - (1-q)e^{-\frac{\sigma q}{N}x}$$

which means that the asymptotic sojourn time can be zero with probability q and it is exponentially distributed with probability $1-q$. As a result, the mean sojourn time

$$E(T) \approx (1-q)N/(\sigma q).$$

From probabilistic reasoning

$$\lambda(1-\kappa_1) = \mu R_1,$$

which means that the mean arrival rate is equal to the mean departure rate and from Little's formula

$$\mu R_1(1-q) = \sigma q \kappa_1$$

should be valid. These formulas can be used for validation of the obtained numerical and asymptotic results. The asymptotic sample examples concerning the sojourn time distribution have not been validated by simulation in the paper.

8.3.2. *M/GI/1 system*

This section deals with the results when the required service times are generally distributed but in the examples the gamma distribution is used due to its useful properties. Namely, it is easy to see that its squared coefficient of variation can be less, equal to or greater than 1 depending on the values of the shape and scale parameters.

8.3.2.1. *Algorithmic approach*

A previous study (Kvach and Nazarov 2015a) deals with the algorithmic approach of how to get the steady-state distribution of the system. The method of supplementary variable technique with residual service time were applied and several numerical examples were treated with gamma distributed service time. The results helped the validation of asymptotic results for the same model.

8.3.2.2. *Stochastic simulation*

Papers of Nazarov *et al.* (2017b,c) are devoted to the asymptotic analysis of the mean total service time, distribution of the sojourn time in the system and the distribution of the number of retrials. It must be noted that the results have not been validated by simulation. Meanwhile, simulations have been carried out, the estimations for the mean and variance of the sojourn time have been obtained, and the distribution of the number of retrials also has been determined. The simulation analysis will be published in the near future.

8.3.2.3. *Asymptotic approach*

In this section, the asymptotic results published in Nazarov *et al.* (2017b,c) are summarized. Before doing that we need some notations, namely

$$B^*(\alpha) = \int_0^\infty e^{-\alpha x} dB(x), \quad \delta(\kappa_1) = \lambda + (\sigma - \lambda)\kappa_1.$$

Then κ_1 can be obtained from

$$\kappa_1 = 1 - \frac{\delta(\kappa_1)}{\lambda} \cdot \frac{B^*(\delta(\kappa_1))}{2 - B^*(\delta(\kappa_1))}, \qquad [8.2]$$

and the distribution of the server's state can be determined by

$$R_0 = \frac{1}{2 - B^*(\delta)}, \quad R_1 = \frac{1 - B^*(\delta)}{2 - B^*(\delta)}.$$

In the case of the residual service time until the successful service completion approach for the busy server's sate, we have

$$R_1(z) = e^{\delta z} \int_0^z e^{-\delta x} \cdot \{\lambda(1-\kappa_1) - \delta R_0 B(x)\}\, dx,$$

and in the case of the elapsed service time approach we get

$$\int_0^\infty R_1(y)\mu(y)dy = \lambda(1-\kappa).$$

Introducing the notations

$$A_1 = \lambda(1-\kappa_1), \qquad R_1^*(\alpha) = -\delta R_0 \left[B^*(\alpha)\right],$$

we obtain

$$\kappa_2 = \frac{A_1\Big(R_0 \cdot B^*(\delta)\left[\delta + A_1\right] - (\delta + A_1 R_0)\Big)}{A_1(\sigma-\lambda)\Big(R_1^*(\delta) - R_1 - R_0(B^*(\delta)-1)\Big) + \delta\Big((\sigma-\lambda)\Big(R_1^*(\delta) - R_0 B^*(\delta)\Big) - \lambda\Big)}.$$

Consequently, the steady-state prelimit distribution of the number of customers in the system can be approximated by a normal distribution with mean $N\kappa_1$ and variance $N\kappa_2$.

It is interesting that equation [8.2] can have one, two or three roots $0 < \kappa_1 < 1$. For example, for the gamma distribution function $B(x)$ with a shape parameter α and scale parameter β with values $\alpha = \beta = 2$, $\lambda = 0.29$, $\sigma = 20$, equation [8.2] has three roots, namely $\kappa_1^{(1)} = 0.031$, $\kappa_1^{(2)} = 0.188$ and $\kappa_1^{(3)} = 0.549$.

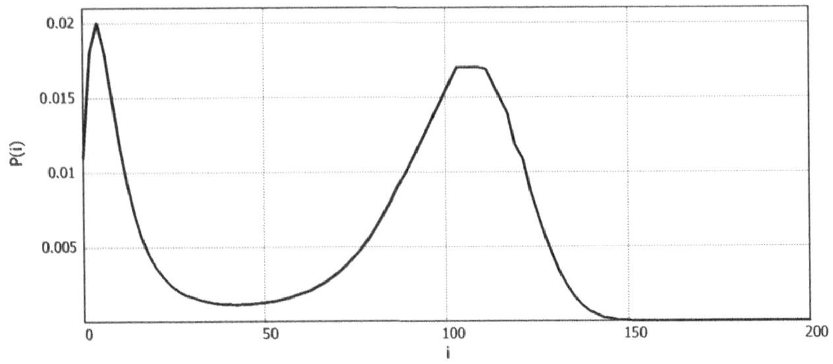

Figure 8.3. *Bimodal probability distribution of the number of customers in the system*

Let us consider such values of the service parameters α, β and the system λ, σ at which equation [8.2] has three roots; the probability distribution $P(i) = P\{i(t) = i\}$ can then be of a bimodal type. In particular, for $\alpha = \beta = 2$, $\lambda = 0.29$, $\sigma = 19.7$ by simulation and numerical methods at $N = 200$ for $P(i)$ we obtain the following graph shown in Figure 8.3.

The reason for this feature is explained by Nazarov *et al.* (2017c).

As a rule, equation [8.2] has three roots in exceptional cases at special values of parameters and such situation arises extremely rare. Therefore, in the following let us consider some properties of the system when the main equation [8.2] has a single root $0 < \kappa_1 < 1$.

In the paper by Nazarov *et al.* (2017b), for the mean sojourn time of the customer in service, we get the following expression:

$$\lim_{N\to\infty} \mathsf{E}(T_S) = \frac{1 - B^*(\delta)}{\delta B^*(\delta)}.$$

Next let us consider the influence of the system parameters on the mean sojourn time $E(T_S)$ of the customer in the server. We will choose $\sigma = 20$ and the values of parameters λ and $\alpha = \beta$ are specified in Table 8.1

As we can see, at $\alpha < 1$ the values of total mean sojourn time $E(T_S)$ of the customer under service takes values less than unity. For $\alpha < 1$, there is a high probability of emergence of small values of service time and this fact undoubtedly influences the total mean sojourn time $E(T_S)$ of the customer at the server and it takes rather small values. Moreover, let us note that with increasing values of parameter λ the values of $E(T_S)$ decrease.

		$\alpha=\beta$							
		0.1	0.3	0.5	0.8	1	2	3	5
λ	0.5	0.324	0.546	0.682	0.857	1	5.038	20.816	153.239
	1	0.204	0.367	0.488	0.727	1	5.576	21.695	154.766
	5	0.067	0.165	0.301	0.640	1	5.937	22.360	155.975
	10	0.046	0.140	0.280	0.632	1	5.979	22.441	156.125
	15	0.039	0.131	0.273	0.629	1	5.993	22.468	156.175

Table 8.1. *Mean sojourn time $E(T_S)$ of the customer under service at various values of λ and $\alpha=\beta$*

In the case of $\alpha > 1$, the values of $E(T_S)$ become greater than unity. Table 8.1 illustrates that with an increase of service parameter α the values of total mean sojourn time $E(T_S)$ of the customer under service considerably increases and reaches very large values. Let us note that in this situation, parameter λ practically does not influence $E(T_S)$ and with an increasing parameter of service α this influence becomes less and less.

Table 8.1 also shows that in case of exponential service time, i.e. $\alpha = \beta = 1$, the values of total mean sojourn time $E(T_S)$ of the customer at the server are equal to the unit.

For the distribution of the number of retrials/transitions of the tagged customer into the orbit, we have the following results.

Let ν be the number of transitions of the tagged customer into the orbit, then

$$\lim_{N\to\infty} \mathsf{E}\, z^{\nu} = \frac{q}{1-(1-q)z},$$

where value of parameter q has a form

$$q = R_0 B^*(\delta).$$

From the proved theorem, it obviously follows that the probability distribution $P\{\nu = n\}$, $n = \overline{0,\infty}$ of the number of transitions of the tagged customer into the orbit is geometric and

$$P\{\nu = n\} = q(1-q)^n, \quad n = \overline{0,\infty}.$$

Consequently, by using the law of total probability for the characteristic function of the sojourn/waiting time W of the tagged customer in the orbit we get

$$\mathsf{E}e^{iuW} \approx q + (1-q)\frac{\sigma q}{\sigma q - iuN}.$$

In the case of $N \to \infty$, the limiting probability distributions of the sojourn time of the customer in the system T and the sojourn time of the customer in the orbit W coincide, namely

$$\lim_{N\to\infty} \mathsf{E}\exp\left\{iu\frac{T}{N}\right\} = \lim_{N\to\infty} \mathsf{E}\exp\left\{iu\frac{W}{N}\right\} = q + (1-q)\frac{\sigma q}{\sigma q - iu}.$$

Let us consider the influence of the system parameters on the values of the mean number of retrials ν. We will choose the same system parameters which have been considered in the previous examples, namely $\sigma = 20$, and the values of parameters λ and $\alpha = \beta$, are specified in Table 8.2.

		$\alpha = \beta$							
		0.1	0.3	0.5	0.8	1	2	3	5
	0.5	0.470	1.131	1.86	3.65	6.32	162.7	793	6091
	1	0.656	2.002	4.19	11.57	21.83	204.1	848	6172
λ	5	1.114	4.192	9.31	22.73	37.07	234.4	891	6236
	10	1.278	4.755	10.28	24.30	39.02	238.1	896	6244
	15	1.355	4.969	10.63	24.83	39.67	239.3	898	6247

Table 8.2. *Mean number of retrials for various values of λ and $\alpha = \beta$*

From Table 8.2, it is shown that the mean number of retrials obviously depends on service parameter α that is the expected and logical result. It should be noted that for small values of parameter α, the influence of parameter λ on the mean number of retrials is significant and obvious. But with the increase in parameter α, this influence decreases and already for great values of α practically disappears.

The following conclusions can be drawn: the mean number of retrials and mean sojourn time of the customer under service is a consequence of the collision of customers and the admissibility of repeated attempts of service by the same customer. Duration of the customer service for repeated attempts has the same probability distribution $B(x)$, but its repeated realization, naturally, accepts various values. If for the distribution $B(x)$, there is a high probability of emergence of small values of service time as in the gamma distribution with the shape parameter $\alpha < 1$, then a small number of retries is sufficient to realize a small value of the service time which will be successful. If the small values of the service time are unlikely for the probability distribution $B(x)$, as in the gamma distribution with the shape parameter $\alpha > 1$, then the number of unsuccessful attempts of service becomes big, as we can see in Table 8.2, the server works without results, the mean sojourn time of the customer under service increases (Table 8.1).

8.4. Systems with an unreliable server

In many practical situations, the server is not reliable and after a random time it can fail and needs repair which also takes a random duration. To deal with these service interruptions, several papers have been published (see, for example, Almási *et al.* (2005); Dragieva (2014); Gharbi and Dutheillet (2011); Gharbi and Ioualalen (2006); Gharbi *et al.* (2014); Krishnamoorthy *et al.* (2014); Roszik (2004); Sztrik *et al.* (2006); Wang *et al.* (2010, 2011); Zhang and Wang (2013)). In the following sections, we summarize our results obtained by different methods.

8.4.1. *M/M/1 system*

8.4.1.1. *Tool-supported approach by MOSEL*

Because of the fact that in many practical situations the state space of the describing Markov chain is very large, it is rather difficult to calculate the system measures in the traditional way of writing down and solving the underlying steady-state equations. To simplify this procedure, several software packages have been developed and effectively used for performance evaluation of complex systems (see, for example, Gharbi and Dutheillet (2011); Gharbi and Ioualalen (2006); Gharbi *et al* (2015); Gharbi *et al.* (2014); Ikhlef *et al.* (2016)). In our investigations, a similar software tool called MOSEL (Modeling, Specification and Evaluation Language) has been used to formulate the model and to obtain the performance measures. A previous study by Bérczes *et al.* (2017) deals with the model formulation, derivation of several performance measures and generation of illustrative examples showing an interesting phenomenon of finite-source retrial queues, that is under specific parameter setup the mean waiting/sojourn time has a maximum as the arrival intensity is increasing. We showed that it still remains in the case of collisions as the following example shows

Case	N	λ/N	γ_0	γ_1	γ_2	σ/N	μ
Figure 8.4 case 1	100	0.01	0.01	0.01	1	0.1	1
Figure 8.4 case 2	100	0.01	0.1	0.1	1	0.1	1
Figure 8.4 case 3	100	0.01	1	1	1	0.1	1
Figure 8.5	100	0.03 - 8.1	0.01	0.01	1	0.1	1
Figure 8.6	100	0.03 - 8.1	0.1	0.1	1	0.1	1

Table 8.3. *Numerical values of model parameters*

Figure 8.4 shows the steady-state distribution of the three investigated cases. In this figure, we can see also the effect of the breakdown of the server. We can see that the mean number of customers increases as the breakdown intensity gets larger. From the shape of the curves, it is clearly visible that the steady-state distribution of the cases are normally distributed.

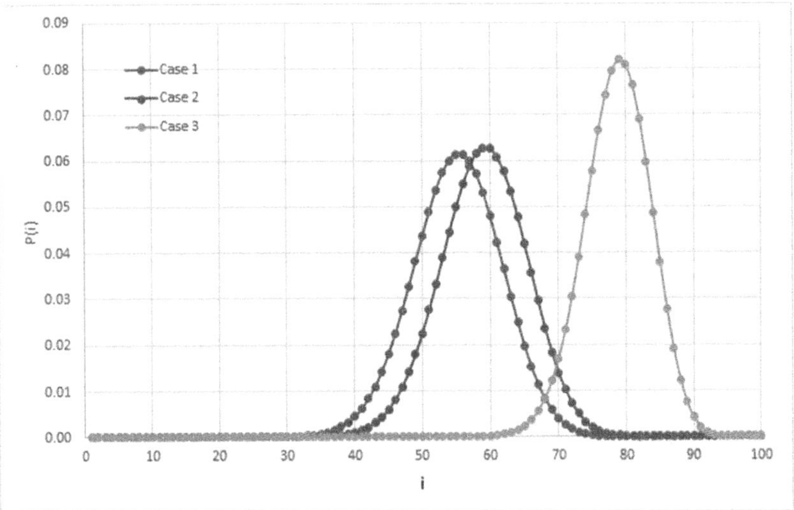

Figure 8.4. *Steady-state distributions. For a color version of this figure, see www.iste.co.uk/anisimov/queueing1.zip*

Figures 8.5 and 8.6 show the mean response time as a function of the customer generation rate. As we can see, the mean response time will be greater as we increase the generation rate, but after λ/N is greater than 1.5 the mean response time starts to decrease.

8.4.1.2. *Algorithmic approach*

The advantage of an algorithmic approach to the general tool supported method is that in this case we do not have a state space explosion and N can be much higher than in case of calculations carried by MOSEL as we will see soon. The publication of the results has been submitted (see Kuki *et al.* (2019)). The calculation has been performed by a spreadsheet program, MS Excel, and N can be arbitrary large. When the calculations are performed by the MOSEL-2 tool (see Bérczes *et al.* (2017)), we run into strict limitations, namely the state space grows extremely fast; consequently the number of sources cannot exceed 200. In Excel, we can go far higher above 200. Other advantages for using this spreadsheet are that the effect of parameter modifications can be seen immediately and the set of the steady-state probabilities, both the two- and one-dimensional ones, are present in separate columns and can be used directly for further investigations.

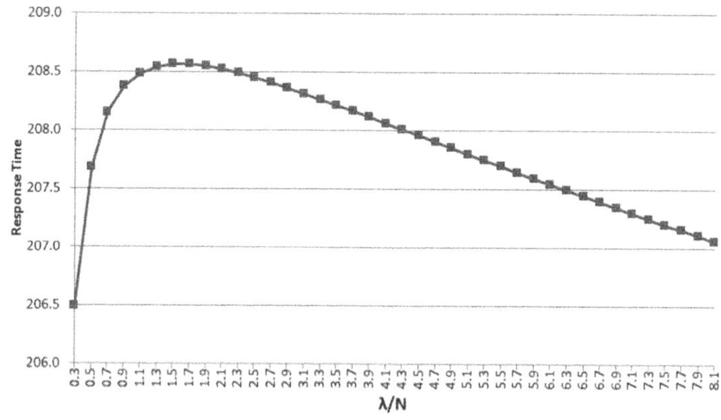

Figure 8.5. *vs* λ/N, $\gamma_0 = \gamma_1 = 0.01$

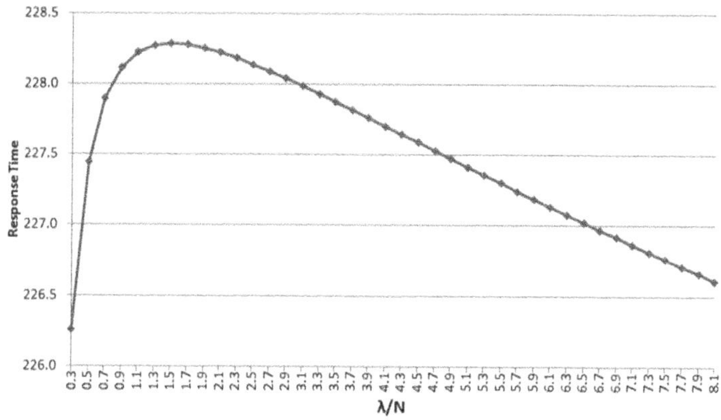

Figure 8.6. *vs* λ/N, $\gamma_0 = \gamma_1 = 0.1$

For illustration, we consider three systems with the following input parameters

$$N = 100, \quad \lambda = 1, \quad \mu = 1, \quad \sigma = 5, \quad \gamma_0 = 0.1, \quad \gamma_1 = 0.1, \quad \gamma_2 = 1.$$

In Figure 8.7, the steady-state probabilities of three models are displayed: a basic reliable system with no conflict, a reliable system with conflict, and an unreliable system with conflict. In the no conflict case, the expectation of states are lower than for the other cases. The probabilities of the states have the greatest mean in the unreliable system as it was expected.

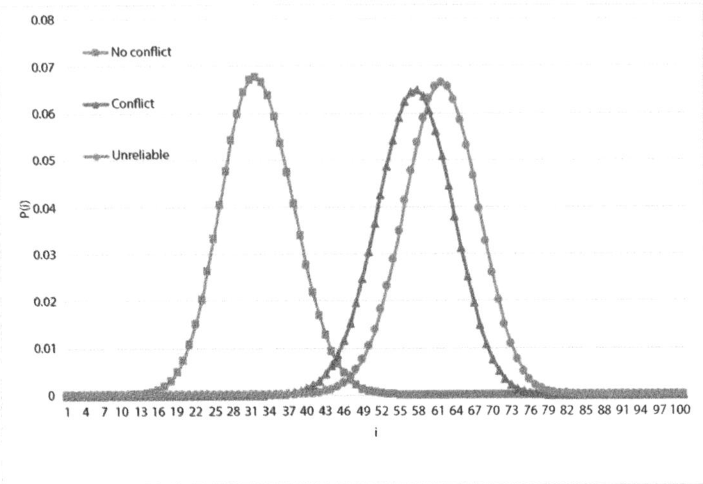

Figure 8.7. *Reliable no conflict, reliable with conflict, and unreliable with conflict. For a color version of this figure, see www.iste.co.uk/anisimov/queueing1.zip*

8.4.1.3. *Stochastic simulation*

To validate the applicability of the asymptotic approach, we need either numerical or simulation results. The correct operation of the simulation software was tested by the numerical sample examples. The investigations carried out by the simulation and asymptotic methods have been submitted for publication (see Nazarov, Sztrik, Kvach and Bérczes (2018) and Nazarov, Sztrik, Kvach and Tóth (2018)).

8.4.1.4. *Asymptotic approach*

First, we deal with the distribution of the number of customers in the system as it has been published by Nazarov *et al.* (2018). The first-order asymptotic results are the following

$$\lim_{N\to\infty} E\exp\left\{iw\frac{J}{N}\right\} = \exp\left\{iw\kappa_1\right\},$$

where κ_1 is the positive solution of the equation

$$(1-\kappa_1)\lambda - \mu R_1(\kappa_1) = 0,$$

where the stationary distributions of probabilities $R_k(\kappa_1)$ of the server state $k = 0, 1, 2$ are obtained as follows

$$R_0(\kappa_1) = \left\{ \frac{\gamma_0 + \gamma_2}{\gamma_2} + \frac{\gamma_1 + \gamma_2}{\gamma_2} \cdot \frac{a(\kappa_1)}{a(\kappa_1) + \gamma_1 + \mu} \right\}^{-1},$$

$$R_1(\kappa_1) = \frac{a(\kappa_1)}{a(\kappa_1) + \gamma_1 + \mu} \cdot R_0(\kappa_1),$$

$$R_2(\kappa_1) = \frac{1}{\gamma_2} [\gamma_0 R_0(\kappa_1) + \gamma_1 R_1(\kappa_1)],$$

here $a(\kappa_1)$ is

$$a(\kappa_1) = (1 - \kappa_1)\lambda + \sigma\kappa_1.$$

The second order asymptotic results are

$$\lim_{N\to\infty} E\exp\left\{ iw \frac{J - \kappa_1 N}{\sqrt{N}} \right\} = \exp\left\{ \frac{(iw)^2}{2} \kappa_2 \right\},$$

where κ_2 is

$$\kappa_2 = \frac{\gamma_2\mu(R_1 - b_1) + (1 - \kappa_1)\lambda \{(\gamma_1 + \gamma_2) b_1 + (1 - \kappa_1)\lambda R_2\}}{(\lambda + \mu b_2)\gamma_2 - (1 - \kappa_1)\lambda(\gamma_1 + \gamma_2) b_2},$$

and

$$b_1 = \frac{(1 - \kappa_1)\lambda}{a + \gamma_1 + \mu} R_0, \qquad b_2 = \frac{(\sigma - \lambda)(R_0 - R_1)}{a + \gamma_1 + \mu}.$$

Consequently, the prelimit distribution of the number of customers in the system can be approximated by a normal distribution with mean $N\kappa_1$ and variance $N\kappa_2$.

Let us determine the accuracy and area of applicability of this approximation.

We have the following input parameters:

$$\mu = 1, \qquad \sigma = 5, \qquad \gamma_0 = 0.1, \qquad \gamma_1 = 0.2, \qquad \gamma_2 = 1$$

and we will provide values of Kolmogorov distance Δ in Table 8.4.

	$N=5$	$N=10$	$N=20$	$N=30$	$N=50$
$\lambda=0.5$	0.095	0.059	0.032	0.023	0.017
$\lambda=1.0$	0.037	0.023	0.017	0.014	0.011
$\lambda=2.0$	0.078	0.046	0.022	0.014	0.013

Table 8.4. *Kolmogorov distance between prelimit distribution $P(i)$ and the asymptotic distribution $P_{as}(i)$ for various values of the parameters N and λ*

Table 8.5 demonstrates the Kolmogorov distance for the following parameters:

$$\lambda = 1, \qquad \mu = 1, \qquad \gamma_0 = 0.1, \qquad \gamma_1 = 0.2, \qquad \gamma_2 = 1$$

and for various values of the parameter N and σ.

	$N=5$	$N=10$	$N=20$	$N=30$	$N=50$
$\sigma=0.2$	0.066	0.035	0.018	0.013	0.010
$\sigma=1.0$	0.033	0.018	0.014	0.014	0.008
$\sigma=5.0$	0.037	0.023	0.017	0.014	0.011

Table 8.5. *Kolmogorov distance between prelimit distribution $P(i)$ and the asymptotic distribution $P_{as}(i)$ for various values of the parameters N and σ*

Assuming an acceptable error $\Delta \leq 0.05$ of the given values in Tables 8.4–8.5, we can conclude that approximation has a large error at $N \leq 10$, at $10 < N < 20$ the acceptability of approximation is doubtful, and at $N \geq 20$ approximation has quite acceptable accuracy.

Let us turn our attention to the analysis of the waiting/sojourn time distribution. Using Little's formula, we have

$$(1-\kappa_1)\,\lambda\,\mathsf{E}\left(\frac{1}{N}T\right) = \kappa_1\,,$$

therefore

$$\mathsf{E}\left(\frac{1}{N}T\right) = \frac{\kappa_1}{(1-\kappa_1)\,\lambda}\,, \qquad [8.3]$$

where κ_1 is the asymptotic average value of the number of customers in the system, and $(1-\kappa_1)\,\lambda$ is the mean arrival intensity of the incoming flow.

However, we should underline that the above equality [8.3] defines only the mean sojourn time of the customer in the system. We would like to carry out a more detailed investigation for T of the tagged customer.

One of the main contributions of Nazarov *et al.* (2018) is that for the limit of the characteristic function of the normalized sojourn time we have

$$\lim_{N\to\infty} \mathsf{E}\exp\left\{iw\frac{T}{N}\right\} = q + (1-q)\frac{\sigma q}{\sigma q - iw},$$

where q is

$$q = \frac{(1-\kappa_1)\lambda}{(1-\kappa_1)\lambda + \sigma\kappa_1}.$$

Consequently, the characteristic function of the sojourn time of the customer in the system in the prelimit situation of finite N can be approximated by

$$\mathsf{E}\, e^{iuT} \approx q + (1-q)\frac{\sigma q}{\sigma q - iuN}. \qquad [8.4]$$

In the following, let us find the average value of the normalized sojourn time of the customer in the system which is

$$\mathsf{E}\left(\frac{1}{N}T\right) = (1-q)\frac{1}{\sigma q} = \frac{1}{\sigma}\cdot\frac{1-q}{q} = \frac{1}{\sigma}\frac{\sigma\kappa_1}{(1-\kappa_1)\lambda} = \frac{\kappa_1}{(1-\kappa_1)\lambda},$$

and that coincides with the result earlier obtained by Little's formula.

For the distribution of the number of transitions/retrials of the tagged customer into the orbit, we get the following results.

Let ν be the number of transitions of the tagged customer into the orbit, then

$$\lim_{N\to\infty} \mathsf{E}\, z^{\nu} = \frac{q}{1-(1-q)z},$$

resulting in the probability distribution $P\{\nu = n\},\ n = \overline{0,\infty}$ of the number of transitions of the tagged customer into the orbit being geometric and having the form

$$P\{\nu = n\} = q(1-q)^n, \quad n = \overline{0,\infty}.$$

Consequently, the prelimit characteristic function of the sojourn/waiting time W of the tagged customer in an orbit can be approximated as

$$\mathsf{E}e^{iuW} \approx q + (1-q)\frac{\sigma q}{\sigma q - iuN}.$$

In the case of $N \to \infty$, the limiting probability distributions of the sojourn time of the customer in the system T and the sojourn time of the customer in an orbit W coincide, namely

$$\lim_{N\to\infty} \mathsf{E}\exp\left\{iu\frac{T}{N}\right\} = \lim_{N\to\infty} \mathsf{E}\exp\left\{iu\frac{W}{N}\right\} = q + (1-q)\frac{\sigma q}{\sigma q - iu}.$$

Previously, we have obtained that the probability distribution of the number of transitions of the tagged customer into the orbit is geometric with parameter q. Let us find out how close the limiting results to the simulation results and at what values N this approximation is admissible.

To do so, let us denote by $P_{as}(\nu = n)$ the asymptotic geometric distribution of probabilities with parameter q and by $P_s(\nu = n)$ the probability distribution of the number of transitions of the tagged customer into the orbit obtained with the help of simulation program. Furthermore, let us determine the accuracy (error) of approximation of distribution by mean of Kolmogorov distance Δ, which for probability distributions $P_{as}(\nu = n)$ and $P_s(\nu = n)$ is defined as

$$\Delta = \max_{0\le i<\infty}\left|\sum_{n=0}^{i}\left(P_{as}(\nu = n) - P_s(\nu = n)\right)\right|.$$

Realizing the simulation program for

$$\lambda = 1, \qquad \mu = 1, \qquad \sigma = 4, \qquad \gamma_2 = 1$$

and applying the approximation we will provide the Kolmogorov distance Δ for various values N and $\gamma = \gamma_0 = \gamma_1$ in Table 8.6.

We can see, as expected, that by increasing N the Kolmogorov distance should decrease, but with this parameter setup there is no essential reduction if $N > 50$.

Let us see the mean number of retrials under the same condition as before.

Again we can observe what was expected: as N increases the mean number of retrials increases since there are more and more customers in the system resulting in more and more collisions. At the same time, it can also be seen how the mean number of retrials increases as the failure rate of the server increases. It should be underlined that each time the limiting values give very good approximations showing the effectiveness of the asymptotic method.

	$N = 20$	$N = 30$	$N = 50$	$N = 100$	$N = 200$
$\gamma = 0.05$	0.026	0.016	0.009	0.005	0.003
$\gamma = 0.1$	0.024	0.015	0.009	0.004	0.002
$\gamma = 0.5$	0.017	0.011	0.006	0.004	0.001

Table 8.6. *Kolmogorov distance between distribution $P_s(i)$ and approximation of the geometric distribution $P_{as}(i)$ for various values of the parameters N and γ*

	$N = 20$	$N = 30$	$N = 50$	$N = 100$	$N = 200$	**Limiting**
$\gamma = 0.05$	5.512	5.727	5.842	5.900	5.930	5.977
$\gamma = 0.1$	6.090	6.233	6.334	6.415	6.442	6.494
$\gamma = 0.5$	10.336	10.501	10.640	10.715	10.777	10.821

Table 8.7. *Mean number of retrials in prelimiting and limiting situations for various values of the parameters N and γ*

8.4.2. *M/GI/1 system*

8.4.2.1. *Stochastic simulation*

In Tóth *et al.* (2017), the required service time is supposed to be gamma distributed and the input parameters of the system are collected in Table 8.8.

Case	N	λ/N	γ_0	γ_1	γ_2	σ/N	α	β
1	100	0.01	0.1	0.1	1	0.01	0.5	0.5
2	100	0.01	0.1	0.1	1	0.01	1	1
3	100	0.01	0.1	0.1	1	0.01	2	2

Table 8.8. *Numerical values of model parameters*

Figure 8.8 shows the steady-state distribution of the three investigated cases. It is observed that the mean number of customers increases as α and β are getting larger. *Case 2* is a special case because when $\alpha = 1$, it represents the exponential distribution. From the shape of the curves, it is clearly visible that the steady-state distribution of the cases are normally distributed. The next table presents the considered performance measures in relation with the different cases (see Table 8.9).

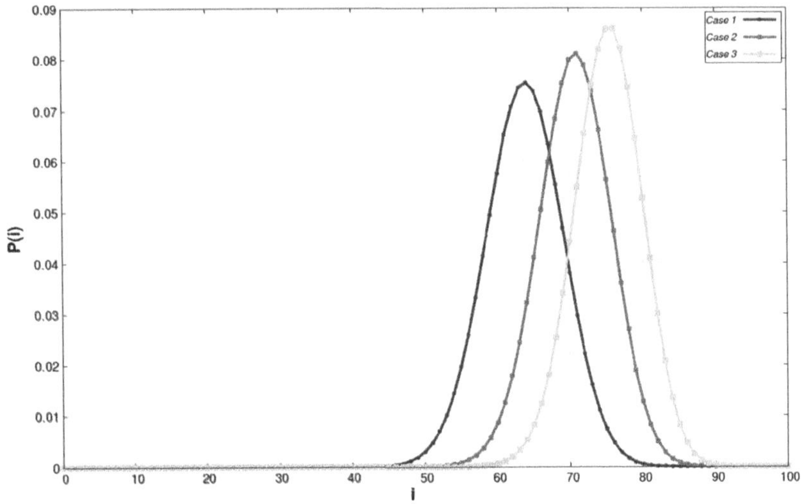

Figure 8.8. *Comparison of steady-state distributions. For a color version of this figure, see www.iste.co.uk/anisimov/queueing1.zip*

In Table 8.9, the notations mean the followings: $E(J)$ and $Var(J)$ are mean number and variance of customers in the system, $E(T)$ and $Var(T)$ are mean and variance of response time, $E(W)$ and $Var(W)$ are mean and variance of waiting time, $E(S)$ and $Var(S)$ are mean and variance of successful service time, and$E(IS)$ is mean interrupted service time.

Case	E(J)	$Var(J)$	E(T)	$Var(T)$	E(W)	$Var(W)$	E(S)	$Var(S)$	E(IS)
1	63.6842	27.9734	175.3073	65,657.3454	174.5884	65,434.6696	0.3147	0.1979	0.4041
2	70.5912	24.3012	239.9734	105,273.4267	238.9734	104,918.6389	0.4784	0.2289	0.5217
3	75.1825	21.2439	302.8106	151,781.1411	301.5377	151,277.6006	0.6472	0.2095	0.6257

Table 8.9. *Simulation results*

Figure 8.9 represents the confirmation of mean waiting time. The same parameters are (see Table 8.9) used as shown in Figure 8.8 but here the running parameter is λ/N. As is expected with the increment of λ/N, mean waiting time increases as well but an interesting phenomenon is noticeable, namely that after λ/N is greater than 0.1 mean waiting time starts to decrease.

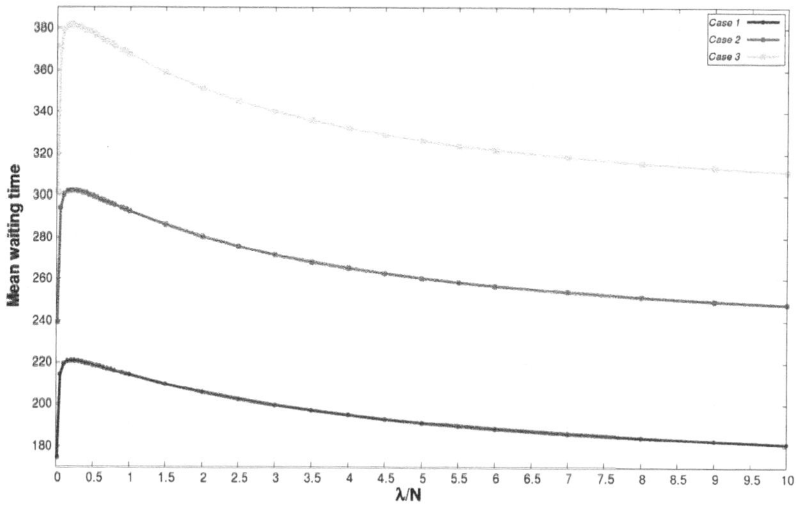

Figure 8.9. *Mean waiting time versus intensity of incoming customers. For a color version of this figure, see www.iste.co.uk/anisimov/queueing1.zip*

8.4.2.2. *Asymptotic approach*

These results have been published by Nazarov *et al.* (2017a) using a supplementary variable technique. The limit of the characteristic function of the scaled number of customers in the systems can be written in the following form

$$\lim_{N\to\infty} \mathsf{E}\exp\left\{iw\frac{J}{N}\right\} = \exp\left\{iw\kappa_1\right\},$$

where κ_1 is the positive solution of the equation

$$(1-\kappa_1)\,\lambda - \delta(\kappa_1)\,[R_0(\kappa_1) - R_1(\kappa_1)] + \gamma_1 R_1(\kappa_1) = 0,$$

here $\delta\,(\kappa_1)$ is

$$\delta\,(\kappa_1) = (1-\kappa_1)\,\lambda + \sigma\kappa_1,$$

and the stationary distributions of probabilities $R_k(\kappa_1)$ of the server's state $k = 0, 1, 2$ are determined as follows:

$$R_0(\kappa_1) = \left\{ \frac{\gamma_0 + \gamma_2}{\gamma_2} + \frac{\gamma_1 + \gamma_2}{\gamma_2} \cdot \frac{\delta(\kappa_1)}{\delta(\kappa_1) + \gamma_1} \left[1 - B^*(\delta(\kappa_1) + \gamma_1)\right] \right\}^{-1},$$

$$R_1(\kappa_1) = R_0(\kappa_1) \frac{\delta(\kappa_1)}{\delta(\kappa_1) + \gamma_1} \cdot \left[1 - B^*(\delta(\kappa_1) + \gamma_1)\right],$$

$$R_2(\kappa_1) = \frac{1}{\gamma_2} \left[\gamma_0 R_0(\kappa_1) + \gamma_1 R_1(\kappa_1)\right].$$

8.4.3. *Stochastic simulation of special systems*

In Tóth *et al.* (2017), systems with not only gamma distributed service times but also gamma distributed interarrival and gamma distributed retrial times have been investigated.

8.4.3.1. *Gamma distributed interarrival times*

Table 8.10 presents the input parameters of this case.

Case	N	α	β	γ_0	γ_1	γ_2	σ/N	α_1	β_1/N
1	100	1	1	0.1	0.1	1	0.01	0.5	0.01
2	100	1	1	0.1	0.1	1	0.01	1	0.01
3	100	1	1	0.1	0.1	1	0.01	2	0.01

Table 8.10. *Numerical values of parameters of gamma distributed interarrival times*

Figure 8.10 displays the steady-state distribution of this case. Now the service time is exponentially distributed ($\alpha = 1$) and the interarrival time is gamma distributed. It is interesting to notice that in these cases, the steady-state distributions are still normally distributed and when α_1 is greater than 1 the mean number of requests in the system is significantly lower than in the other cases. In Table 8.11, the estimations for the main performance measures can be found.

Case	E(J)	$Var(J)$	E(T)	$Var(T)$	E(W)	$Var(W)$	E(S)	$Var(S)$	E(IS)
1	84.3609	14.1827	270.0351	128,831.5059	269.0354	128,420.4087	0.4502	0.2039	0.5495
2	70.5912	24.3012	239.9734	105,273.4267	238.9734	104,918.6389	0.4784	0.2289	0.5217
3	47.7859	34.2376	183.0164	69,830.9728	182.0164	69,573.992	0.5462	0.2982	0.4538

Table 8.11. *Simulation results*

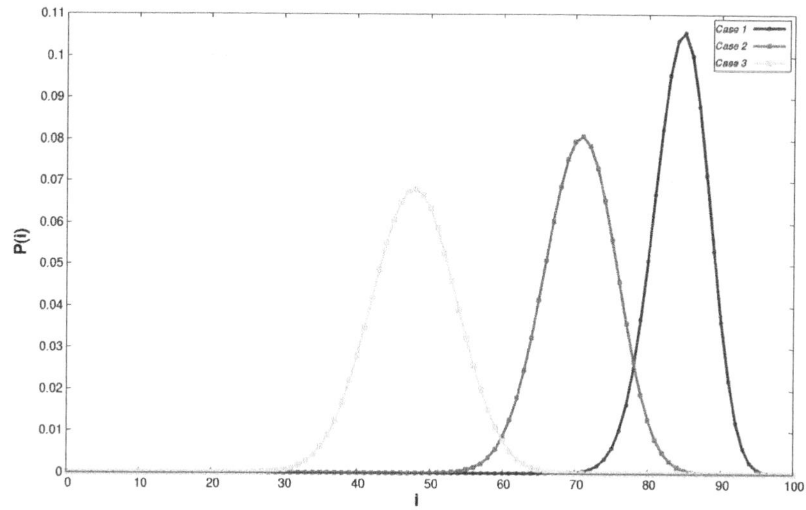

Figure 8.10. *Steady-state distributions of gamma distributed interarrival times. For a color version of this figure, see www.iste.co.uk/anisimov/queueing1.zip*

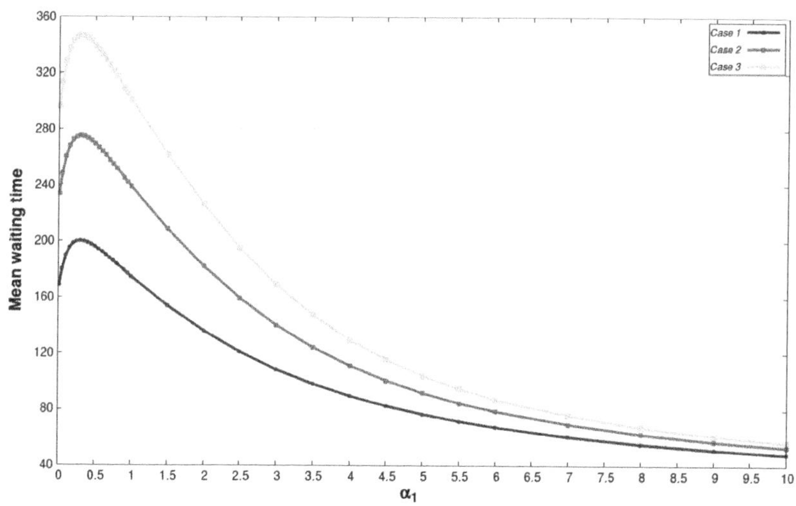

Figure 8.11. *Mean waiting time versus shape parameter. For a color version of this figure, see www.iste.co.uk/anisimov/queueing1.zip*

In Figures 8.11 and 8.13, the service time distribution of the *cases* has the following parameters: $\alpha = \beta = 0.5, \quad 1, \quad 2$, respectively, and all the other parameters are the same as in Table 8.11. The running parameter is α_1, so in this way the impact of different distributions on the various performance measures can be discovered. First the mean waiting time (Figure 8.11), after an initial jump mean waiting time, starts to monotonically decrease resulting in α_1 getting bigger the less time the customers spend in the system. At the end, the values of separate cases are almost equal. As a result (Figure 8.11), it is not surprising how the mean interrupted service time changes as a function of α_1. It is interesting that around $\alpha_1 \approx 4.5$ we observe that the curves intersect each other.

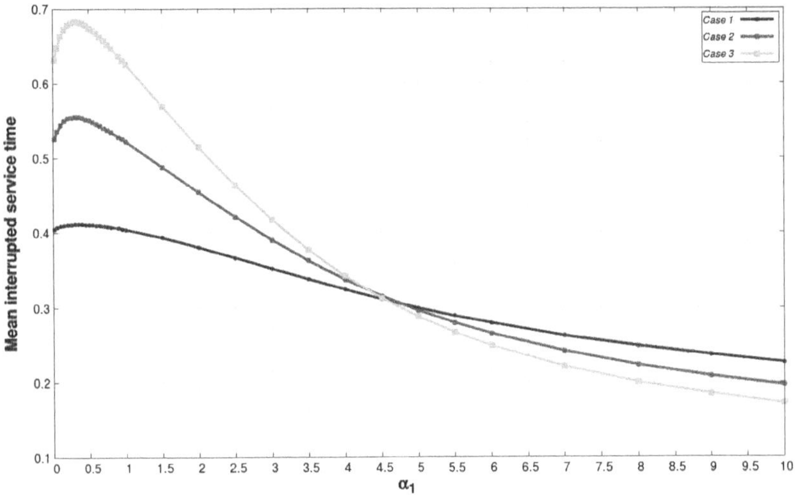

Figure 8.12. *Mean interrupted service time versus shape parameter. For a color version of this figure, see www.iste.co.uk/anisimov/queueing1.zip*

8.4.4. *Gamma distributed retrial times*

In this case not just the interarrival but also the retrial time are gamma distributed with the same parameters. Table 8.12 shows the input parameters of this system.

Case	N	α	β	γ_0	γ_1	γ_2	α_1	β_1/N
1	100	1	1	0.1	0.1	1	0.5	0.01
2	100	1	1	0.1	0.1	1	1	0.01
3	100	1	1	0.1	0.1	1	2	0.01

Table 8.12. *Numerical values of parameters of gamma distributed retrial times*

Table 8.13 contains the main performance measures in connection with the cases.

Case	E(J)	$Var(J)$	E(T)	$Var(T)$	E(W)	$Var(W)$	E(S)	$Var(S)$	E(IS)
1	81.5384	16.9599	220.8314	82,735.764	219.8314	82,377.8317	0.3398	0.1183	0.6602
2	70.5912	24.3012	239.9734	105,273.4267	238.9734	104,918.6389	0.4784	0.2289	0.5217
3	56.6635	30.1636	261.483	146,264.4778	260.4827	145,919.0907	0.626	0.3915	0.3743

Table 8.13. *Simulation results*

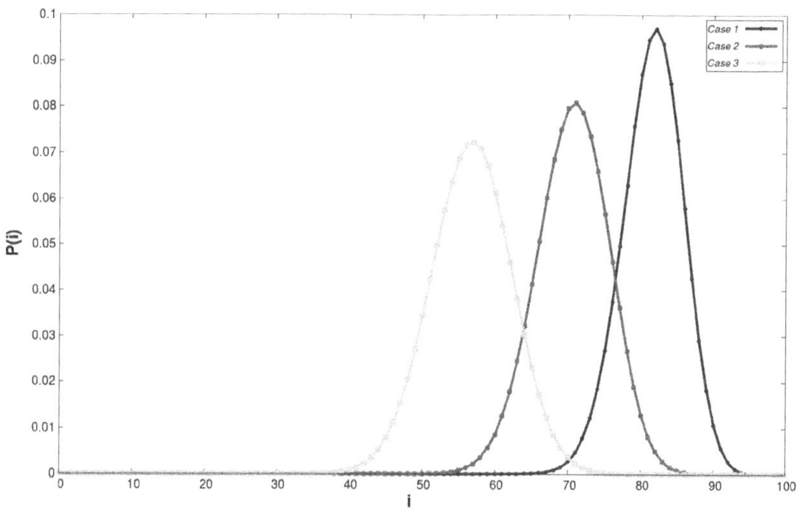

Figure 8.13. *Steady-state distributions of gamma distributed retrial times. For a color version of this figure, see www.iste.co.uk/anisimov/queueing1.zip*

Let us notice that under this parameter setup this modification has no significant effect on the steady-state distribution (see Figure 8.13). Of course, the mean customers in the system is quite disparate but the distribution remains normal. Also when α_1 is less than 1, it results in a higher mean number of customers in the system compared to when it is more than one.

8.4.5. *The effect of breakdowns disciplines*

In Tóth *et al.* (2019), the $M/G/1//N$ and $G/M/1//N$ systems were investigated with exponentially distributed operating and repair times. In cases of server failure, two operation modes are considered:

– the interrupted request gets into the orbit instantaneously (mode 1);

– the service of the interrupted request is suspended and it continues after repairing the server (mode 2).

8.4.5.1. *Scenario A, gamma distributed service time*

Table 8.14 shows the input parameters of scenario A.

Case	N	λ/N	γ_0	γ_1	γ_2	σ/N	α	β
1	100	0.01	0.1	0.1	1	0.01	0.5	0.5
2	100	0.01	0.1	0.1	1	0.01	1	1
3	100	0.01	0.1	0.1	1	0.01	2	2

Table 8.14. *Numerical values of model parameters*

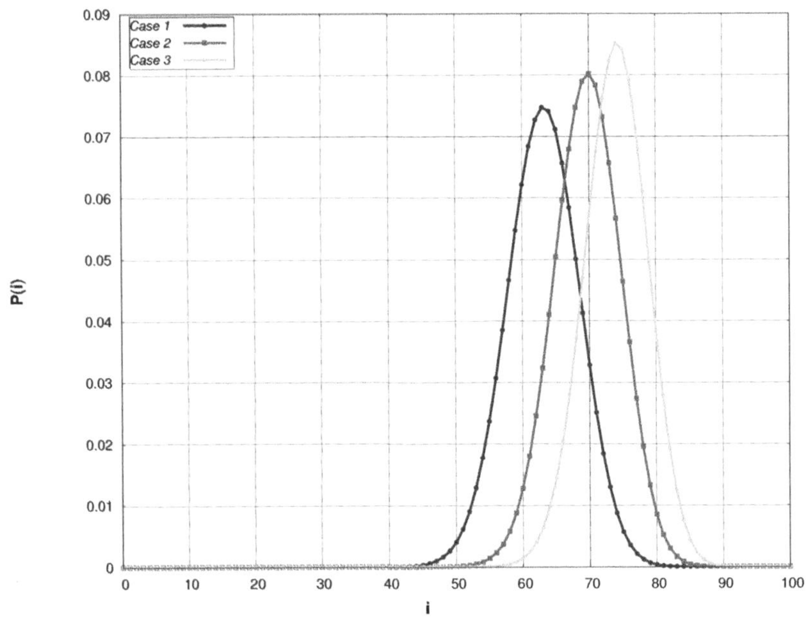

Figure 8.14. *Comparison of steady-state distributions for mode 2. For a color version of this figure, see www.iste.co.uk/anisimov/queueing1.zip*

Figure 8.14 shows the steady-state distribution of the investigated cases for operation mode 2. *Case 2* is a special case because when parameter α is equal to 1, it results in the exponential distribution. The mean number of customers increases with the increment of parameters α and β and, taking a closer look at the shape of the curves, steady-state distribution of the cases follow normal distribution. Table 8.15 presents the considered performance measures in relation to the different cases.

Figures 8.15–8.17 compare the mean waiting time of the two different operation modes of the investigated cases. Operation mode 1 reflects the mode when the interrupted requests get into the orbit instantaneously, and under operation mode 2, we consider the mode when the service of the interrupted request is suspended and it continues after repairing the server. In all cases, the results confirm the expectation that applying operation mode 2 results in lower mean waiting time. When the values of parameter α and β are higher, the difference between the modes is higher as well. With increasing arrival intensity, we should expect higher waiting times but after λ/N reaches 0.15 it starts to monotonically decrease, which is an interesting phenomenon.

Case	E(J)	$Var(J)$	E(T)	$Var(T)$	E(W)	$Var(W)$	E(S)	$Var(S)$
1	63.0526	28.3198	170.5618	63,092.8264	169.7553	62,855.9266	0.3357	0.2255
2	69.6114	24.6949	228.9834	97,974.6274	227.8834	97,613.9701	0.5022	0.2523
3	73.9099	21.9244	283.1452	136,396.9409	281.7728	135,909.2373	0.6688	0.2238

Table 8.15. *Numerical results of scenario A*

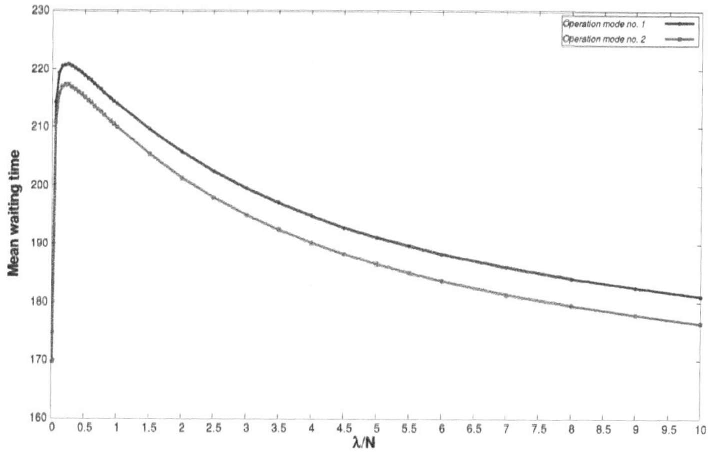

Figure 8.15. *Mean waiting time versus intensity of incoming customers of* Case 1. *For a color version of this figure, see www.iste.co.uk/anisimov/queueing1.zip*

Figure 8.15 shows the mean waiting time as a function of failure rate. As it is expected, the mean waiting time increases with the higher failure rate in all cases. Comparing the operation modes, the difference is quite obvious. While using operation mode number 2, the growth seems linear, and in the case of operation mode number 1, the rise is quite significant especially in *Case 3*.

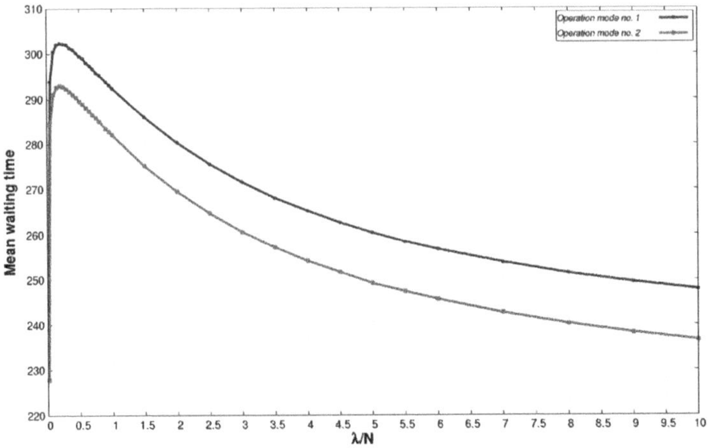

Figure 8.16. *Mean waiting time versus intensity of incoming customers of* Case 2. *For a color version of this figure, see www.iste.co.uk/anisimov/queueing1.zip*

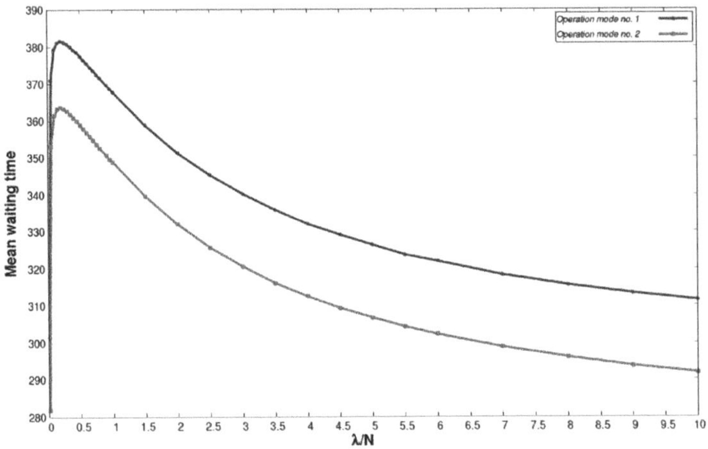

Figure 8.17. *Mean waiting time versus intensity of incoming customers of* Case 3. *For a color version of this figure, see www.iste.co.uk/anisimov/queueing1.zip*

8.4.5.2. *Scenario B, gamma distributed interarrival time*

In scenario B, the distribution of interarrival times of the customers is gamma distributed with parameter λ_1 and β_1. Table 8.16 presents the numerical values of parameters of scenario B.

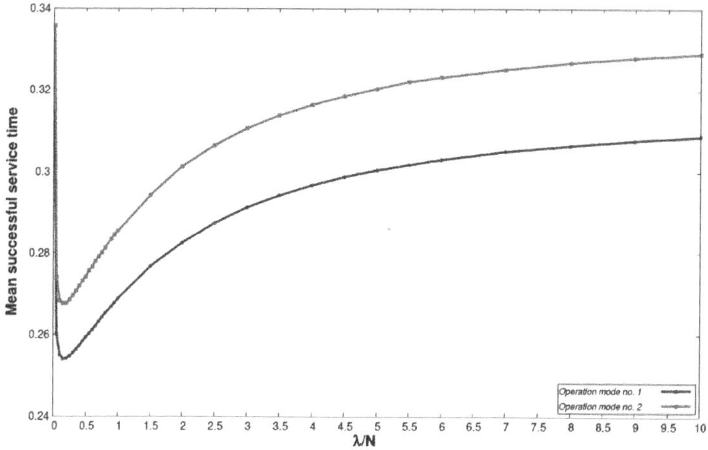

Figure 8.18. *Mean successful service time versus intensity of incoming customers of* Case 1. *For a color version of this figure, see www.iste.co.uk/anisimov/queueing1.zip*

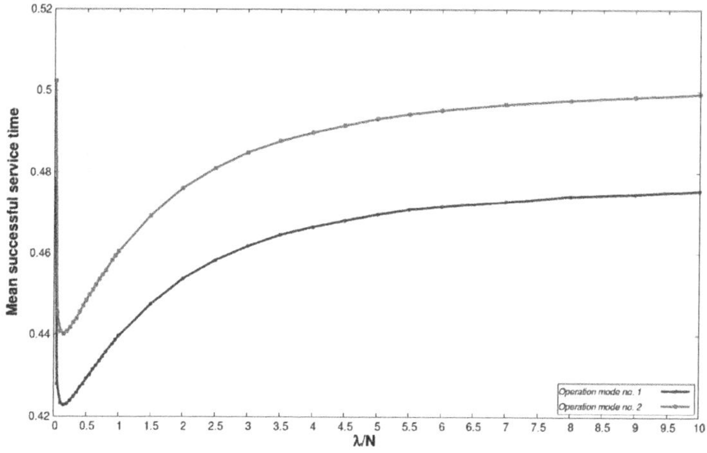

Figure 8.19. *Mean successful service time versus intensity of incoming customers of* Case 2. *For a color version of this figure, see www.iste.co.uk/anisimov/queueing1.zip*

Case	N	α	β	γ_0	γ_1	γ_2	σ/N	α_1	β_1/N
1	100	1	1	0.1	0.1	1	0.01	0.5	0.01
2	100	1	1	0.1	0.1	1	0.01	1	0.01
3	100	1	1	0.1	0.1	1	0.01	2	0.01

Table 8.16. *Numerical values of parameters of scenario B*

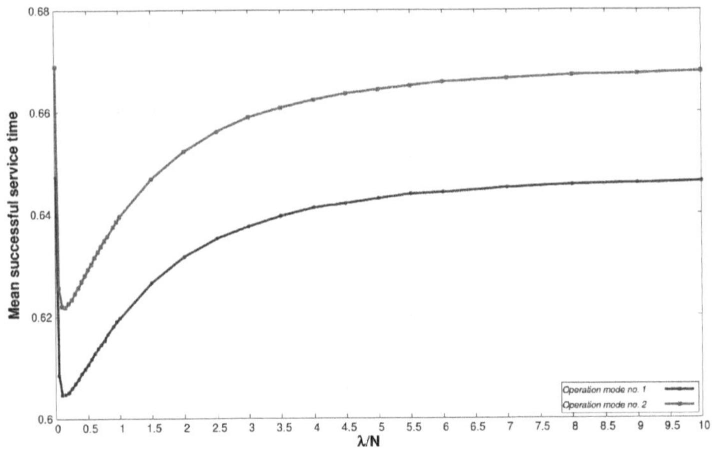

Figure 8.20. *Mean successful service time versus intensity of incoming customers of* Case 3. *For a color version of this figure, see www.iste.co.uk/anisimov/queueing1.zip*

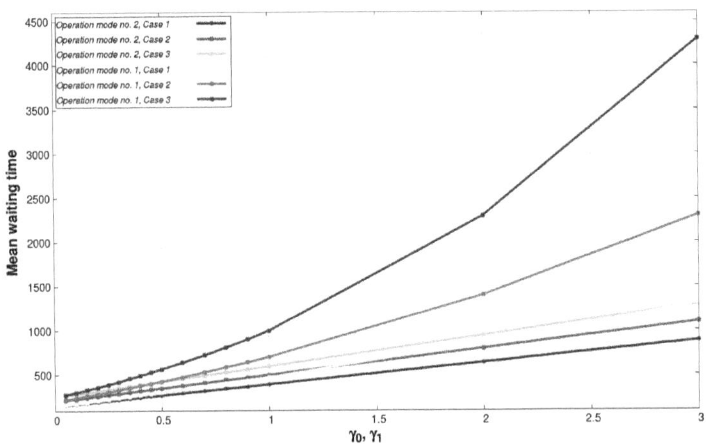

Figure 8.21. *Mean waiting time versus intensity of failure rate. For a color version of this figure, see www.iste.co.uk/anisimov/queueing1.zip*

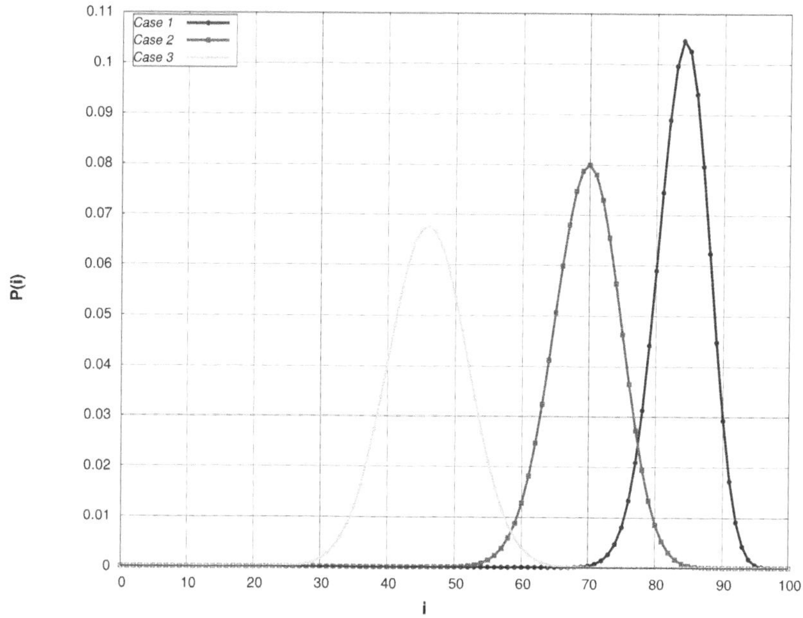

Figure 8.22. *Steady-state distributions of scenario B. For a color version of this figure, see www.iste.co.uk/anisimov/queueing1.zip*

Case	E(J)	$Var(J)$	E(T)	$Var(T)$	E(W)	$Var(W)$	E(S)	$Var(S)$
1	83.856	14.4707	259.7716	121,328.2493	258.6719	120,904.8333	0.4707	0.2232
2	69.5984	24.7715	228.9076	97,997.6219	227.8074	97,637.2906	0.5025	0.2525
3	45.9508	34.3607	170.0134	62,628.4067	168.9133	62,379.0441	0.5806	0.337

Table 8.17. *Numerical results of scenario B*

Figure 8.22 displays the steady-state distributions of Scenario B. Now, the service time is supposed to be exponentially distributed ($\alpha = 1$). With this modification compared to scenario A, the steady-state distributions still follow a normal distribution and as the value of α_1 is increasing, the mean number of requests in the system is decreasing. In *Case 3,* the mean number of customers in the system is significantly lower among the cases. In Table 8.17, an estimation for the main performance measures can be found in connection with the cases.

The running parameter is α_1 and helps to discover the impact of different distributions on the various examined performance measures. Figures 8.23–8.25 show the comparison of the mean waiting time between the two operation modes of the investigated cases. As in scenario A, using operation mode number 2, when the interrupted requests stay at the server in cases of server failure and their services continue after the server is ready to process jobs again, ensures lower mean waiting times. In all cases, the mean waiting time starts to increase till α_1 reaches 0.3, then it monotonically decreases. With higher values of α and β, the difference between the operation modes are higher as well.

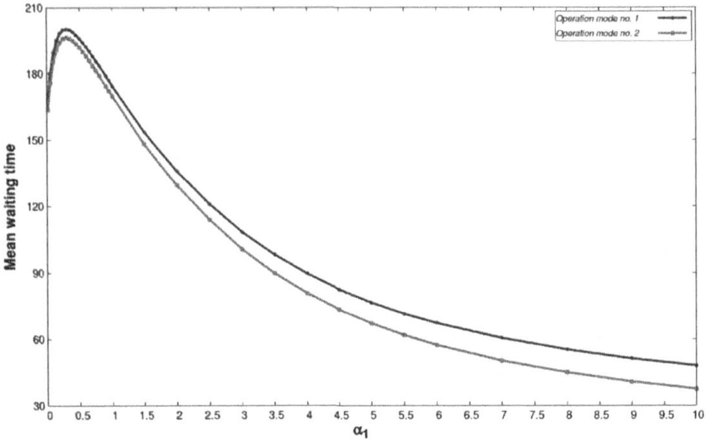

Figure 8.23. *Mean waiting time versus shape parameter, $\alpha = \beta = 0.5$. For a color version of this figure, see www.iste.co.uk/anisimov/queueing1.zip*

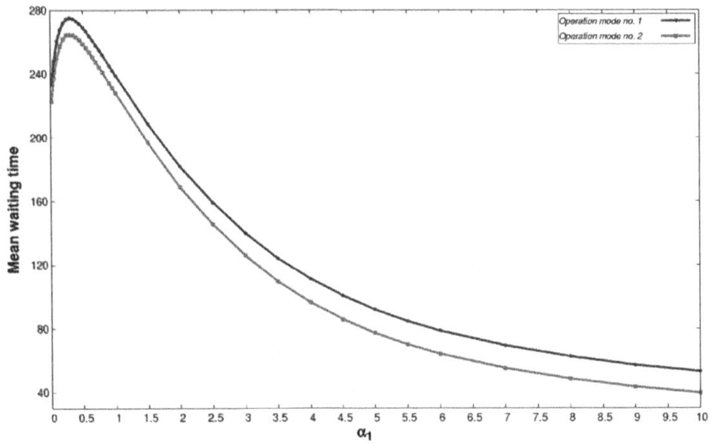

Figure 8.24. *Mean waiting time versus shape parameter, $\alpha = \beta = 1$. For a color version of this figure, see www.iste.co.uk/anisimov/queueing1.zip*

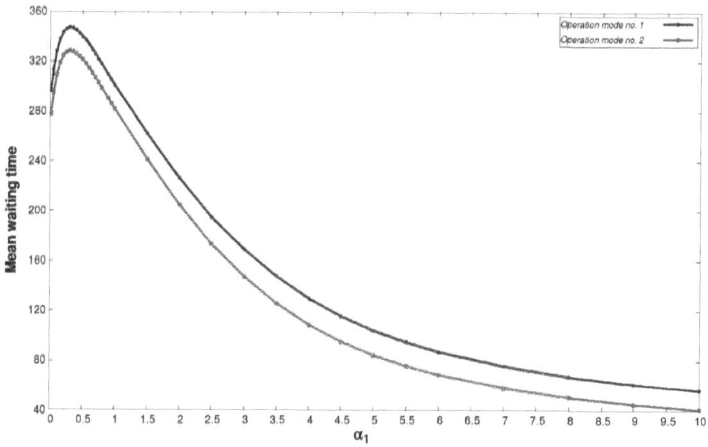

Figure 8.25. *Mean waiting time versus shape parameter, $\alpha = \beta = 2$. For a color version of this figure, see www.iste.co.uk/anisimov/queueing1.zip*

Figures 8.26–8.28 show the mean successful service time versus the shape parameter of the interarrival time using both operation mode. As we can see in scenario A, we get what we expected as the usage of operation mode number 2 provides higher values of mean successful service time. The difference is quite high between the applied operation modes in all cases especially when α and β is equal to 0.5. The mean successful service time behaves in reverse compared to the mean waiting time because when the mean waiting time increases the mean successful service time decreases and vice versa.

8.5. Conclusion

In this chapter, tool-supported, numerical, simulation, and asymptotic methods were considered under the condition of an unlimited growing number of sources in a finite-source retrial queue with collisions of customers and an unreliable server. During the survey, several cases and examples were treated and the results of different approaches were compared to each other showing the advantages and disadvantages of the given method. Tables and figures used in this chapter illustrate some special features of these systems. In the near future, the two research groups would like to continue their investigations in this direction including systems with impatient customers, systems embedded in a random environment and systems with two-way communications, just to mention some alternative generalizations.

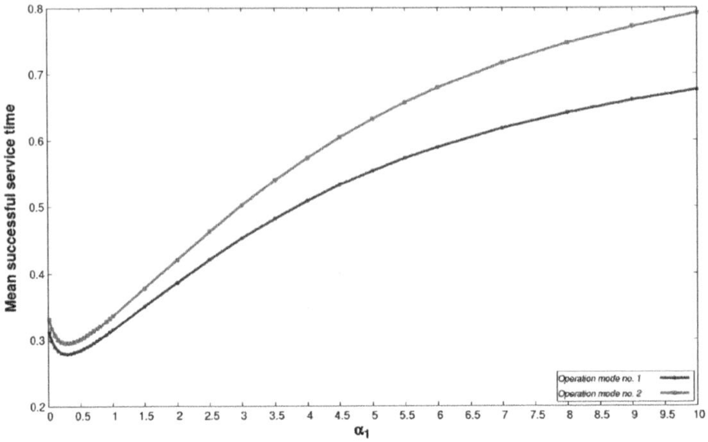

Figure 8.26. *Mean successful service time versus shape parameter, $\alpha = \beta = 0.5$. For a color version of this figure, see www.iste.co.uk/anisimov/queueing1.zip*

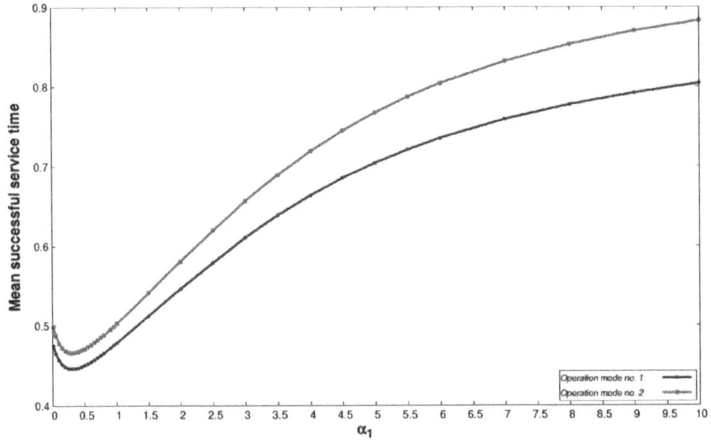

Figure 8.27. *Mean successful service time versus shape parameter, $\alpha = \beta = 1$. For a color version of this figure, see www.iste.co.uk/anisimov/queueing1.zip*

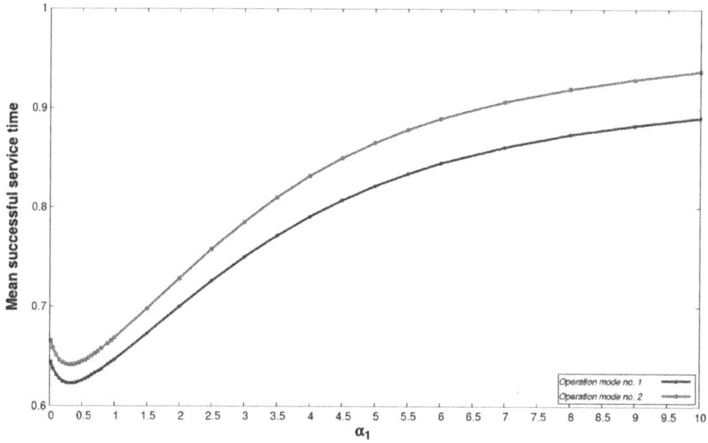

Figure 8.28. *Mean successful service time versus shape parameter, $\alpha = \beta = 2$. For a color version of this figure, see www.iste.co.uk/anisimov/queueing1.zip*

8.6. Acknowledgments

The authors are very grateful to the reviewer for their comments and recommendations, which improved the presentation of the chapter.

The work/publication of J. Sztrik is supported by the EFOP-3.6.1-16-2016-00022 project. The project is co-financed by the European Union and the European Social Fund.

8.7. References

Ali, A.-A., Wei, S. (2015). Modeling of coupled collision and congestion in finite source wireless access systems. *Wireless Communications and Networking Conference (WCNC)*, IEEE, New Orleans, 1113–1118.

Almási, B., Roszik, J., Sztrik, J. (2005). Homogeneous finite-source retrial queues with server subject to breakdowns and repairs. *Math. Comput. Model.*, 42(5–6), 673–682.

Anisimov, V. (1999). Averaging methods for transient regimes in overloading retrial queueing systems. *Math. Comput. Model.*, 30(3–4), 65–78.

Anisimov, V., Sztrik, J. (1989). Asymptotic analysis of some complex renewable systems operating in random environments. *European Journal of Operational Research*, 41(2), 162–168.

Anisimov, V., Artalejo, J. R. (2001). Analysis of Markov multiserver retrial queues with negative arrivals. *Queueing Systems*, 39(2-3), 157–182.

Artalejo, J., Corral, A.G. (2008). *Retrial Queueing Systems: A Computational Approach*. Springer, Berlin.

Bérczes, T., Sztrik, J., Tóth, Á., Nazarov, A. (2017). Performance modeling of finite-source retrial queueing systems with collisions and non-reliable server using MOSEL. *International Conference on Distributed Computer and Communication Networks*, Springer, Berlin, 248–258.

Bhat, U. N. (2015). *An Introduction to Queueing Theory. Modeling and Analysis in Applications*. Birkhäuser, Boston.

Bossel, H. (2013). *Modeling and Simulation*. Springer-Verlag, Berlin.

Cao, Y., Khosla, D., Chen, Y., Huber, D. J. (2018). System and method for real-time collision detection. US Patent 9,934,437.

Choi, B.D., Shin, Y.W., Ahn, W.C. (1992). Retrial queues with collision arising from unslotted CSMA/CD protocol. *Queueing Syst.*, 11(4), 335–356.

Dragieva, V. I. (2014). Number of retrials in a finite source retrial queue with unreliable server. *Asia-Pac. J. Oper. Res.*, 31(2), 23.

Falin, G., Artalejo, J. (1998). A finite source retrial queue. *European Journal of Operational Research*, 108, 409–424.

Falin, G., Templeton, J.G.C. (1997), *Retrial Queues*. Chapman and Hall, London.

Fiems, D., Phung-Duc, T. (2017), Light-traffic analysis of random access systems without collisions. *Annals of Operations Research*, 277(2), 311–327

Gharbi, N., Dutheillet, C. (2011). An algorithmic approach for analysis of finite-source retrial systems with unreliable servers. *Computers & Mathematics with Applications*, 62(6), 2535–2546.

Gharbi, N., Ioualalen, M. (2006). GSPN analysis of retrial systems with servers breakdowns and repairs. *Applied Mathematics and Computation*, 174(2), 1151–1168.

Gharbi, N., Mokdad, L., Ben-Othman, J. (2015). A performance study of next generation cellular networks with base stations channels vacations. *Global Communications Conference (GLOBECOM)*, IEEE, 1–6.

Gharbi, N., Nemmouchi, B., Mokdad, L., Ben-Othman, J. (2014). The impact of breakdowns disciplines and repeated attempts on performances of small cell networks. *Journal of Computational Science*, 5(4), 633–644.

Gómez-Corral, A., Phung-Duc, T. (2016). Retrial queues and related models. *Annals of Operations Research*, 247(1), 1–2.

Harchol-Balter, M. (2013). *Performance Modeling and Design of Computer Systems: Queueing Theory in Action*. Cambridge University Press, Cambridge.

Ikhlef, L., Lekadir, O., Aïssani, D. (2016). MRSPN analysis of semi-Markovian finite source retrial queues. *Annals of Operations Research*, 247(1), 141–167.

Jinsoo, A., Kim, Y., Kwak, J., Son, J. (2018). Wireless communication method for multi-user transmission scheduling, and wireless communication terminal using same. US Patent App. 15/736,968.

Kim, J., Kim, B. (2016). A survey of retrial queueing systems. *Annals of Operations Research*, 247(1), 3–36.

Kim, J.S. (2010). Retrial queueing system with collision and impatience. *Communications of the Korean Mathematical Society*, 25(4), 647–653.

Kobayashi, H., Mark, B. L. (2009). *System Modeling and Analysis: Foundations of System Performance Evaluation*. Pearson Education India, Chennai.

Krishnamoorthy, A., Pramod, P.K., Chakravarthy, S.R. (2014). Queues with interruptions: a survey. *TOP*, 22(1), 290–320.

Kuki, A., Bérczes, T., Sztrik, J., Kvach, A. (2019). Numerical analysis of retrial queueing systems with conflict of customers. *Journal of Mathematical Sciences*, 237, 637–683.

Kulkarni, V.G. (2016). *Modeling and Analysis of Stochastic Systems*. CRC Press, Boca Raton.

Kumar, B.K., Vijayalakshmi, G., Krishnamoorthy, A., Basha, S.S. (2010). A single server feedback retrial queue with collisions. *Computers & Operations Research*, 37(7), 1247–1255.

Kvach, A. (2014). Numerical research of a Markov closed retrial queueing system without collisions and with the collision of the customers. *Proceedings of Tomsk State University*. TSU Publishing House, Tomsk, 105–112 (in Russian).

Kvach, A., Nazarov, A. (2015a). Numerical research of a closed retrial queueing system M/GI/1//N with collision of the customers. *Proceedings of Tomsk State University*. TSU Publishing House, Tomsk, 65–70 (in Russian).

Kvach, A., Nazarov, A. (2015b). *Sojourn Time Analysis of Finite Source Markov Retrial Queuing System with Collision*. Springer International Publishing, Cham.

Kvach, A., Nazarov, A. (2015c). The research of a closed RQ-system M/GI/1//N with collision of the customers in the condition of an unlimited increasing number of sources. *Probability Theory, Random Processes, Mathematical Statistics and Applications*, 65–70 (in Russian).

Kwak, B.-J., Rhee, J.-K., Kim, J., Kyounghye, K. (2018). Random access method and terminal supporting the same. US Patent 9,954,754.

Lakaour, L., Aïssani, D., Adel-Aissanou, K., Barkaoui, K. (2018). $M/M/1$ retrial queue with collisions and transmission errors. *Methodology and Computing in Applied Probability*, 1–12.

Lakatos, L., Szeidl, L., Telek, M. (2013). *Introduction to Queueing Systems with Telecommunication Applications*. Springer, New York.

Law, A.M., Kelton, W.D. (1991). *Simulation Modeling and Analysis*. McGraw-Hill, New York.

Nazarov, A., Kvach, A., Yampolsky, V. (2014). *Asymptotic Analysis of Closed Markov Retrial Queuing System with Collision*. Springer International Publishing, Cham.

Nazarov, A., Moiseeva, S.P. (2006). *Methods of Asymptotic Analysis in Queueing Theory*. NTL Publishing House of Tomsk University, Tomsk (in Russian).

Nazarov, A., Sudyko, E. (2010). Method of asymptotic semi-invariants for studying a mathematical model of a random access network. *Probl. Inf. Transm.*, 46(1), 86–102.

Nazarov, A., Sztrik, J., Kvach, A. (2017a). Comparative analysis of methods of residual and elapsed service time in the study of the closed retrial queuing system $M/GI/1//N$ with collision of the customers and unreliable server. *International Conference on Information Technologies and Mathematical Modelling*, 97–110.

Nazarov, A., Sztrik, J., Kvach, A. (2017b). Some features of a finite-source $M/GI/1$ retrial queuing system with collisions of customers. *International Conference on Distributed Computer and Communication Networks*, 186–200.

Nazarov, A., Sztrik, J., Kvach, A. (2017c). Some features of a finite-source $M/GI/1$ retrial queuing system with collisions of customers. *International Conference on Distributed Computer and Communication Networks*, 79–86.

Nazarov, A., Sztrik, J., Kvach, A. (2018). A survey of recent results in finite-source retrial queues with collisions. In *Information Technologies and Mathematical Modelling. Queueing Theory and Applications*, Dudin, A., Nazarov, A., Moiseev, A. (eds). Springer, Berlin.

Nazarov, A., Sztrik, J., Kvach, A., Bérczes, T. (2018). Asymptotic analysis of finite-source M/M/1 retrial queueing system with collisions and server subject to breakdowns and repairs. *Annals of Operations Research*, 277(2), 213–229.

Nazarov, A., Sztrik, J., Kvach, A., Tóth, A. (2018). Asymptotic sojourn time analysis of Markov finite-source M/M/1 retrial queueing system with collisions and server subject to breakdowns and repairs. *Probability in the Engineering and Informational Sciences*, 33(3), 387–403.

Nazarov, A., Terpugov, A. (2004). *Theory of Mass Service*. NTL Publishing House of Tomsk University, Tomsk (in Russian).

Peng, Y., Liu, Z., Wu, J. (2014). An $\mathrm{M/G/1}$ retrial G-queue with preemptive resume priority and collisions subject to the server breakdowns and delayed repairs. *J. Appl. Math. Comput.*, 44(1–2), 187–213.

Reith III, H. C. (2017). System and method for collision detection and avoidance for network communications. US Patent 9,553,828.

Roszik, J. (2004). Homogeneous finite-source retrial queues with server and sources subject to breakdowns and repairs. *Ann. Univ. Sci. Budap. Rolando Eötvös, Sect. Comput.*, 23, 213–227.

Rubinstein, R. Y., Kroese, D. P. (2016). *Simulation and the Monte Carlo Method*. John Wiley & Sons, Hoboken.

Stewart, W. J. (2009). *Probability, Markov Chains, Queues, and Simulation: The Mathematical Basis of Performance Modeling*. Princeton University Press, Princeton.

Sztrik, J. (2005). Tool supported performance modelling of finite-source retrial queues with breakdowns. *Publicationes Mathematicae*, 66, 197–211.

Sztrik, J., Almási, B., Roszik, J. (2006). Heterogeneous finite-source retrial queues with server subject to breakdowns and repairs. *Journal of Mathematical Sciences*, 132, 677–685.

Takeda, T., Yoshihiro, T. (2017). A distributed scheduling through queue-length exchange in CSMA-based wireless mesh networks. *Journal of Information Processing*, 25, 174–181.

Tóth, Á., Bérczes, T., Sztrik, J., Kuki, A. (2019). Comparison of two operation modes of finite-source retrial queueing systems with collisions and non-reliable server by using simulation. *Journal of Mathematical Sciences*, 237, 846–857.

Tóth, Á., Bérczes, T., Sztrik, J., Kvach, A. (2017). Simulation of finite-source retrial queueing systems with collisions and a non-reliable server. *International Conference on Distributed Computer and Communication Networks*, 146–158. Springer.

Wang, J., Zhao, L., Zhang, F. (2010). Performance analysis of the finite source retrial queue with server breakdowns and repairs. *Proceedings of the 5th International Conference on Queueing Theory and Network Applications*, ACM, 169–176.

Wang, J., Zhao, L., Zhang, F. (2011). Analysis of the finite source retrial queues with server breakdowns and repairs. *Journal of Industrial and Management Optimization*, 7(3), 655–676.

Wentink, M. (2017). Collision avoidance systems and methods. US Patent App. 15/401,606.

Wüchner, P., Sztrik, J., de Meer, H. (2010). Finite-source retrial queues with applications. *Proceedings of 8th International Conference on Applied Informatics*, Eger, Hungary, 275–285.

Yao, J. (2016). *Asymptotic Analysis of Service Systems with Congestion-Sensitive Customers*. Columbia University, New York City.

Yeo, G. M., Kim, Y. I., Park, D. G., Song, S. Y., Lee, Y. T., Lee, H. W. (2017). Control method and apparatus for collision avoidance in low-power wireless sensor communication. US Patent App. 15/361,194.

Zhang, F., Wang, J. (2013). Performance analysis of the retrial queues with finite number of sources and service interruptions. *Journal of the Korean Statistical Society*, 42(1), 117–131.

9

Strong Stability of Queueing Systems and Networks: a Survey and Perspectives

Boualem RABTA, Ouiza LEKADIR and Djamil AÏSSANI
Research Unit LaMOS (Modeling and Optimization of Systems), University of Béjaïa, Algeria

The analysis of the stability of queueing models aims at determining the conditions under which the mathematical model is a good representation of the real system despite approximation and estimation errors. A system is stable if small perturbations in its parameters generate at most a bounded deviation in its characteristics. The strong stability method has been used for the study of the sensitivity of diverse types of queues and queueing networks. In addition to the qualitative affirmation of stability, quantitative estimates of the perturbation error have been obtained in most cases. In this chapter, we review the application of the strong stability method to queues and queueing networks and provide directions for future research.

9.1. Introduction

Queueing models are useful for modeling and analysis of numerous systems such as communication systems, computer networks as well as production and manufacturing lines. The analysis of queueing models aims at evaluating a set of performance measures such as the utilization of resources, throughput and response time. However, real systems are usually very complicated and their representation by

Queueing Theory 1,
coordinated by Vladimir ANISIMOV, Nikolaos LIMNIOS.

mathematical models is performed only through approximations. Hence, the real system is often replaced by another one, which is close to it in some sense but simpler in structure and/or components. This is necessary in order to obtain a model that is analytically tractable or can be solved by numerical methods. Moreover, the parameters of the model as well as the underlying probability distributions are estimated from empirical data by means of statistical methods. Usually, those approximations are performed with care in order to insure that the constructed model is sufficiently robust to resist the perturbations in its structure and parameters and that it remains a reliable representation of the real system. It is therefore very important to justify these approximations and estimate the resulting error.

In addition to the qualitative properties of the model, it is also important to have an estimation of the deviation in the characteristics (output) resulting from the perturbation of the parameters (input). Sensitivity analysis is a very important step in validating mathematical models.

Different mathematical methods have been elaborated in the study of the qualitative properties of stochastic systems, especially their stability. Here, we use the term "stability" to designate the ability of the system to resist perturbations (robustness, insensitivity). The first results were obtained by Rossberg (1965), Gnedenko (1970), Franken (1970) and Kennedy (1972). Kalashnikov and Tsitsiachvili (1972) proposed the method of test functions inspired by the classical Liapunov method initially applied to investigate the stability of differential equations (Kalashnikov and Tsitsiachvili 1972; Kalashnikov 1978). Stoyan (1977) investigated continuity properties of queueing models based on the weak convergence theory (see also Stoyan (1984)). Zolotariev (1975) and Rachev (1989) considered the stability problem as a continuity problem, which appears when applying some metric spaces in other spaces. Borovkov (1984) obtained theorems of ergodicity and stability with minimal conditions using renewal theory. Cao (1998) presented an approach based on Maclaurin series expansions of the stationary distribution of Markov chains to study the effect of parameter perturbation (see also Heidergott and Hordijk (2003)). Using operator-theoretic and probabilistic methods, Anisimov (1988) expressed the bounds for general Markov chains in terms of ergodicity coefficients of the iterated transition kernel, which are difficult to compute for infinite state spaces.

The strong stability method (Aïssani and Kartashov 1983a,b; Kartashov 1996) can be used to investigate the stability of a Harris recurrent Markov chain in general state space when the perturbation of its transition kernel is small with respect to a certain norm. A Markov chain is said to be strongly stable when small perturbations in the inputs (transition kernel) can lead to at most a bounded deviation of the outputs (stationary measure). Under this condition, approximations and parameter estimation errors result in a controlled deviation in the characteristics of the system in some sense. In addition to the qualitative affirmation of the stability (robustness) of the considered Markov chain, the strong stability method also allows the derivation of upper bounds on the deviation of the stationary characteristics resulting from the perturbation (approximation errors).

Queueing systems are among the first and most studied stochastic systems in the context of the strong stability theory. Many types of queues and queueing networks have been analyzed and their stability established. In most cases, quantitative estimates (perturbation bounds) are also obtained. This chapter focuses on the applicability of the strong stability method to queueing systems, reviews previous results and discusses future perspectives.

The remainder of this paper is organized as follows. In the first section, we introduce the notations and the basic definitions and theorems of the strong stability theory. Next, we review the application of the strong stability method to single queues, queueing networks as well as the use of non-parametric density estimation method in the study of those systems. The chapter will be finalized by a general conclusion with some future perspectives.

9.2. Preliminary and notations

Let $X = (X_t, t \geq 0)$, a homogeneous Markov chain with values in a measurable space $(\mathbb{E}, \mathcal{E})$ (where we assume that the σ-algebra $\mathcal{E}$ is countably generated), given by a regular transition kernel $P(x, A)$, $x \in \mathbb{E}$, $A \in \mathcal{E}$ and having a unique invariant measure π.

Denote by $m\mathcal{E}$ ($m\mathcal{E}^+$) the space of finite (nonnegative) measures on $\mathcal{E}$, $f\mathcal{E}$ ($f\mathcal{E}^+$) the space of bounded (nonnegative) measurable functions on $\mathbb{E}$.

Consider in the space $m\mathcal{E}$, the Banach space $\mathcal{M} = \{\mu \in m\mathcal{E} : ||\mu|| < \infty\}$ with norm $||.||$ compatible with the structural order in $m\mathcal{E}$, i.e.:

$$||\mu_1|| \leq ||\mu_1 + \mu_2|| \text{ for } \mu_i \in \mathcal{M}^+, i = 1, 2.; \quad [9.1]$$

$$||\mu_1|| \leq ||\mu_1 - \mu_2|| \text{ for } \mu_i \in \mathcal{M}^+, i = 1, 2 \text{ and } \mu_1 \perp \mu_2; \quad [9.2]$$

$$|\mu|(\mathbb{E}) \leq k||\mu|| \text{ for } \mu \in \mathcal{M}; \quad [9.3]$$

where $|\mu|$ is the variation of the measure μ, k is a finite positive constant and $\mathcal{M}^+ = \mathcal{M} \cap (m\mathcal{E}^+)$.

We introduce in $m\mathcal{E}$, the special family of norms:

$$||\mu||_v = \int_{\mathbb{E}} v(x)|\mu|(dx), \forall \mu \in m\mathcal{E}, \quad [9.4]$$

where v is a measurable function bounded from below by a positive constant, (not necessary finite) on $\mathbb{E}$. Therefore, the induced norms on $f\mathcal{E}$ and $\mathcal{M}$ will have the following forms:

$$\|P\|_v = \sup\{\|\mu P\|_v, \|\mu\|_v \leq 1\} = \sup_{x\in\mathbb{E}} (v(x))^{-1} \int_{\mathbb{E}} |P(x, dy)|v(y), \qquad [9.5]$$

$$\|f\|_v = \sup\{|\mu f|, \|\mu\|_v \leq 1\} \quad = \sup_{x\in\mathbb{E}} (v(x))^{-1} |f(x)|. \qquad [9.6]$$

We associate to every transition kernel $P(x, A)$ in the space of bounded linear operators, the linear mappings $\mathcal{L}_P : m\mathcal{E} \to m\mathcal{E}$ and $\mathcal{L}_P^* : f\mathcal{E} \to f\mathcal{E}$, the values of which for $\mu \in m\mathcal{E}$ and $f \in f\mathcal{E}$ are, respectively:

$$\mu P(A) = \mathcal{L}_P(\mu)(A) = \int_{\mathbb{E}} \mu(dx)P(x, A), \ \ \forall A \in \mathcal{E},$$

$$Pf(x) = \mathcal{L}_P^*(f)(x) = \int_{\mathbb{E}} P(x, dy)f(y), \ \ \forall x \in \mathbb{E},$$

and with every function $f \in f\mathcal{E}$, we associate the linear functional $f : \mu \to \mu f$ such that:

$$\mu f = \int_{\mathbb{E}} \mu(dx)f(x).$$

For $\mu \in m\mathcal{E}$ and $f \in f\mathcal{E}$, $f \circ \mu$ is the transition kernel having the form:

$$f(x)\mu(A), x \in \mathbb{E}, A \in \mathcal{E},$$

where $\circ$ denotes the convolution between a measure and a function.

DEFINITION 9.1.– *The Markov chain X verifying $\|P\|_v < \infty$ is strongly $v-$stable, if every stochastic kernel Q in the neighborhood $\{Q : \|Q - P\|_v < \epsilon\}$ admits a unique stationary measure ν and:*

$$\|\nu - \pi\|_v \longrightarrow 0 \ \textit{when} \ \|Q - P\|_v \longrightarrow 0.$$

The following result (see Aïssani and Kartashov (1983a)) gives sufficient conditions for the strong $v-$stability of a Harris recurrent Markov chain.

THEOREM 9.1.– The Harris recurrent Markov chain X verifying $\|P\|_v < \infty$ is strongly v-stable, if the following conditions are satisfied:

1) $\exists \alpha \in \mathcal{M}^{+}, \exists h \in f\mathcal{E}^{+}$ such that : $\pi h > 0$, $\alpha \mathbb{I} = 1$, $\alpha h > 0$;

2) $T = P - h \circ \alpha$ is a non-negative kernel;

3) $\exists\ \rho < 1$ such that, $Tv(x) \ \leq\ \rho\, v(x), \forall x \in \mathbb{E}$;

where $\mathbb{I}$ is the function identically equal to 1.

One important feature of the strong stability method is the possibility of obtaining quantitative estimates. The following theorem (see Kartashov (1981, 1986c)) allows us to obtain an upper bound to the norm of the deviation of the stationary distribution of the strongly stable Markov chain X.

THEOREM 9.2.– Under the conditions of theorem 9.1 and for $\Delta = (Q - P)$ verifying the condition $\|\Delta\|_v < C^{-1}(1 - \rho)$, we have:

$$\|\nu - \pi\|_v \leq \|\Delta\|_v \|\pi\|_v C \left(1 - \rho - C\|\Delta\|_v\right)^{-1}, \qquad [9.7]$$

where

$$C = 1 + \|\mathbb{I}\|_v \|\pi\|_v \quad \text{and} \quad \|\pi\|_v \leq (\alpha v)(1 - \rho)^{-1}(\pi h).$$

Proofs of the above theorems, further conditions for the strong stability of homogeneous Markov chains and perturbation bounds as well as extensive results on the strong stability theory can be found in various studies (Aïssani 1990; Aïssani and Kartashov 1983a; Kartashov 1985, 1986a,b,c, 1996; Rabta and Aïssani 2008; Mouhoubi and Aïssani 2014; Rabta and Aïssani 2018).

Finally, we denote by $\mathbb{Z}$ the set of integer numbers and by $\mathbb{R}$ the set of real numbers. $\mathbb{A}^{+}$ denotes the non-negative part of the set $\mathbb{A}$ and $\mathbb{A}^{*} = \mathbb{A}/\{0\}$. Note that the notations introduced in each section are independent from those of the rest of the paper and might be redefined in another section.

9.3. Strong stability of queueing systems

Queueing models might have a large number of parameters that are subject to perturbations given that in practice those are unknown and must be approximated and/or fitted from empirical data. Obviously, the interarrival time and service time distributions are the first candidates. We distinguish the following kinds of perturbations.

– Parameter perturbations: only perturbations to individual parameters of the system are considered. For instance, in an $M/M/1$ queueing system, arrivals are Poisson distributed but the arrival rate has to be estimated. Considering the perturbation of the arrival rate while using a distribution of the same form (exponential) for the interarrival times, the perturbed system is also of type $M/M/1$.

– Distributional perturbations: here, the distribution of a constituent random variable is unknown and/or approximated by another distribution of a given (but different) form. In an $M/G/1$ queueing system, the general distribution of service times is unknown. Under certain conditions (distance), it might be replaced by an exponential distribution with the same mean. The resultant system is of type $M/M/1$.

– Non-parametric perturbations: we put in this category the applications of the strong stability method where the unknown distribution is fitted from data using non-parametric statistical methods such as kernel density estimation.

In all of the above kinds of perturbations, the same question is asked. The estimated parameters and/or distributions are imprecise, and hence, the mathematical representation (model) differs from the original system. We need to make sure that the estimation errors will not greatly impact the performance of our model and estimate the deviation in the performance measures between the original system and its mathematical representation, namely, the perturbation errors.

9.3.1. $M/M/1$ *queue*

The $M/M/1$ queueing system is the simplest queueing model. Customers arrive according to a Poisson process with rate λ and wait for service in front of a single server. We denote by E_λ the exponential distribution of the interarrival times. The queue capacity is infinite, the service durations follow an exponential distribution E_μ with rate μ and the service discipline is first-in-first-out (FIFO). The performance measures of this model are calculated in a closed form. In this section, we clarify the conditions under which the $M/M/1$ queueing model can be used as a good approximation in the case where the interarrival time distribution or the service distribution is different (but sufficiently close in some sense) from the exponential distribution.

9.3.1.1. *Perturbation of the interarrival time distribution*

Consider the queueing system $M/M/1\ (FIFO, \infty)$ as described above. Let X_n be the number of customers in the queue just before the nth arrival. $X = \{X_n : n \geq 0\}$ is a homogeneous Markov chain with states in $\mathbb{Z}^+$ and transition matrix $P = (P_{ij})_{ij \geq 0}$ where

$$P_{ij} = \begin{cases} d_{i+1-j} = \frac{\lambda \mu^{i+1-j}}{(\lambda+\mu)^{i+2-j}} & \text{if } 1 \leq j \leq i+1, \\ 1 - \sum_{k=0}^{i} d_k = \left(\frac{\mu}{\mu+\lambda}\right)^{i} & \text{if } j = 0, \\ 0 & \text{otherwise.} \end{cases} \qquad [9.8]$$

Provided that the utilization $\frac{\lambda}{\mu} < 1$, the Markov chain X admits a unique stationary vector $\pi = (\pi_k)_{k \geq 0}$. Consider, on the other hand, a $GI/M/1(FIFO, \infty)$ queueing system where interarrival times are independent and identically distributed according to a general distribution H. Service times are distributed according to an exponential distribution E_μ. Let X_n^* be the number of customers in the queue just before the nth arrival. It is very easy to show that $X^* = \{X_n^* : n \geq 0\}$ is a homogeneous Markov chain with states in $\mathbb{Z}^+$ and a transition matrix $P^* = (P_{ij}^*)_{i,j \geq 0}$ where

$$P_{ij}^* = \begin{cases} d_{i+1-j}^* = \int_0^\infty \frac{1}{(i+1-j)!} e^{-\mu t} (\mu t)^{i+1-j} \, dH(t) & \text{if } 1 \leq j \leq i+1, \\ 1 - \sum_{k=0}^{i} d_k^* & \text{if } j = 0, \\ 0 & \text{otherwise.} \end{cases} \qquad [9.9]$$

Again, a unique stationary vector $\pi^* = (\pi_k^*)_{k \geq 0}$ of the Markov chain X^* exists provided that the utilization is lower than 1. In order for the first system to be a good approximation of the latter, the interarrival time distributions of the two systems must be sufficiently close to each other. We measure the distance between the two distributions with the following metric:

$$w = w(H, E_\lambda) = \int_0^\infty |H - E_\lambda| \, (dt), \qquad [9.10]$$

where $|a|$ is the variation of measure a.

THEOREM 9.3.– Suppose that the utilization of the $M/M/1$ queue $\frac{\lambda}{\mu} < 1$. Then, for each β such that $1 < \beta < \frac{\mu}{\lambda}$, the Markov chain X is stongly v-stable (with respect to the perturbation of the interarrival time distribution) with $v(k) = \beta^k$.

To prove the strong v-stability of the Markov chain X with respect to the function $v(k) = \beta^k$, $\beta > 1$, we check the conditions of theorem 9.1. Suppose that $\lambda/\mu < 1$ and let

$$h_i = \left(\frac{\mu}{\lambda + \mu} \right)^i \text{ for } i \geq 0 \quad \text{and} \quad \alpha_j = \begin{cases} 0 \text{ if } j \geq 1, \\ 1 \text{ if } j = 0. \end{cases}$$

The calculation is straightforward. In particular,

$$\rho = \frac{\beta \lambda}{\mu - \frac{\mu}{\beta} + \lambda} < 1 \text{ for } 1 < \beta < \mu/\lambda. \qquad [9.11]$$

The following result provides an estimation of the distance (in the norm $\|.\|_v$) between the transition matrices of the two Markov chains. This could tell us how close the real system ($G/M/1$) is to its mathematical representation ($M/M/1$).

THEOREM 9.4.– For each β such that $1 < \beta < \mu/\lambda$, we have

$$\|P^* - P\|_v \leq (1+\beta)\, w,$$

where w is given by [9.10].

Now, we are ready to estimate the deviation of stationary distribution that results from the considered perturbation of the interarrival time distribution. The following theorem gives an upper bound to the difference between the stationary distributions π and π^* with respect to the norm $\|.\|_v$.

THEOREM 9.5.– Under the conditions of theorem 9.3 and for each distribution H satisfying

$$w < \frac{(1-\rho)\,(\mu - \lambda\beta)}{(1+\beta)\,(2\mu - \lambda\,(1+\beta))},$$

we have

$$\|\pi^* - \pi\|_v \leq \frac{(1+\beta)\,(2\mu - \lambda\,(1+\beta))\,(\mu - \lambda)\, w}{\frac{(\beta-1)(\mu-\lambda\beta)^3}{(\beta-1)\mu+\lambda\beta} - (2\mu - \lambda\,(1+\beta))\,(1+\beta)\,(\mu - \lambda\beta)\, w},$$

where ρ is given by [9.11].

The proof of this result is based on theorem 9.2. For detailed calculations, see Bouallouche and Aïssani (2006a).

9.3.1.2. *Perturbation of the service time distribution*

Consider again the queueing system of type $M/M/1$ $(FIFO, \infty)$ described above. This time, we consider the random variable X_n representing the number of customers in the queue just after the nth departure. $X = \{X_n : n \geq 1\}$ is a homogeneous Markov chain with states in $\mathbb{Z}^+$ and transition matrix $P = (P_{ij})_{ij \geq 0}$ where

$$P_{ij} = \begin{cases} f_j & \text{if } i = 0 \\ f_{j-i+1} & \text{if } 1 \leq i \leq j+1 \\ 0 & \text{otherwise,} \end{cases} \qquad [9.12]$$

where

$$f_k = \frac{\mu\lambda^k}{(\lambda+\mu)^{k+1}}.$$

Under the usual utilization condition $\frac{\lambda}{\mu} < 1$, X is irreducible and aperiodic. Therefore, it admits a unique stationary vector π.

Consider the $M/G/1$ $(FIFO, \infty)$ queueing system obtained by replacing the service time distribution in the previous model by a general distribution F with the same mean. Let X_n^* be the number of customers in the queue in this new model just after the nth departure. $X^* = \{X_n^* : n \geq 1\}$ is a homogeneous Markov chain with states in $\mathbb{Z}^+$ and transition matrix $P^* = (P_{ij}^*)_{ij \geq 0}$ where

$$P_{ij}^* = \begin{cases} f_j^* & \text{if } i = 0, \\ f_{j-i+1}^* & \text{if } 1 \leq i \leq j+1, \\ 0 & \text{otherwise.} \end{cases} \qquad [9.13]$$

with

$$f_k^* = \int_0^\infty e^{-\lambda t} \frac{(\lambda t)^k}{k!} dF(t).$$

Suppose that the distribution of service durations in the $M/G/1$ system is close to the exponential service time distribution of the $M/M/1$ system. We measure the distance between the two distributions by

$$w = w(F, E_\mu) = \int_0^\infty |F - E_\mu| (dt). \qquad [9.14]$$

First, we prove the following result.

THEOREM 9.6.– Suppose that the utilization of the $M/M/1$ system $\lambda/\mu < 1$. Then, for every β such that $1 < \beta < \mu/\lambda$, the Markov chain X is strongly v-stable for a test function $v(k) = \beta^k$ with respect to the perturbation of the service time distribution.

To prove this result, we choose

$$h(i) = \delta_{i0} = \begin{cases} 1 \text{ if } i = 0, \\ 0 \text{ if } i > 0, \end{cases}$$

and

$$\alpha_j = f_j.$$

Then, we check the conditions of theorem 9.1. In particular, it emerges that if $\beta < \mu/\lambda$, then

$$\rho = \frac{\mu}{\beta(\lambda + \mu - \beta\lambda)} < 1. \qquad [9.15]$$

This intermediary result is necessary to prove the next theorems.

LEMMA 9.1.– Under the conditions

$$\int_0^\infty t^2 |F - E_\mu| (dt) < +\infty.$$

and

$$\int_0^\infty t |F - E_\mu| (dt) < w/\lambda,$$

where w is given by [9.14], there exists $\beta > 1$ such that

$$\int_0^\infty e^{\lambda(\beta-1)t} |F - E_\mu| (dt) < \beta w.$$

The distance between the transition matrices of the two systems is estimated by the following result.

THEOREM 9.7.– Under the conditions of lemma 9.1, we have

$$\|P - P^*\|_v \leq \beta_o w,$$

with $v(k) = \beta^k$ and

$$\beta_o = \max\left(\beta \;:\; 1 < \beta < \frac{\mu}{\lambda} \text{ and } \int_0^\infty e^{\lambda(\beta-1)t} |F - E_\mu| (dt) < \beta w\right). \qquad [9.16]$$

Now, we apply theorem 9.2 to estimate the deviation of the stationary vector with respect to the perturbation of the service time distribution.

THEOREM 9.8.– Under the conditions of theorem 9.6 and lemma 9.1, for every β such that $1 < \beta < \mu/\lambda$ and if

$$w \leq \frac{(1-\rho)}{C\beta_o},$$

we have

$$\|\pi - \pi^*\|_v \leq \beta_o wCC' (1 - \rho - \beta_o wC)^{-1} = e(\beta).$$

with $v(k) = \beta^k$, β_o is given by [9.16] and ρ is given by [9.15],

$$C' = \frac{\mu - \lambda}{\mu - \lambda\beta} \text{ and } C = \frac{2\mu - \lambda(1+\beta)}{\mu - \lambda\beta}.$$

Additionally to the stationary probabilities, the deviation of the other performance measures can be obtained. For example, let N_s (respectively, N_s^*) be the average number of customers in the system, N_q (respectively, N_q^*) be the average number of customers in the queue and T_s (respectively, T_s^*) be the average response time while T_q (respectively, T_q^*) is the average waiting time in the $M/M/1$ (respectively, $M/G/1$) system.

THEOREM 9.9.– Under the same conditions as theorem 9.8, we have the inequalities:

$$|N_s - N_s^*| = |N_q - N_q^*| < e(\beta)/\ln(\beta), \qquad [9.17]$$

$$|T_s - T_s^*| = |T_q - T_q^*| < e(\beta)/(\lambda \ln(\beta)), \qquad [9.18]$$

with $e(\beta) = \beta_o wCC' (1 - \rho - \beta_o wC)^{-1}$.

The detailed proofs of the results of this section can be found in Bouallouche and Aïssani (2006b).

9.3.2. $PH/M/1$ and $M/PH/1$ queues

In the previous section, the interarrival time distribution (respectively, the service time distribution) is approximated by an exponential one having the same mean. This is a rather strict condition because in practice not many distributions can be approximated by the exponential distribution at an acceptable precision level. It would be very helpful if we could find a family of distributions that can approximate a large set of general distributions with a good level of precision and still have properties to allow analytical solutions of the underlying queueing model. This is precisely what the family of phase type (PH) distributions can provide us. Many PH queueing problems can be analytically solved using matrix–geometric techniques (Latouche and Ramaswami 1999). Additionally, the set of PH distributions is dense in the set of positive distributions allowing one (in theory) to approximate any positive distribution by a PH distribution at any desired precision level (Asmussen 2003).

In Djabali *et al.* (2018), the $M/PH/1$ queueing model is used to represent an $M/G/1$ system by approximating the general service time distribution of the latter by a PH distribution. This is achieved by matching the first two moments of both distributions. The matched PH distribution is chosen among the family of hyperexponential or hypoexponential distributions depending on the value of variability coefficient of the original distribution. The confirmation of the strong stability of the underlying Markov chain as well as quantitative estimates of the perturbation error are obtained in each case. Similar results (Djabali *et al.* 2015) exist for the $PH/M/1$ queue when perturbing the interarrival time distribution.

9.3.3. $G/M/1$ and $M/G/1$ queues

This section is concerned with the perturbation of the service time (respectively, interarrival time) distribution in a $G/M/1$ (respectively, $M/G/1$) queueing system. The exponential distribution is replaced by a general distribution with the same mean and the result of this perturbation is a $G/G/1$ queueing model. The conditions of the strong stability of the underlying Markov chain are clarified in each case and upper bounds to the deviation of the stationary vectors are obtained.

9.3.3.1. *Strong stability in the $G/M/1$ queueing system*

Consider a $G/G/1$ queueing system with a general service times distribution G and a general interarrival times probability distribution F. The following notations are used: θ_n (the arrival time of the nth demand), ω_n (the departure time of the nth customer), γ_n (the time interval from θ_n to the departure of the next customer) and $V_n = V(\theta_n - 0)$ (the number of customers found in the system immediately prior to θ_n).

Let us denote by $\nu_{\theta_n} = \min\{m > 0, \omega_m \geq \theta_n\}$. Then, $\gamma_n = \omega_{\nu_{\theta_n}} - \theta_n$.

Define recursively the following sequence,

$$\begin{cases} T_0 = \omega_{\nu_{\theta_n}} - (\theta_n + \gamma_n) = 0, \\ T_k = T_{k-1} + \omega_{\nu_{\theta_n}+k}, \forall k > 0. \end{cases} \qquad [9.19]$$

The sequence $\{T_k\}_{k\in\mathbb{Z}^+}$ describes the departure process after θ_n.

Let us also consider a $G/M/1$ system with exponentially distributed service times with parameter μ and with the same distribution of the interarrival times as the $G/G/1$ one. We introduce the corresponding following notation: $\bar{\theta}_n, \bar{\omega}_n, \bar{\gamma}_n$ and $\bar{V}_n = \bar{V}(\bar{\theta}_n - 0)$ defined as above. We also define the process $\{\bar{T}_n\}_{n\in\mathbb{N}}$ as the sequence $\{T_n\}$.

In what follows, when no domain of integration is indicated, an integral is extended over $\mathbb{R}^+$.

LEMMA 9.2.– The sequence $X_n = (V_n, \gamma_n)$ forms a homogeneous Markov chain with state space $\mathbb{Z}^+ \times \mathbb{R}^+$ and transition operator $Q = (Q_{ij})_{i,j\geq 0}$ defined by

$$Q_{ij}(x, dy) = P(V_{n+1} = j, \gamma_{n+1} \in dy / V_n = i, \gamma_n = x),$$
$$= \begin{cases} q_{i-j}(x, dy) & \text{for } 1 \leq j \leq i,\ i \geq 1, \\ \sum_{k\geq i} q_k(x, dy) & \text{for } j = 0,\ i \geq 0, \\ p(x, dy) & \text{for } j = i+1,\ i \geq 0, \\ 0 & \text{for } j > i+1,\ i \geq 0. \end{cases}$$

where

$$\begin{cases} q_k(x, dy) = \int_x^\infty P(T_k \leq u - x < T_{k+1}, T_{k+1} - (u-x) \in dy) dF(u), \\ p(x, dy) = \int_0^x P(x - u \in dy) dF(u). \end{cases} \quad [9.20]$$

LEMMA 9.3.– The sequence $\bar{X}_n = (\bar{V}_n, \bar{\gamma}_n)$ forms a homogeneous Markov chain with state space $\mathbb{Z}^+ \times \mathbb{R}^+$ and transition operator $\bar{Q} = (\bar{Q}_{ij})_{i,j\geq 0}$ having the same form as Q (Lemma 9.2), where

$$\bar{q}_k(x) = \int_x^\infty e^{-\mu(u-x)} \frac{[\mu(u-x)]^k}{k!} dF(u). \quad [9.21]$$

REMARK 9.1.– The assumption $\bar{\tau}\mu > 1$, where $\bar{\tau}$ is a mean time between arrivals in the $G/M/1$ queueing system, implies the existence of a stationary distribution $\bar{\pi}$ for the embedded Markov chain $\bar{X}$. This distribution has the following form:

$$\bar{\pi}(\{k\}, A) = \bar{\pi}_k(A) = p_k E_\mu(A), \qquad \forall\ \{k\} \subset \mathbb{Z}^+ \ and\ A \subset \mathbb{R}^+, \quad [9.22]$$

where $p_k = \lim_{n\to\infty} P(\bar{V}_n = k)$ is given by the following relation:

$$p_k = (1-\sigma)\sigma^k, \quad k = 0, 1, 2, \ldots \quad [9.23]$$

σ is the unique solution of the equation

$$\sigma = F^*(\mu - \mu\sigma) = \int_0^\infty e^{-(\mu-\mu\sigma)x} dF(x), \quad [9.24]$$

F^* is the Laplace transform of the probability density function of the demands' inter-arrival times. We can show that $0 < \sigma < 1$ (Kleinrock 1975).

Otherwise, note that

$$\lim_{t\to\infty} P(X(t)=k) = \frac{1}{\bar{\tau}\mu} p_{k-1},\ k=1,2,\ldots \ \ and \ \ \lim_{t\to\infty} P(X(t)=0) = 1-\frac{1}{\bar{\tau}\mu}, \tag{9.25}$$

where $X(t)$ represents the size of the $G/M/1$ system at time t.

The formulas [9.23] and [9.25] permit us to compute the stationary distribution of the queue length in a $G/M/1$ system. Unfortunately, for the $G/G/1$ system, these exact formulas are not known. So, if we suppose that the $G/G/1$ system is close to the $G/M/1$ system, then we can use the formulas [9.23] and [9.25] to approximate the $G/G/1$ system characteristics with prior estimation of the corresponding approximation error.

Suppose that the service time distribution of the $G/G/1$ system is close to the exponential one with parameter μ. This proximity is characterized by the distance of variation,

$$W^* = W^*(G, E_\mu) = \int e^{\delta t}|G - E_\mu|(dt)\ , \qquad \text{where} \quad \delta > 0. \tag{9.26}$$

Let also consider the following deviation:

$$W_0 = W_0(G, E_\mu) = \int |G - E_\mu|(dt). \tag{9.27}$$

We apply theorem 9.1 to the imbedded Markov chain $\bar{X}$. Consider the test function

$$\begin{aligned} v : \mathbb{Z}^+ \times \mathbb{R}^+ \to & \quad \mathbb{R}^+ \\ (k, x) & \mapsto v(k,x) = \beta^k e^{\delta x} \end{aligned} \tag{9.28}$$

where $1 < \beta < 1/\sigma$ and $0 < \delta = \mu - \frac{\mu}{\beta} < \mu$ and σ is given by relation [9.24].

Let α be a measure defined as follows, for $\{j\} \times dy \in \mathcal{E}_\mu$, we have,

$$\alpha(\{j\}, dy) = \alpha_j(dy) = \begin{cases} E_\mu(dy) & \text{for } j = 0, \\ 0 & \text{for } j \neq 0. \end{cases} \tag{9.29}$$

And the measurable function

$$\begin{array}{rl} h : \mathbb{Z}^+ \times \mathbb{R}^+ \to & \mathbb{R} \\ (i,x) & \mapsto h_i(x) = h(i,x) = \sum_{k \geq i} \bar{q}_k(x) \end{array} \quad [9.30]$$

where $\bar{q}_k(x)$ is defined by the relation [9.21].

Now, we can state the following result.

THEOREM 9.10.– Suppose that the geometric ergodicity condition $\mu\bar{\tau} > 1$ holds. Then, $\forall \beta \in \mathbb{R}^+$ such that, $1 < \beta < \frac{1}{\sigma}$, the Markov chain $\bar{X}$ is strongly v-stable for a function $v(k,x) = \beta^k e^{\delta x}$ where, $0 < \delta = \mu - \frac{\mu}{\beta} < \mu$.

To prove this result, we check that all the conditions of theorem 9.1 are satisfied (Benaouicha and Aïssani 2005). In particular,

$$\rho = \beta F^*(\mu - \frac{\mu}{\beta}) < 1.$$

The following result gives us the quantitative estimation of the deviation of the norm of transition operator in the $G/M/1$ system, after perturbation of the service time distribution. The proof is based on a series of lemmas demonstrated in Benaouicha and Aïssani (2005).

THEOREM 9.11.– Let Q and $\bar{Q}$ be the transition kernels of the Markov chains X and $\bar{X}$, respectively. Suppose that, for each β such that $1 < \beta < 1/\sigma$, the following conditions hold:

1) $G^* = \int e^{\delta t} G(dt) < +\infty$;

2) $\exists a > 0$ such that $\int e^{au} dF(u) = N < +\infty$;

3) $W_0 = \int |G - E_\mu|(dt) < \frac{a}{a+\mu}$;

4) the geometric ergodicity condition $\mu\bar{\tau} > 1$.

Then, the following inequality holds:

$$\|Q - \bar{Q}\|_v \leq W^*(1 + \mu\bar{\tau}) + W_0 G^* \frac{N + \mu M}{1 - C_0},$$

where $C_0 = W_0 + \frac{\mu}{a+\mu} < 1$ and $M = \int u e^{au} dF(u) < +\infty$.

The upper bound on the norm of the deviation of the stationary measure of the Markov chain $\bar{X}$ is given by the following theorem.

THEOREM 9.12.– Let π *and* $\bar{\pi}$ be the stationary measures of X and $\bar{X}$, respectively. If

$$W^* = W^*(G, E_\mu) < \frac{1-\rho}{2C(1+\mu\bar{\tau}+C_1)},$$

and if

$$W_0 < \frac{a}{a+\mu},$$

then the following inequality holds:

$$\|\pi - \bar{\pi}\|_v \leq 2[(1+\mu\bar{\tau})W^* + C_1 W_0]\frac{C(C-1)}{1-\rho},$$

where

$$C = 1 + \|\bar{\pi}\|_v = \frac{1+\beta(1-2\sigma)}{1-\beta\sigma}, \qquad C_1 = \frac{N+\mu M}{1-C_0}G^*. \qquad [9.31]$$

The proof of this theorem is based on theorem 9.2 and uses a series of intermediary results. Further details can be found in previous studies (Aïssani 1982, 1987a,b; Benaouicha and Aïssani 2005).

9.3.3.2. *Strong stability of the $M/G/1$ queue*

Consider a $G/G/1$ (FIFO,∞) queueing model. We denote by τ'_n the interarrival time between the arrival dates of the $(n-1)$th and the nth customers. The sequence of independent and identically distributed service times is denoted by $\{\xi_n\}$. Let $\theta'_n = \tau'_1 + \cdots + \tau'_n$ be the arrival moment of the $(n+1)$th customer with $\theta'_0 = 0$, and consider the distributions $G(t) = P(\tau'_n < t)$ and $F(t) = P(\xi_n < t)$ with $m = \mathrm{E}\xi_n$.

Let q_n be the number of customers in the system at the end of service of the nth customer, γ_n the (residual) time until the next arrival. The sequence $X_n = (q_n, \gamma_n)$ is therefore a homogeneous Markov chain with states in $\mathbb{Z}^+ \times \mathbb{R}^+$ and a transition kernel

$$Qf(n,x) = \mathrm{E}(f(n-1,\ x-\xi_1),\ \xi_1 < x) +$$

$$+\sum_{k=1}^{\infty} \mathrm{E}(f(n-1+k,\ \theta'_k + x - \xi_1),\ \theta'_{k-1} \leq \xi_1 - x < \theta'_k), \qquad [9.32]$$

where $n > 0$ and $Qf(n,x) = \mathrm{E}Qf(1,\tau'_1)$. Q can be obtained from the distributions G and F.

On the other hand, consider a $M/G/1$ (FIFO,∞) queueing model with (i.i.d) exponential interarrival times $\{\theta_n\}_{n\geq 1}$, $\theta_0 = 0$ of mean $1/\lambda$. The service time distribution is F. Let P be the transition kernel of the Markov chain $\{X_n\}$ in the $M/G/1$ system.

In what follows, we assume that the following conditions are satisfied:

$$\lambda \, \mathrm{E} \, \xi_1 < 1 \, , \; \mathrm{E} \, exp \, (a\xi_1) < +\infty, \tag{9.33}$$

for a certain $a > 0$.

Suppose that the interarrival time distribution G in the $G/G/1$ system is close to the exponential interarrival time distribution E_λ of the $M/G/1$ model. The distance between the two distributions is measured by

$$w(G, E_\lambda) = \int \exp(ct)|G - E_\lambda|(dt), \tag{9.34}$$

where $0 < c < \lambda$ is a fixed parameter.

Consider the measurable test function $v(k, x)$ on $\mathbb{Z}^+ \times \mathbb{R}^+$ defined by

$$v(k, x) = \beta^k \exp{(\varepsilon x)},$$

and the corresponding $v-$norm $||.||_{\beta\varepsilon}$.

It is easy to prove that $||P||_{\beta\varepsilon} < \infty$ under the condition $\lambda/\mu < 1$ and for every $0 < \varepsilon < \lambda$ and $1 < \beta < 1 + a\lambda^{-1}$. Additionally, we have the following result:

LEMMA 9.4.– Suppose that the condition $\lambda/\mu < 1$ is satisfied and let $0 < c < \lambda$ be a fixed constant. Then, there exist $\beta_1 = \beta_1(\lambda, c, F) > 1$ and $L_1(\beta) < \infty$ such that $||Q - P||_{\beta c} \leq L_1(\beta)w(G, E_\lambda)$ for every $1 < \beta < \beta_1$ and every distribution G such that $w(G, E_\lambda)$ is sufficiently small.

Let $v(n, x) = \beta^n(exp\,(cx) + bexp(-\delta x))$. We show that the Markov chain X is strongly stable with respect to the norm $||.||_v$. By the fact that $b^{-1}||T||_{\beta c} \leq ||T||_v \leq (1 + b)||T||_{\beta c}$, the chain X_n is strongly stable with respect to the norm $||.||_{\beta c}$.

Let us check the conditions of the strong stability of the Markov chain X.

– it is easy to check that $||P||_v < \infty$;

– define the measure α and the measurable function h on $I \times \mathbb{R}_+$ as follows:

$$h(n, x) = 0 \text{ for } n > 0, h(0, x) = 1,$$

$$\alpha(\{n\} \times A) = P(\theta_n \leq \xi_1 < \theta_{n+1}, \; \theta_{n+1} - \xi_1 \in A).$$

We may easily show that for the operator $T = P - ho\alpha$, $Tf(n,x) = Pf(n,x)$ for $n > 0$ and $Tf(0,x) = 0$. Clearly, T is non-negative.

Observe that for $n > 0$ and $v(k,x) = \beta^k \, exp\,(-\delta x)$,

$$\begin{aligned} Tv(n,x) &= \beta^{n-1}\mathrm{E}(exp\,(\delta\xi_1 - \delta x),\ \xi_1 < x) + \sum_{k\geq 1} \beta^{n-1+k} \times \\ &\times \mathrm{E}(exp(-\delta\theta_k - \delta x + \delta\xi_1),\ \theta_{k-1} \leq \xi_1 - x < \theta_k) = \beta^{n-1} exp(-\delta x) \times \\ &\times \mathrm{E}(exp(\delta\xi_1 - \delta x),\ \xi_1 < x) + \lambda(\lambda+\delta)^{-1}\beta^n \mathrm{E}(exp(\lambda(\beta-1)(\xi_1 - x), (\xi_1 \geq x). \end{aligned} \quad [9.35]$$

For $\delta = \lambda(\beta - 1)$, we obtain $Tv(n,x) = \rho_0 v(n,x)$, where $\rho_0 = \beta^{-1}\mathrm{E}\, exp(\lambda(\beta - 1)\xi_1)$. Given that $\lambda/\mu < 1$, it is easy to establish that $\rho_0 < 1$ for $\beta < 1$ if sufficiently small.

By choosing β and setting $\delta = \lambda(\beta - 1)$ in the expression of $v(n,x)$, we obtain for $n > 0$

$$Tv(n,x) \leq (b\rho_0 + \lambda(\lambda - c)^{-1}\beta)\, \beta^n \, exp\,(-\delta x) + \beta^{n-1} exp(cx) \leq \rho V(n,x),$$

where $\rho = \rho_0 + b^{-1}\lambda(\lambda - C)^{-1}\beta$. In addition, $Tv(0,x) = 0 < \rho v(0,x)$.

It is sufficient to choose $\rho < 1$. This is possible because $\rho_0 < 1$ and the constant b can be taken sufficiently large.

9.3.4. *Other queues*

In the same way, a multitude of queueing systems have been studied. Hence, strong stability results and perturbation bounds were also obtained for the following queueing systems:

– group arrival queues: perturbation of the distribution of the size of the group (Boukir *et al.* 2009);

– retrial queues: perturbation of the retrial rate (Berdjoudj and Aïssani 2003);

– multiple server $M/M/m$ queue: perturbation of the interarrival time distribution (Issaadi *et al.* 2016);

– queues with unreliable server: perturbation of failure rate (Abbas and Aïssani 2010a,c);

– queues with server vacation: perturbation of vacation rate (Rahmoune and Aïssani 2008, 2014);

– $GI/M/\infty$ queue: perturbation of the size of the system (Aïssani 1992a,b; Bareche *et al.* 2016);

– $M_2/G_2/1$ queue with priority: perturbation of the rate of priority arrivals (Aïssani 1991; Bouallouche and Aïssani 2008; Hamadouche and Aïssani 2011);

– $GI/M/1$ queue with negative customers: perturbation of the negative arrival rate (Abbas and Aïssani 2010b).

9.3.5. *Queueing networks*

The application of the strong stability method to queueing networks poses numerous challenges. Except from Jackson networks, the performance measures cannot be obtained in closed form. The approximation of general queueing networks by Jackson networks also poses problems. First, the dimension of the underlying Markov chain might be high. Additionally, the strong stability method supposes that the perturbed process is also Markovian. The interconnection of the network nodes and the complex dynamics of such systems make it very difficult to maintain the Markov property or to define an embedded Markov chain in the perturbed system.

9.3.5.1. *Jackson networks with two tandem stations*

Consider the following Jackson network with two tandem queues $[M/M/1 \rightarrow M/M/1]$. Customers arrive at the first station according to a Poisson process with rate λ. Denote by E_λ the exponential distribution of the interarrival times. Service durations follow an exponential distribution with rate μ in the first station and an exponential distribution with rate μ_1 in the second one. The performance measures of this model can be calculated in the closed form by exploiting the product form property (Jackson 1957). In this section, we clarify the conditions under which this type of queueing network can be used as a good approximation in the case where the service distribution in the first station is different (but sufficiently close in some sense) from the exponential distribution.

The state of the tandem network above is completely described by the Markovian bidimensional process $\overline{V}(t) = (\overline{X}(t), \overline{Y}(t))$, where $\overline{X}(t)$ is the number of customers in the first station at time t, and $\overline{Y}(t)$ is the number of customers in the second station at the same time t.

Denote by $\overline{d}_n^+$ the moment just after the departure of the nth customer from the first station. Then, the embedded sequence $(\overline{V}_n)_{n\geq 0}$ of random variables where $\overline{V}_n =$

$\overline{V}(\overline{d}_n^+)$, $\overline{V}_0 = 0$ is a Markov chain. Under the condition $\lambda \leq \min(\mu_1, \mu)$, the transition probabilities of the Markov chain $\overline{V}_n$ are given by:

$$\overline{Q}_{ij}(k,l) = \begin{cases} \overline{P}_j \overline{q}_{kl}, & \text{if } i = 0, j \geq 0, 1 \leq l \leq k+1, k \geq 0, \\ \overline{P}_j \overline{q}_{k0}, & \text{if } i = 0, j \geq 0, l = 0, k \geq 0, \\ \overline{P}_{j-i+1} \overline{q}_{kl}, & \text{if } 1 \leq i \leq j+1, j \geq 0, 1 \leq l \leq k+1, k \geq 0, \\ \overline{P}_{j-i+1} \overline{q}_{k0}, & \text{if } 1 \leq i \leq j+1, j \geq 0, l = 0, k \geq 0, \\ 0, & \text{otherwise} \end{cases} \quad [9.36]$$

where:

$$\overline{P}_r = \int_0^\infty \exp(-\lambda x) \frac{(\lambda x)^r}{r!} dE_{\mu_1}(x), \forall r \in \mathbb{N},$$

$$\overline{q}_{kl} = \int_0^\infty \exp(-\mu x) \frac{(\mu x)^{k+1-l}}{(k+1-l)!} dE_\lambda(x),$$

$$\overline{q}_{ko} = 1 - \sum_{l=1}^{k+1} \overline{q}_{kl}.$$

On the other hand, consider the two stations tandem network $[M/G/1 \to \cdot/M/1]$. Arrivals to the first station are Poisson distributed with the same rate λ as before while the service times distribution H is general. Consequently, the arrival process to the second station is not Poissonian anymore. Denote by F the distribution of interarrival times to this station. Service times at the second station are exponentially distributed having common distribution E_μ with rate μ. Let $V(t) = (X(t), Y(t))$ be the two-dimensional process, where $X(t)$ is the number of customers in the first station at time t, $Y(t)$ is the number of customers in the second station at time t. $V(t)$ completely describes the state of the considered network $[M/G/1 \to ./M/1]$. However, $V(t) = (X(t), Y(t))$ is not a Markovian process. Denote by d_n^+ the moment just after the departure of the nth customer from the first station. Then, the sequence of random variables $V_n = (\overline{X}_n, \overline{Y}_n)$ at these times is an embedded Markov chain that describes the state of the network $[M/G/1 \to \cdot/M/1]$ at those specific dates. Under the condition $\lambda < \min(\mu_1, \mu_2)$ the transition probabilities of the chain V_n write:

$$Q_{ij}(k,l) = \begin{cases} P_j q_{kl}, & \text{if } i = 0, j \geq 0, 1 \leq l \leq k+1, k \geq 0, \\ P_j q_{k0}, & \text{if } i = 0, j \geq 0, l = 0, k \geq 0, \\ P_{j-i+1} q_{kl}, & \text{if } 1 \leq i \leq j+1, j \geq 0, 1 \leq l \leq k+1, k \geq 0, \\ P_{j-i+1} q_{k0}, & \text{if } 1 \leq i \leq j+1, j \geq 0, l = 0, k \geq 0, \\ 0, & \text{otherwise} \end{cases} \quad [9.37]$$

where

$$P_r = \int_0^{\infty} \exp(-\lambda x) \frac{(\lambda x)^r}{r!} dH(x), \forall r \in \mathbb{N},$$

$$q_{kl} = \int_0^{\infty} \exp(-\mu x) \frac{(\mu x)^{k+1-l}}{(k+1-l)!} dE_{\lambda}(x),$$

$$q_{ko} = 1 - \sum_{l=1}^{k+1} q_{kl}.$$

Let α be the measure defined as follows:

$$\alpha(\{i\}, \{j\}) = \begin{cases} P_i q_{kj} & \text{if } 0 \leq l \leq k+1, \\ 0 & \text{otherwise.} \end{cases} \quad [9.38]$$

Consider as well the measurable function h defined by:

$$h(i,k) = h_i(k) = \mathbf{1}_{\{i=0\}}. \quad [9.39]$$

Using them in theorem 9.1, the following result can be proved.

THEOREM 9.13.– Under the condition $\lambda < \min(\mu_1, \mu)$, the Markov chain $\{V_n\}_{n \geq 0} = \{X_n, Y_n\}_{n \geq 0}$ is strongly v-stable with respect to the test function $v(i,j) = \gamma^i \beta^j$, for every γ, β such that

$$1 < \gamma < \frac{\mu_1}{\lambda}, \text{and} 1 < \beta < \gamma \Big(\frac{\lambda + \mu_1 - \gamma\lambda}{\mu_1} \Big).$$

9.3.5.2. *Tandem queues with constant retrials*

Consider the tandem queueing network $[M/G/1 \to ./G/1/1]$ with blocking after service, consisting of a sequence of two service stations without intermediate queue. Customers arrive to the first station according to a Poisson process with intensity λ. Each customer receives service at station 1 and then proceeds to station 2 for an additional service. Since there is no intermediate waiting room, a customer whose service in the station 1 is completed cannot proceed to the second station if the latter is busy. Instead, the customer remains at station 1 that is blocked until station 2 becomes empty. The arriving customer who finds station 1 busy or blocked behaves like a retrial customer, i.e., he does not join a queue but instead he is placed in a hypothetical retrial queue (orbit) of infinite capacity and retries for service under the constant retrial policy. According to this policy, the parameter of the exponential time of each customer in the retrial group is $\frac{\mu}{n}$, where n is the size of the retrial group. Thus, the total intensity is μ. If the server of station 1 is free at the time of an attempt, the customer at the head of the retrial group receives service immediately. Otherwise, they repeat their demand later.

Service times at stations 1 and 2 are independent and arbitrarily distributed random variables with probability density functions $b_i(x)$, distribution functions $B_i(x)$ and finite mean values $1/\mu_i$, for $i = 1, 2$, respectively.

Let $X(t)$ represent the number of customers in the orbit at time t, and for $l = 1, 2$:

$$\xi^l(t) = \begin{cases} 0 & \text{if the } l^{\text{th}} \text{ server is idle at time } t, \\ 1 & \text{if the } l^{\text{th}} \text{ server is working at time } t, \\ 2 & \text{if the } l^{\text{th}} \text{ server is blocked at time } t. \end{cases}$$

The considered model is completely described by the regenerative process:

$$V(t) = \left(X(t), \xi^1(t), \xi^2(t)\right).$$

However, this process is not Markovian. We denote by $d_n, n \in \mathrm{N}$, the instant of the nth departure from station 1. We assume, without loss of generality, that $d_0 = 0$, and we note that:

$$V_n = V(d_n + 0) = \left(X(d_n + 0), \xi^1(d_n + 0), \xi^2(d_n + 0)\right) = (X_n, 0, 0).$$

Thus, V_n is a semiregenerative process with embedded Markov renewal process $(X, D) = \{X_n, d_n : n \in \mathrm{N}\}$. The process $\{X_n\}$ is an homogeneous, irreductible and aperiodic Markov chain with the transition matrix $\mathbf{P} = \{p_{ij}\}_{i,j \geq 0}$, where:

$$p_{ij} = \begin{cases} \int_0^{+\infty} \frac{(\lambda t)^j}{j!} e^{-\lambda t} f_0(t) dt, & \text{for } i = 0, \\ \int_0^{+\infty} \frac{(\lambda t)^{j-i}}{(j-i)!} e^{-\lambda t} f_1(t) dt \\ \quad + \int_0^{+\infty} \frac{(\lambda t)^{j-i+1}}{(j-i+1)!} e^{-\lambda t} f_2(t) dt, & \text{for } 1 \leq i < j+1, \\ \int_0^{+\infty} e^{-\lambda t} f_2(t) dt & \text{for } i = j+1, \\ 0, & \text{otherwise,} \end{cases} \qquad [9.40]$$

$$\text{with: } \begin{cases} f_0(t) = \int_0^{+\infty} \lambda e^{-\lambda w} \frac{d}{dt} \left(B_1(t) B_2(t+w)\right) dw; \\ f_1(t) = \int_0^{+\infty} \lambda e^{-(\lambda+\mu) w} \frac{d}{dt} \left(B_1(t) B_2(t+w)\right) dw; \\ f_2(t) = \int_0^{+\infty} \mu e^{-(\lambda+\mu) w} \frac{d}{dt} \left(B_1(t) B_2(t+w)\right) dw. \end{cases}$$

Let $\psi_u(s)$ be the function defined as:

$$\psi_u(s) = \int_0^{+\infty} ue^{-uw}dw \int_0^{+\infty} e^{-sx}d_x\left(B_1(x)B_2(x+w)\right). \qquad [9.41]$$

Assume that the mean retrial rate in the above tandem queues tends to infinity, i.e. the customers in the retrial group try continuously to find a position for service and they become ordinary customers. It means that if $\mu \to +\infty$, the tandem network with constant retrials becomes similar to the classical model of two queues in tandem without intermediate room.

Now, let $\overline{X}(t)$ denote the number of customers in the first queue of the classical tandem network at time t and for $l = 1, 2$, we consider:

$$\overline{\xi}^l(t) = \begin{cases} 0 & \text{if the } l^{\text{th}} \text{ server is idle at time } t, \\ 1 & \text{if the } l^{\text{th}} \text{ server is working at time } t, \\ 2 & \text{if the } l^{\text{th}} \text{ server is blocked at time } t. \end{cases}$$

The state of the ordinary tandem network $[M/G/1 \to ./G/1]$ is completely described by the process $\overline{V}(t) = \left(\overline{X}(t), \overline{\xi}^1(t), \overline{\xi}^2(t)\right)$. It is clear that $\overline{X}_n = (\overline{X}, D) = \{\overline{X}_n, \overline{d}_n, n \geq 0\}$ is the embedded Markov renewal process of the semiregenerative process $\left(\overline{X}(t), \overline{\xi}^1(t), \overline{\xi}^2(t)\right)$ at the instant $\overline{d}_n$ of the nth departure from station 1.

Now, suppose that $\rho = \lim\limits_{\mu \to +\infty} \rho^* < 1$. Then $\overline{X}_n$ is an irreductible and homogeneous, aperiodic, positive recurrent Markov chain with transition matrix $\overline{\mathbf{P}} = \{\overline{p}_{ij}\}_{i,j \geq 0}$:

$$\overline{p}_{ij} = \begin{cases} \int_0^{+\infty} \frac{(\lambda t)^j}{j!} e^{-\lambda t} f_0(t)dt, & i = 0, \\ \\ \int_0^{+\infty} \frac{(\lambda t)^{j-k+1}}{(j-k+1)!} e^{-\lambda t} d_t\left(B_1(t)B_2(t)\right), & i \in [i, j+1], \\ \\ 0, & \text{otherwise}. \end{cases} \qquad [9.42]$$

It is easy to show that:

$$\lim_{\mu \to +\infty} \psi_{\lambda+\mu}(s) = \psi(s) = \int_0^{+\infty} e^{-st}d_t\left(B_1(t)B_2(t)\right); \quad \lim_{\mu \to +\infty} v_{\lambda+\mu} = \frac{\rho}{\lambda};$$

$$\lim_{\mu \to +\infty} \rho^* = \rho = -\lambda \frac{d\psi(s)}{ds}\Big|_{s=0} = \lambda \int_0^{+\infty} t d_t\left(B_1(t)B_2(t)\right).$$

Suppose that the mean retrial rate in the tandem network with constant retrials tends to infinity. The distance between the two distribution's interarrivals is measured by: $W = \int_0^{+\infty} |f_2(t) - \frac{d}{dt}(B_1(t)B_2(t))|dt$.

Consider the test function $v(j) = \beta^j, \beta > 1$, the measure α and the measurable function h defined by:

$$\alpha(\{j\}) = \alpha_j = \overline{p}_{0j}, \qquad h(i) = \begin{cases} 1, & \text{if } i = 0, \\ 0, & \text{otherwise.} \end{cases}$$

By using them in theorem 9.1, and based on a series of lemmas demonstrated in a previous study (Lekadir and Aïssani 2008a), the following theorems can be proved.

THEOREM 9.14.– In the two stations' tandem network with blocking $[M/G/1 \to ./G/1/1]$, the Markov chain $\overline{X}_n$ representing the number of customers in the first station at the instant of the nth departure from the first station is strongly v-stable with respect to the function $v(k) = \beta^k$ for all β such that $1 < \beta \leq \beta_0$, where β_0 is given by: $\beta_0 = \sup\left\{\beta : \gamma(\beta) = \frac{\psi(\lambda - \lambda\beta)}{\beta} < 1\right\}$.

THEOREM 9.15.– Let π (respectively, $\overline{\pi}$) be the stationary distribution of the Markov chain X_n (resp. $\overline{X}_n$). For β such that $1 < \beta < \beta_0$, we have:

$$\|\pi - \overline{\pi}\|_v \leq c_0(1 + c_0)\|\Delta\|_v\left(1 - \gamma - (1 + c_0)\|\Delta\|_v\right)^{-1},$$

where: $c_0 = \frac{\psi_\lambda(\lambda\beta - \lambda) - \gamma}{1 - \gamma}$.

9.3.5.3. *Tandem queues with non-preemptive priority*

Consider the tandem queueing network $[M_2/G_2/1 \to ./G/1/1]$ with non-preemptive priority. Customers arrive to the first station according to a Poisson process. Denote by $(\theta\lambda)$ (respectively, (λ)) the arrival rate of the priority (respectively, the non-priority) customers. The distribution function of the service times of the priority (respectively, non-priority) customers in the first station is $C_1(x)$ (respectively, $C_2(x)$)) and the corresponding density function is $c_1(x)$ (respectively, $c_2(x)$). Moreover, service times in the second station are independent having distribution function $D(x)$ and density function $d(x)$ common for both priority classes. Now, let $\{(X^i(t)\}_{i=1,2}$ represent the number of customers in station i at the departure time of the nth customer. The state of the considered model is completely described by the bidimentional process $V(t) = \left(X^1(t), X^2(t)\right)$. Suppose that θ tends to zero, the part of the tandem queue related to priority customers can be interpreted as a perturbation of the network $[M/G/1 \to ./G/1/1]$ of two stations in tandem with only non-priority customers. The Markov chain $\left(X_n^{(1)}, X_n^{(2)}\right)_{n \geq 0}$ has transition probabilities $P_{k,l}(i,j,\theta)$ defined by:

$$\begin{cases} \int_0^\infty \left(\frac{(\lambda\theta x)^{i-k+1}}{(i-k+1)!}\right)\left(\frac{(\lambda x)^{j-l}}{(j-l)!}\right) e^{-\lambda(\theta+1)x} d\left[C_1(x)D(x)\right]; \text{if } k \neq 0, l \geq 0; i \geq k-1, j \geq l, \\ \int_0^\infty \left(\frac{(\lambda\theta x)^{i}}{i!}\right)\left(\frac{(\lambda x)^{j-l+1}}{(j-l+1)!}\right) e^{-\lambda(1+\theta)x} \left[\int_0^\infty \lambda e^{-\lambda t} d\left(C_2(x)D(x+t)\right) dt\right]; \\ \qquad \text{if } k = 0, l \neq 0; i \geq 0, j \geq l-1, \\ \frac{\theta}{1+\theta}\int_0^\infty \left(\frac{(\lambda\theta x)^{i}}{i!}\right)\left(\frac{(\lambda x)^{j}}{j!}\right) e^{-\lambda(1+\theta)x} \left[\int_0^\infty \lambda e^{-\lambda t} d\left(C_1(x)D(x+t)\right) dt\right] + \\ \quad + \frac{1}{1+\theta}\int_0^\infty \left(\frac{(\lambda\theta x)^{i}}{i!}\right)\left(\frac{(\lambda x)^{j}}{j!}\right) e^{-\lambda(1+\theta)x} \left[\int_0^\infty \lambda e^{-\lambda t} d\left(C_2(x)D(x+t)\right) dt\right]; \\ \qquad \text{if } k = 0, l = 0; i \geq 0, j \geq 0, \\ 0; \qquad \text{otherwise.} \end{cases}$$

The transition probabilities $P_{k,l}(i,j,0)$ of the Markov chain $\left(\overline{X}_n^{(1)}, \overline{X}_n^{(2)}\right)_{n\geq 0}$ describing the state of the tandem network $[M/G/1 \rightarrow ./G/1/1]$ are given by:

$$\begin{cases} \int_0^\infty \left(\frac{(\lambda x)^{j-l}}{(j-l)!}\right) e^{-\lambda x} d\left[C_1(x)D(x)\right]; & \text{if } k \neq 0, l \geq 0; i = k-1, j \geq l, \\ \int_0^\infty \left(\frac{(\lambda x)^{j-l+1}}{(j-l+1)!}\right) e^{-\lambda x} \left[\int_0^\infty \lambda e^{-\lambda t} d\left(C_2(x)D(x+t)\right) dt\right]; & \text{if } k = 0, l \neq 0; i = 0, j \geq l-1, \\ \int_0^\infty \left(\frac{(\lambda x)^{j}}{j!}\right) e^{-\lambda x} \left[\int_0^\infty \lambda e^{-\lambda t} d\left(C_2(x)D(x+t)\right) dt\right]; & \text{if } k = 0, l = 0; i = 0, j \geq 0, \\ 0; & \text{otherwise.} \end{cases}$$

The conditions of the v-stability of the network $[M/G/1 \rightarrow ./G/1/1]$ are clarified by the following theorem:

THEOREM 9.16.– Suppose that in the tandem network $[M/G/1 \rightarrow ./G/1/1]$, the following the assumptions holds:

– $\lambda E(\xi_2) < 1$ (geometric ergodicity condition), ξ_i is the random variable representing the service times of customers with priority i;

– $\exists a > 0 / E(e^{a\xi_2}) < \infty$ (Cramer condition).

Then, for all β such that $1 < \beta < \beta_0$ and $\delta > 1$, the Markov chain $\left(\overline{X}_n^{(1)}, \overline{X}_n^{(2)}\right)_{n \geq 0}$ is v-strongly stable with respect to the function $v(i,j) = \delta^i \beta^j$ with , $\delta = \frac{E(e^{x\xi_1})}{\gamma_2}$, where $\gamma_2 = \frac{E(e^{x\xi_2})}{\beta}$ and $\beta_0 = \sup\{\beta / E(e^{x\xi_2}) < \beta\}$.

To prove theorem 9.16, it is sufficient to check the conditions of theorem 9.1 using the test function:

$$
\begin{aligned}
v : \mathbb{N} \times \mathbb{N} &\longrightarrow \mathbb{R}_+^* \\
(i,\ j) &\longrightarrow V(i,\ j) = \delta^i \times \beta^j;\ \text{with } \delta > 1,\ \beta > 1.
\end{aligned}
$$

The measure α and the measurable function h are defined by:

$$
\begin{aligned}
\alpha : \sigma(N) \times \sigma(N) &\longrightarrow \mathbb{R}^+ \\
(\{i\}, \{j\}) &\longrightarrow P_{0,0}(i,j,0)
\end{aligned}
\qquad
\begin{aligned}
h : \mathbb{N} \times \mathbb{N} &\longrightarrow \mathbb{R} \\
(k,\ l) &\longrightarrow 1_{\{k=0,\ l=0\}}.
\end{aligned}
$$

Note $f_i(x) = \displaystyle\int_0^{+\infty} \lambda e^{-\lambda t} \frac{d}{dx}\Big(C_i(x) D(x+t)\Big) dt$, $i = \overline{1,2}$, then:

$$
\alpha(\{i\}, \{j\}) = \begin{cases} P_{0,0}(0,j,0) = \displaystyle\int_0^{+\infty} \frac{(\lambda x)^j}{j!} e^{-\lambda x} f_2(x) dx, & \text{if } i = 0; \\ P_{0,0}(i,j,0) = 0, \text{ if } i \neq 0. \end{cases}
$$

From the above theorem, the tandem network $[M/G/1 \to ./G/1/1]$ is strongly v-stable. It means that its characteristics can approximate the tandem network $[M_2/G_2/1 \to ./G/1/1]$ under the condition that the arrival rate of priority customers θ tends to zero. To characterize this proximity, it is essential to estimate the deviation between the stationary distributions of the chains X_n and $\overline{X}_n$. To do so, we first estimate the deviation $\|\Delta\|_v$ of the transition operator $\mathbf{P}$.

LEMMA 9.5.–

$$
\|\Delta\|_v = \|\mathbf{P} - \overline{\mathbf{P}}\|_v = \sup_{k \geq 0} \sup_{l \geq 0} \frac{1}{\delta^k \beta^l} \sum_{i \geq 0} \sum_{j \geq 0} \delta^i \beta^j \left|\Delta_{kl}(i,j,\theta)\right|,
$$

$$
\text{with} \ : \ \left|\Delta_{kl}(i,j,\theta)\right| = \left|P_{kl}(i,j,\theta) - P_{kl}(i,j,0)\right|.
$$

Denote by $\Theta = \sup\{E_{00}, E_{0l}, E_{kl}\}$. Then:

$$
\|\Delta\|_v \leq \Theta; \qquad [9.43]
$$

where:

$$E_{00} \leq \phi_2(\lambda - \lambda\beta) + \frac{1}{1+\theta}\phi_2(\lambda\theta + \lambda - \lambda\beta - \lambda\delta\theta)$$
$$- \left(\frac{1}{1+\theta} + 1\right)\phi_2(\lambda\theta + \lambda - \lambda\beta) + \frac{\theta}{1+\theta}\phi_1(\lambda\theta + \lambda - \lambda\beta - \lambda\theta\delta);$$
$$E_{0l} = \frac{1}{\beta}\left[\psi_2(\lambda - \lambda\beta) + \psi_2(\lambda + \lambda\theta - \lambda\beta - \lambda\delta\theta) - 2\psi_2(\lambda + \lambda\theta - \lambda\beta)\right];$$
$$E_{kl} = \frac{1}{\delta}\left[\psi_1(\lambda + \lambda\theta - \lambda\theta\delta - \lambda\beta) + \psi_1(\lambda - \lambda\beta) - 2\psi_1(\lambda + \lambda\theta - \lambda\beta)\right];$$
$$\phi_i(s) = \int_0^\infty \lambda e^{-\lambda t} dt \int_0^\infty e^{-sx} d[C_i(x)D(x+t)],$$
$$\psi_i(s) = \int_0^\infty e^{-sx} d[C_i(x)D(x)].$$

The following lemma is also needed to prove the next theorem.

LEMMA 9.6.– Let us consider the constant ϖ defined by $\varpi = \beta P_0\left(\frac{\gamma}{1-\gamma}\right)$, where: $P_0 = \frac{1+\lambda\psi_1'(0)+\theta\lambda\phi_1'(0)}{1+\lambda\psi_1'(0)+\theta\lambda\psi_2'(0)-\lambda\phi_1'(0)-\theta\lambda\phi_1'(0)}$, then $\|\overline{\pi}\|_v \leq \varpi$.

The upper bound on the norm of the deviation of the stationary measure of the Markov chains of the tandem queues $[M/G/1 \rightarrow ./M/1/1]$ and $[M_2/G_2/1 \rightarrow ./M/1/1]$ is given by the following theorem.

THEOREM 9.17.– Let $\overline{\pi}$ (respectively, π) be the stationary distribution of $[M/G/1 \rightarrow ./M/1/1]$ (respectively, $[M_2/G_2/1 \rightarrow ./M/1/1]$). Under the condition $\Theta < \frac{1-\gamma}{C}$, we have:

$$\|\pi - \overline{\pi}\|_v \leq \frac{\Theta\varpi(1+\varpi)}{1-\gamma-(1+\varpi)\Theta} \qquad [9.44]$$

where γ, Θ, ϖ and P_0 are defined previously. $1 < \beta < \beta_0$ where β_0 is defined by: $\beta_0 = \sup\left\{\beta : \frac{\psi_2(\lambda\beta-\lambda)}{\beta} < 1\right\}, 1 < \beta_0 < +\infty$.

The proof of this theorem is based on theorem 9.2 and uses a series of intermediary results (for more details, see Lekadir and Aïssani (2011)).

9.3.5.4. *Tandem queues with blocking*

A two-station tandem network of type $[M/M/1/\infty \leftrightarrow M/M/1/N]$ with blocking was considered in a previous study (Adel-Aissanou *et al.* 2012). The infinite buffer of the first station is truncated by rejecting arrivals when the queue length reaches a given level Q. It is expected that such a truncation well approximates the original model as the truncation level (or size) becomes large. The conditions that guarantee that the steady-state joint queue length distribution of the original tandem queueing system is well approximated by the finite buffer truncation are clarified and error bounds on the stationary queue length distributions are obtained.

9.3.6. *Non-parametric perturbation*

The above results are concerned with the use of parametric distributions with known or unknown form. In other works, the authors explore the use of non-parametric distributions such as kernel density functions fitted from empirical data. Combining statistics theory results with the strong stability method, one is able to estimate the impact of the use of such technique.

Note that, in practice, all model parameters are imprecisely known because they are obtained by means of statistical methods. That is why the strong stability inequalities will allow us to numerically estimate the uncertainty shown during this analysis.

In this sense, one aspect which is of interest is when a distribution governing a queueing system is unknown and we resort to non-parametric methods to estimate its density function. For instance, if one had real data, then one could apply the kernel density method to estimate the related density function.

Moreover, it is very often the case that the natural domain of definition of a density to be estimated is not the whole real line but an interval bounded on one or both sides. For example, when the data are measurements of positive quantities, it will be preferable to obtain density estimates that take the value zero for all negative numbers.

Indeed, the strong stability method states that the perturbation must be small in the sense that the general law G of arrivals (respectively, service times) must be close but not equal to the Poisson (respectively, exponential) law. Consequently, the density function of the law G must be close to the exponential density, which is defined on a bounded support. Thus, the boundary effects must be taken into consideration when using the kernel density method.

By combining the techniques of correction of boundary effects with the calculation of the variation distance characterizing the proximity of the quoted systems, one will be able to check whether this density is sufficiently close to that of the Poisson law (or that of the exponential law), and then apply the strong stability method to approximate the characteristics of the real system with those of a classical method (Bareche and Aïssani 2008, 2011, 2014a,b).

9.4. Conclusion and further directions

The chapter reviews the evolution of the applications of the strong stability method to queueing systems and their networks. Many types of queues and networks have been already studied and both qualitative (stability) and quantitative (perturbation bounds) are obtained in most cases. However, queueing systems may contain a large number of parameters and the perturbation of each might have a completely different effect from the others. Therefore, it is necessary to extend the stability analysis to those unexplored models and perturbation types, for instance, perturbation of the collusion probability in the $M/M/1$ retrial queue with collisions and transmission errors (modeling the access method in the channel of IEEE 802.11 (Lakaour *et al.* 2018)), as well as the perturbation of the parameter characterizing the impatient customers in the $M/G/c/k$ queueing systems with impatient customers (modeling the Cloud Data Center (Outamazirt *et al.* 2018)).

The application to general queueing networks might be challenging because of the high dimension of the embedded Markov chain and the complex dynamics of those models. Additionally, perturbations in one node result in the loss of the Markov property in many other (connected) nodes or in the whole network. However, particular networks (e.g. tandem networks) can be studied.

The theory of strong stability is rich in results. Conditions for the stability and quantitative estimates might be obtained with various techniques. For instance, one might use generalized inverses, ergodicity coefficients, eigenvalues, etc. Applying other strong stability results to queues is essential in order to further sharpen the perturbation bounds.

Finally, comparison with other methods demonstrates the usefulness of the theory (see, e.g., Abbas and Heidergott (2010)).

9.5. References

Abbas, K., Aïssani, D. (2010a). Structural perturbation analysis of a single server queue with breakdowns. *Stoch. Models.*, 26(1), 78–97.

Abbas, K., Aïssani, D. (2010b). Strong stability of the embedded Markov chain in an $GI/M/1$ queue with negative customers. *Appl. Math. Modell.*, 34(10), 2806–2812.

Abbas, K., Aïssani, D. (2010c). Approximation of performance measures in an $M/G/1$ queue with breakdowns. *Quality Technology and Quantitative Management*, 7(4), 353–363.

Abbas, K., Heidergott, B., Aïssani, D. (2010). A Taylor series approach to the numerical analysis of the $M/D/1/N$ queue. *Procedia Comput. Sci.*, 1(1), 1547–1554.

Adel-Aissanou, K., Abbas, K., Aïssani, D. (2012). Strong truncation approximation in tandem queues with blocking. *Math. Prob. Eng.*, 2012, 906486.

Aïssani, D. (1982). Estimate of the strong stability in an $M/G/1$ system. *VINITI N 4119–82, R. Journal Matematika*, 83, 1–33.

Aïssani, D. (1987a). Estimation quantitative de la norme de déviation de l'opérateur de transition d'un système $M/G/1$. *Cahiers Mathématiques*, 2, 125–133.

Aïssani, D. (1987b). Sur les conditions d'approximation d'un système $G/G/1$ par un système $M/G/1$. *Journal of Technology*, 4, 96–107.

Aïssani, D. (1990). Ergodicité uniforme et stabilité forte des chaînes de Markov: application aux systèmes de files d'attente. *Séminaire Math. Rouen*, 167, 115–121.

Aïssani, D. (1991). Application of the operator methods to obtain inequalities of stability in the $M2/G2/1$ system with relative priority. *Annales Maghrèbines de l'Ingénieur*, 2, 701–709.

Aïssani, D. (1992a). Stabilité forte de la chaîne de Markov incluse dans un système $G/M/\infty$. *Technologies Avancées*, 2(1), 28–33.

Aïssani, D. (1992b). Estimate of the strong stability in the system $G/M/\infty$. *Technologies Avancées*, 2(2), 29–33.

Aïssani, D., Kartashov, N.V. (1983a). Ergodicity and stability of Markov chains with respect to operator topology in the space of transition kernels. *Doklady Akademii Nauk Ukrainskoi SSR*, A11, 3–5.

Aïssani, D., Kartashov, N.V. (1983b). Strong stability of an imbedded Markov chain in the $M/G/1$ queueing system, *Theor. Probab. Math. Statist.*, 29(1984), 1–5.

Asmussen, S. (2003). *Applied Probability and Queues*. Springer, New York.

Anisimov, V. (1988). Estimates for the deviations of the transition characteristics of nonhomogeneous Markov processes. *Ukrainian Math. J.*, 40(6), 588–592.

Bareche, A., Aïssani, D. (2008). Kernel density in the study of the strong stability of the $M/M/1$ queueing system. *Oper. Res. Lett.*, 36(5), 535–538.

Bareche, A., Aïssani, D. (2011). Statistical techniques for numerical evaluation of the proximity of $G/G/1$ and $G/M/1$ queueing systems. *Comput. Math. Appl.*, 61(5), 1296–1304.

Bareche, A., Aïssani, D. (2014a). Interest of boundary kernel density techniques in evaluating an approximation error of queueing systems characteristics. *Int. J. Math. Math. Sci.*, 2014, 871357.

Bareche, A., Aïssani, D., (2014b). Statistical Methodology for Approximating $G/G/1$ Queues by the Strong Stability Technique. *Proceedings of the 3rd International Conference on Operations Research and Enterprise Systems*, 241–248.

Bareche, A., Cherfaoui, M., Aïssani, D. (2016). Approximate analysis of an $GI/M/\infty$ queue using the strong stability method. *IFAC-PapersOnLine*, 49(12), 863–868.

Benaouicha, M., Aïssani, D. (2005). Strong stability in a $G/M/1$ queueing system. *Theory Probab. Math. Statist.*, 71, 25–36.

Berdjoudj, L., Aïssani, D. (2003). Strong stability in retrial queues. *Theory Probab. Math. Statist.*, 68, 11–17.

Berdjoudj, L., Benaouicha, M., Aïssani, D. (2012). Measure of performance of the strong stability method. *Math. Comput. Modell.*, 56(11–12), 241–246.

Borovkov, A.A. (ed.). (1984). *Asymptotic Methods in Queueing Theory*. Wiley, New York.

Bouallouche, L., Aïssani, D. (2006a). Measurement and performance of the strong stability method. *Theory Probab. Math. Statist.*, 72, 1–9.

Bouallouche, L., Aïssani, D. (2006b). Performance analysis approximation in a queueing system of Type $M/G/1$. *Math. Method. Oper. Res.*, 63(2), 341–356.

Bouallouche, L., Aïssani, D. (2008). Quantitative estimates in an $M2/G2/1$ queue with non-preemptive priority customers: the method of strong stability. *Stoch. Models*, 24(4), 626–646.

Boukir, L., Bouallouche-Medjkoune, L., Aïssani, D. (2009). Strong stability of the batch arrival queueing systems. *Stoch. Anal. Appl.*, 28(1), 8–25.

Cao, X.R. (1998). The Maclaurin series for performance functions of Markov chains. *Adv. Appl. Probab.*, 30(3), 676–692.

Djabali, Y., Rabta, B., and Aïssani, D. (2015). Strong stability of $PH/M/1$ queues. Unpublished.

Djabali, Y., Rabta, B., and Aïssani, D. (2018). Approximating service-time distributions by phase-type distributions in single-server queues: A strong stability approach. *International Journal of Mathematics and Operation Research*, 12(4), 507–531.

Franken, P. (1970). Ein stetigkeitssatz für Verlustsusteme. *Operations-forschung Math Stat.*, 11, 1–23.

Gnedenko, B.V. (1970). Sur certains problèmes non résolus de la théorie de files d'attente. *Six International Telegraphic Congress*, Munich.

Hamadouche, N., Aïssani, D. (2011). Approximation in the $M_2/G_2/1$ queue with preemptive priority. *Methodol. Comput. Appl.*, 13(3), 563–581.

Heidergott, B., Hordijk, A. (2003). Taylor series expansions for stationary Markov chains. *Adv. Appl. Probab.*, 35(4), 1046–1070.

Issaadi, B., Abbas, K., Aïssani, D. (2016). Perturbation analysis of the $GI/M/s$ queue. *Methodol. Comput. Appl.*, 19(3), 819–841.

Jackson, J.R. (1957). Networks of waiting lines. *Oper. Res.*, 5, 518–521.

Kalashnikov, V.V., Tsitsiachvili, G.C. (1972). Sur la stabilité des systèmes de files d'attente relativement à leurs fonctions de répartition perturbées. *J. Izv AN USSR Technique Cybernétique*, (2), 41–49.

Kalashnikov, V.V. (1978). *Qualitative Analysis of Behavior of Complex Systems Using the Method of Test Functions*. Nauka, Moscow (in Russian)

Kartashov, N.V. (1981). Strong stability of Markov chains. *VNISSI, Vsesayouzni Seminar on Stability Problems for Stochastic Models*, Moscow, 54–59.

Kartashov, N.V. (1985). Criteria for ergodicity and stability for Markov chains with common phase space. *Theory Probab. Math. Statist.*, 30, 71–89.

Kartashov, N.V. (1986a). Inequalities in theorems of ergodicity and stability for Markov chains with common phase space I. *Theor. Probab. Appl.*, 30(2), 247–259.

Kartashov, N.V. (1986a). Inequalities in theorems of ergodicity and stability for Markov chains with common phase space II. *Theor. Probab. Appl.*, 30(3), 507–515.

Kartashov, N.V. (1986c). Strong stability of Markov chains. *J. Sov. Math.*, 34(2), 1493–1498.

Kartashov, N.V. (1996). *Strong Stable Markov Chains*. VSP, Utrecht.

Kennedy, D. (1972). The continuity of the single server queue. *J. Appl. Probab.*, 9(3), 370–381.

Kleinrock, L. (1975). *Queueing Systems. Volume 1: Theory*. John Wiley & Sons, New York.

Lakaour, L., Aïssani, D., Adel, K., Barkaoui, K. (2018). $M/M/1$ retrial queue with collisions and transmission errors. *Methodol. Comput. Appl.*. https://doi.org/10.1007/s11009-018-9680-x

Latouche, G., Ramaswami, V. (1999). *Introduction to Matrix Analytic Methods in Stochastic Modelling*. ASA-SIAM, Philadelphia.

Lekadir, O., Aïssani, D. (2008a). Stability of two-stage queues with blocking. In *Modelling, Computation and Optimization in Information Systems and Management Sciences. MCO 2008*. Le Thi, H.A., Bouvry, P., Pham Dinh, T. (eds). Springer, Berlin, 526–535.

Lekadir, O., Aïssani, D. (2008b). Strong stability in a Jackson queueing network. *Theory Probab. Math. Statist.*, 77, 107–119.

Lekadir, O., Aïssani, D. (2011). Error bound on practical approximations for two tandem queues with non-preemptive priority. *Comput. Math. Appl.*, 61(7), 1810–1822.

Mouhoubi, Z., Aïssani, D. (2014). Uniform ergodicity and strong stability of homogeneous Markov chains. *J. Math. Sci.*, 200(4), 452–461.

Outamazirt, A., Barkaoui, K., Aïssani, D. (2018). Maximizing profit in cloud computing using $M/G/c/k$ queuing model. *13th International Symposium on Programming and System*, Zeralda, Algiers.

Rabta, B., Aïssani, D. (2008). Strong stability and perturbation bounds for discrete Markov chains. *Linear Algebra Appl.*, 428(8–9), 1921–1927.

Rabta, B., Aïssani, D. (2018). Perturbation bounds for Markov chains with general states space. *J. Math. Sci.*, 228(5), 510–521.

Rachev, S.T. (1989). The problem of stability in queueing theory. *Queueing Syst.*, 4(4), 287–318.

Rahmoune, F., Aïssani, D. (2008). Strong stability of queues with multiple vacations of the server. *Stoch. Anal. Appl.*, 26(3), 665–678.

Rahmoune, F., Aïssani, D. (2014). Quantitative stability estimates in queues with server vacation. *J. Math. Sci.*, 200(4), 480–485.

Rossberg, H.J. (1965). Über die Verteilung von Wartezeiten. *Math. Nachr.*, 20(1/2), 1–16.

Stoyan, D. (1977). Ein stetigkeitssatz für einlinige Wartemodelle der Bedienungstheorie. *Math. Operationsforsch. Stat.*, 3(2), 103–111.

Stoyan, D. (1984). *Comparison Methods for Queues and Other Stochastic Models.* Wiley, New York.

Zolotariev, V.M. (1975). On the continuity of stochastic sequences generated by recurrence procedures. *Theor. Probab. Appl.*, 20(4), 834–847.

10

Time-varying Queues: a Two-time-scale Approach

George YIN[1], Hanqin ZHANG[2] and Qing ZHANG[3]
[1] *Wayne State University, Detroit, USA*
[2] *School of Business, National University of Singapore, Singapore*
[3] *University of Georgia, Athens, USA*

This chapter is concerned with time-varying queues with customer abandonment and finite waiting rooms. A two-time-scale Markov chain approach is adopted to study the performance measures of the system including the queue length, the probabilities of customer reneging and the abandonment for such time-varying queues. Both moderate deviations and large deviations are considered when time scale separation is in effect. In the study of large deviations, the calculation of rate function is crucial. In this chapter, general expression of the rate function is provided. At the same time, a numerical example is also included to illustrate a related rate function calculation.

10.1. Introduction

Many queueing systems, that arise in applications are time varying in nature. These queueing systems include service systems such as customer contact centers, patient

This research was supported in part by the Army Research Office under grant W911NF-19-1-0176, the Social Science Research Council (Singapore), MOE2016-SSRTG-059 and the Simons Foundation under 235179.

Queueing Theory 1,
coordinated by Vladimir ANISIMOV, Nikolaos LIMNIOS. © ISTE Editions 2020.

flows in emergency departments and ward management in hospitals. In these systems, either customer arrival rates, service rates or hazard rates of customer abandonment are time dependent. There are several empirical studies that have been conducted on such time inhomogeneity (see, for instance, Armony *et al.* (2015); Batt and Terwiesch (2015); Yom-Tov and Mandelbaum (2014), among others).

By and large, there are a few ways to treat these time-varying queues. One way is to apply time-invariant (or time-homogeneous or stationary) queues to approximate the time-inhomogeneous queues. When building the models for the time-stationary queues, usually we can establish three different stationary models based on the various managerial motivation. One of the models is to assume that the customer arrival rate and service rate are time dependent but change sufficiently slowly. Thus, the system performance measures such as queue length and waiting time at running time t can be approximated by that of the steady-state performance measures of the time stationary queue with the customer arrival rate and service rate being fixed at time t. This approximation is called a pointwise-stationary approximation in the queueing literature (see Green and Kolesar (1991); Whitt (1991)). Another two time-stationary models are constructed by considering the quality of service at all times. For example, to analyze the time-varying queue over time interval $[0, T]$, we use either the maximum of the customer arrival rates over $[0, T]$ or the average of the customer arrival rates over $[0, T]$ instead of time-varying customer arrival rates to construct the time-stationary queue, as it is referred to in queueing literature. Apart from applying time-stationary queues, another way to study the time-varying queues is to use the many-server queue idea. In this setting, the customer arrival rate and the number of servers are assumed to be large. Then the system performance measures can be approximated by a Gaussian process or a function of a Gaussian process (see Liu (2018); Liu and Whitt (2014)). A survey on the vast literature on time-varying queues has been provided by Whitt (2018).

In contrast with traditional approaches, we develop a two-time-scale Markov chain approach to analyze the time-varying queues in this chapter. Treating time-inhomogeneous Markov chains with two-time scales, Yin and Zhang (2013) developed the approximation schemes based on time-scale separation. The central theme is the use of the so-called quasi-stationary (or quasi-steady state) distributions (see Chapter 2 of Yin and Zhang (2013) for a definition). A quasi-steady-state distribution is roughly like a steady-state distribution. However, the distribution can be time dependent. The time dependence arises from the two-time-scale formulation, in which one has two time scales. One of them is real running time t, and the other is a "stretched time" t/ε. Holding t as a "constant" and letting $\varepsilon \to 0$ yields a limit, which is t dependent. For further details, we refer the reader to Yin and Zhang (2013) (see also Yin *et al.* (2013) and references therein). Such an approximation result is rather convenient for us to carry out the study in the time-varying queues. Here, we use the phrase "quasi" to reflect the time dependence involved. We model the queue length process as a continuous-time Markov chain. When the customer arrival rate and the number of servers get larger and larger, the two-time-scale feature in the Markov chain emerges. Using a small parameter $\varepsilon > 0$ to highlight the time scale separation, the aforementioned feature is pronounced when ε is sufficiently small. One of the central issues in the approximation of the time-varying queues with two-time scales is to closely characterize the approximation by looking at the deviation of the original occupation measures from that of the quasi-steady-state measures. A closed caption of this property can be done by using large deviations analysis. A related work on diffusion approximation can be found in the work of Anisimov (1995).

Because of the time-scale separation, within a short period of time, the queueing system will reach its quasi-steady state. One important question concerns the probability of the rates of convergence to the quasi-steady distribution. The answer to such a question is beyond the usual normal deviations range of the errors. A pertinent analysis is based on larger than normal deviations. In this chapter, we use the large deviations principal for two-time-scale non-homogeneous Markov chains to analyze the queue length process and associated probabilities, which characterizes the quality of service and the efficiency of the systems. In addition, in this chapter, we also characterize the quality of approximation by examining deviations that are larger than the usual normal deviations but are not quite in the large deviation range. This is the study of moderate deviations.

The rest of the chapter is organized as follows. In section 10.2, we give a mathematical description for our time-varying queues. Section 10.3 demonstrates how to use the large deviation principal to analyze the time-varying queues. The chapter concludes in section 10.4 with some further remarks.

10.2. Time-varying queues

We consider the $M_t/M_t/s_t + M_t/N$ model, in which the customer arrival process is a non-stationary Poisson process with rate $\lambda(t)$; the system has s_t identical servers, and their service rate is $\mu(t)$ at time t. The arriving customers are impatient and their patient times are independent and follow exponential distribution with parameter $\gamma(t)$ at time t. Here, N is the system capacity, that is, the system is a loss system and arriving customers will be forced to balk when the system reaches its capacity, that is, when there are N customers in the system. Let $Q(t)$ represent the number of customers in the system at time t. By the memoryless property of the customer arrival process, service time and patience time, we know that $\{Q(t) : t \geq 0\}$ is a Markov process with the state space $\{0, 1, \dots, N\}$. At time t, the system service completion rate is

$$\mu(t) \cdot \min\{s_t, Q(t)\},$$

and the customer abandonment rate is

$$\gamma(t) \cdot (Q(t) - s_t)^+.$$

Based on the above observation, we know that the generator $\mathcal{G}(t)$ of the queue length process $\{Q(t) : t \geq 0\}$ is given by

$$\mathcal{G}(t) = \begin{pmatrix} -q_0(t) & \lambda(t) & & & \\ d_1(t) & -q_1(t) & \lambda(t) & & \\ & \ddots & \ddots & \ddots & \\ & & d_{N-1}(t) & -q_{N-1}(t) & \lambda(t) \\ & & & d_N(t) & -q_N(t) \end{pmatrix}, \qquad [10.1]$$

where

$$q_i(t) = \begin{cases} \lambda(t) + i \cdot \mu(t), & \text{if } 0 \leq i \leq s_t, \\ \lambda(t) + s_t \cdot \mu(t) + (i - s_t) \cdot \gamma(t), & \text{if } s_t + 1 \leq i \leq N; \end{cases}$$

$$d_i(t) = \begin{cases} i \cdot \mu(t), & \text{if } 1 \leq i \leq s_t, \\ s_t \cdot \mu(t) + (i - s_t) \cdot \gamma(t), & \text{if } s_t + 1 \leq i \leq N. \end{cases}$$

The above formulation regarding the number of servers at time t in the system is very general and can capture many interesting features from practical considerations. Under such a setup, we can consider the following special situations.

PROBLEM 10.1.– First, we consider a dynamic staffing problem. In particular, there are k thresholds related to the number of customers in the system to dynamically adjust the staffing levels. Namely, for given k thresholds

$$0 = n_0 < n_1 < n_2 < \cdots < n_k = N,$$

and k staffing levels

$$0 \leq s_1 < s_2 < \cdots < s_k,$$

when the number of customers in the system at time t is strictly larger than n_ℓ but less than or equal to $n_{\ell+1}$ with $\ell = 0, \ldots, k-1$, we use $s_{\ell+1}$ servers to handle the customer service. With this staffing policy, we have that

$$q_0(t) = \lambda(t),$$

and for $i = n_\ell + 1, \ldots, n_{\ell+1}$ with $0 \le \ell \le k-1$,

$$q_i(t) = \lambda(t) + \mu(t) \cdot \min\{s_\ell, i\} + (i - s_\ell)^+ \cdot \gamma(t),$$
$$d_i(t) = \mu(t) \cdot \min\{s_\ell, i\} + (i - s_\ell)^+ \cdot \gamma(t).$$

To evaluate the efficiency of the staffing policy for this problem, we need to analyze the probability distribution for $t \ge 0$,

$$n_{s_1}(t) = \int_0^t \sum_{j=0}^{n_1} \mathcal{I}_{\{Q(y)=j\}} \mathrm{d}y, \qquad [10.2]$$

$$n_{s_i}(t) = \int_0^t \sum_{j=n_{i-1}+1}^{n_i} \mathcal{I}_{\{Q(y)=j\}} \mathrm{d}y, \quad i = 2, \ldots, k. \qquad [10.3]$$

$n_{s_i}(t)$ indicates the cumulative time that the system uses s_i servers to handle customer service during time interval $[0, t]$.

PROBLEM 10.2.– Next, consider a time-dependent staffing policy. Formally, the time interval $[0, \infty)$ is divided into a few subintervals,

$$[t_0, t_1), [t_1, t_2), \ldots, [t_{k-1}, t_k), \quad t_0 = 0 \text{ and } t_k = \infty,$$

and the staffing level at interval $[t_\ell, t_{\ell+1})$ is given by $s = s_{\ell+1}$ for $\ell = 0, \ldots, k-1$. Thus, for $t \in [t_{\ell-1}, t_\ell)$ with $\ell = 1, \ldots, k$,

$$q_i(t) = \lambda(t) + \mu(t) \cdot \min\{s_\ell, i\} + (i - s_\ell)^+ \cdot \gamma(t),$$
$$d_i(t) = \mu(t) \cdot \min\{s_\ell, i\} + (i - s_\ell)^+ \cdot \gamma(t).$$

Similarly, to analyze the customer loss probability, we introduce

$$L(t) = \int_0^t \mathcal{I}_{\{Q(y)=N\}} \mathrm{d}y, \quad t \ge 0. \qquad [10.4]$$

PROBLEM 10.3.– Finally, we consider the staffing problem based on the ratio of the customer arrival rate to service rate. At time t, the staffing level s_t is chosen such that $\frac{\lambda(t)}{s_t \cdot \mu(t)} \le \rho$ holds with the smallest positive integer s_t. Here, ρ is determined in advance; usually it shows us the system service quality. Thus, for any time t, with the customer arrival rate $\lambda(t)$ and each individual server's service rate $\mu(t)$, we have

$$s_t = \left\lceil \frac{\lambda(t)}{\rho \cdot \mu(t)} \right\rceil.$$

Let

$$s_m(t) = \sup_{0\le x\le t} s_x = \sup_{0\le x\le t} \left\lceil \frac{\lambda(x)}{\rho\cdot\mu(x)} \right\rceil.$$

We consider the probability distribution

$$M(t) = \int_0^t \mathcal{I}_{\{Q(y)=s_m(y)\}} \mathrm{d}y, \quad t \ge 0. \qquad [10.5]$$

This shows the cumulative time that the system takes to handle the customer service during time interval $[0,t]$ using the maximum number of servers.

With the setting of the three problems given above, in what follows, we develop a two-time-scale approach to analyze the probability distributions of $n_{s_i}(t)$, $M(t)$, and $L(t)$. A crucial point is to let the generator of the queue length process depend on a small parameter. As the small parameter goes to 0, the system may be replaced by an average one.

10.3. Main results

We divide this section into several subsections. We first consider large deviations of the two-time-scale queues and then provide some discussion on a relevant computational method. This is followed by applications in various queues. Additional considerations on moderate deviations will also be discussed.

10.3.1. *Large deviations of two-time-scale queues*

Consider a general finite-state continuous-time Markov chain $\{X^\varepsilon(t) : t \ge 0\}$ depending on a small parameter $\varepsilon > 0$ with a generator $\mathcal{G}^\varepsilon(t)$ such that

$$\mathcal{G}^\varepsilon(t) = \frac{1}{\varepsilon}\mathcal{G}(t), \qquad [10.6]$$

where $\mathcal{G}(t)$ is given by [10.1], is irreducible and is twice continuously differentiable with respect to t. Note that although [10.6] only has a fast varying part, the system is still a two-time scale queue. As alluded to previously, there is natural time t and also stretched time t/ε. Let $\mathcal{S} = \{0, 1, \ldots, N\}$ denote its state space, and

$$f(\cdot) = (f_1(\cdot), \ldots, f_k(\cdot)) : \mathcal{S} \longrightarrow \mathbb{R}^k. \qquad [10.7]$$

For a fixed $T > 0$, consider a k-dimensional random variable

$$\alpha^{\varepsilon}(T) = \int_0^T f(X^{\varepsilon}(t))\mathrm{d}t. \tag{10.8}$$

Then using the occupation measure, it is readily seen that

$$\begin{aligned}\alpha^{\varepsilon}(T) &= \int_0^T \sum_{i=0}^{N} f(i)\mathcal{I}_{\{X^{\varepsilon}(t)=i\}}\mathrm{d}t \\ \to \alpha(T) &= \int_0^T \sum_{i=0}^{N} f(i)\nu_i(t)\mathrm{d}t \\ &\text{in probability as } \varepsilon \to 0,\end{aligned}$$

where $\nu(t) = (\nu_0(t), \ldots, \nu_N(t))$ is the quasi-stationary distribution. Note that the random variable $\alpha^{\varepsilon}(T)$ defined above is used to capture the performance measures in problems 10.1–10.3.

For example, consider problem 10.1. Let $k = 1$ in [10.8] and

$$f_1(\ell) = \begin{cases} 1, \text{ if } \ell = 0, \ldots, n_1, \\ 0, \text{ otherwise.} \end{cases} \tag{10.9}$$

Then we have [10.2] with $t = T$ in problem 10.1. That is,

$$n^{\varepsilon}_{s_1}(T) = \alpha^{\varepsilon}(T). \tag{10.10}$$

In general, consider $n^{\varepsilon}_{s_i}(T)$ for $i = 2, \ldots, k$. Let $k = 1$ and

$$f_1(\ell) = \begin{cases} 1, \text{ if } \ell = n_{i-1}+1, \ldots, n_i, \\ 0, \text{ otherwise.} \end{cases} \tag{10.11}$$

Then we have [10.3] with $t = T$ in problem 10.1. That is,

$$n^{\varepsilon}_{s_i}(T) = \alpha^{\varepsilon}(T). \tag{10.12}$$

Consider problem 10.2. Let $k = 1$ in [10.8] and

$$f_1(\ell) = \begin{cases} 1, \text{ if } \ell = N, \\ 0, \text{ otherwise.} \end{cases} \tag{10.13}$$

We have that $L^{\varepsilon}(T) = \alpha^{\varepsilon}(T)$ in [10.4].

Similarly for problem 10.3, let

$$I(t) = \Big(I_{jk}(t)\Big)_{(N+1)\times(N+1)} \tag{10.14}$$

with

$$I_{jk}(t) = \begin{cases} 0, \text{ if } j \neq k, \\ 1, \text{ if } j = k \text{ and } \left\lceil \frac{\lambda(t)}{\rho\cdot\mu(t)} \right\rceil = \sup_{0\le x\le t} \left\lceil \frac{\lambda(x)}{\rho\cdot\mu(x)} \right\rceil, \\ 0, \text{ if } j = k \text{ but } \left\lceil \frac{\lambda(t)}{\rho\cdot\mu(t)} \right\rceil < \sup_{0\le x\le t} \left\lceil \frac{\lambda(x)}{\rho\cdot\mu(x)} \right\rceil. \end{cases}$$

Furthermore, let $k = N+1$ in [10.8] and for $\ell \in \mathcal{S}$,

$$(f_1(\ell,t),\ldots,f_N(\ell,t)) = \Big(\mathcal{I}_{\{\ell=0\}},\ldots,\mathcal{I}_{\{\ell=N\}}\Big) \times I(t). \tag{10.15}$$

Then $M^{\varepsilon}(T) = \alpha^{\varepsilon}(T)$ in [10.5]. Here, we note that $f(\cdot)$ given by [10.7] has been modified to be dependent on t.

Now we give a systematic way to estimate the probability distribution of random variable $\alpha^{\varepsilon}(T)$ given in [10.8]. From theorem 4.2 in He *et al.* (2011), we have the following result.

THEOREM 10.1.– For any $B \subseteq \mathbb{R}^k$, let B^o and $\bar{B}$ be the interior and closure of B in $\mathbb{R}^k$, respectively. Then

$$-\inf_{x\in B^o} H(x) \le \liminf_{\varepsilon\to 0} \varepsilon \log \mathsf{Pr}\Big(\alpha^{\varepsilon}(T) \in B\Big),$$

$$-\inf_{x\in \bar{B}} H(x) \ge \limsup_{\varepsilon\to 0} \varepsilon \log \mathsf{Pr}\Big(\alpha^{\varepsilon}(T) \in B\Big),$$

where

$$H(x) = \sup_{\gamma\in\mathbb{R}^k} \Big\{\sum_{i=1}^{k} x_i\gamma_i - \int_0^T H(\gamma,t)\mathrm{d}t\Big\}$$

with for $y \in \mathbb{R}^k$ and $t \ge 0$,

$$H(y,t) = \lim_{\delta\to 0} \frac{1}{\delta}\Big(\lim_{\varepsilon\to 0} \varepsilon \log \mathsf{E}\Big[\exp\Big(\frac{1}{\varepsilon}\int_t^{t+\delta} \sum_{i=1}^{k} y_i f_i(X^{\varepsilon}(v))\mathrm{d}v\Big)\Big| X^{\varepsilon}(0) = i\Big]\Big).$$

According to theorem 10.1, to get the probabilistic estimate of $\alpha^\varepsilon(T)$ in [10.8], we need to find the function $H(\cdot)$. Although the analytic form of $H(\cdot)$ is available and given above, the calculation in general is not straightforward. Nevertheless, in what follows, we demonstrate how to compute $H(\cdot)$ in certain cases.

10.3.2. *Computation of* $H(y,t)$

In this section, we consider a special case of $\mathcal{G}(t)$ and demonstrate how to compute $H(y,t)$. We consider $\{X^\varepsilon(t) : t \geq 0\}$ as a birth–death process with the state space $\mathcal{S}$ and the generator $\mathcal{G}(t)/\varepsilon$ with $\mathcal{G}(t)$ given by [10.1].

For $i = 0, 1, \ldots, N$, let

$$M(t,i) = \mathsf{E}\left[\exp \frac{1}{\varepsilon}\int_0^t g(X^\varepsilon(u))\mathrm{d}u \middle| X^\varepsilon(0) = i\right],$$

where $g(x) = \sum_{j=1}^k y_j f_j(x)$ and $(f_1(x), \ldots, f_k(x))$ is given in [10.7].

If $\mathcal{G}(t)$ is a piecewise constant matrix, then we can assume it is right continuous with left-hand limit. Otherwise, we can modify it by taking its left-hand limit and use $\mathcal{G}_m(t)$ instead

$$\mathcal{G}_m(t) = \lim_{\Delta t \to 0+} \mathcal{G}(t + \Delta t).$$

In order to evaluate the function $H(y,t)$, we only have to examine the case with constant $\mathcal{G} = \mathcal{G}(t)$. We focus on constant $\mathcal{G}$ in this section. Let $M(t) = (M(t,0), M(t,1), \ldots, M(t,N))'$ and let $A = \text{diag}g(0), g(1), \ldots, g(N) + \mathcal{G}$. Then, the Feynman–Kac formula yields the differential equation

$$\frac{\mathrm{d}M(t)}{\mathrm{d}t} = \frac{1}{\varepsilon}AM(t) \text{ with } M(0) = \mathbf{1} = (1, \ldots, 1)'.$$

Its solution is given by

$$M(t) = \left[\exp A\left(\frac{t}{\varepsilon}\right)\right]\mathbf{1}.$$

EXAMPLE 10.1.– To illustrate a possible way to compute $H(y,t)$, we specialize further and consider a 3×3 constant matrix $\mathcal{G} = \mathcal{G}(t)$ in which the matrix A admits three distinct real eigenvalues $\eta_0 > \eta_1 > \eta_2$. Then there exists an invertible matrix S such that

$$A = S \cdot \operatorname{diag}(\eta_0, \eta_1, \eta_2) \cdot S^{-1}.$$

It follows that

$$M(t) = S \cdot \operatorname{diag}\left(e^{\eta_0 t/\varepsilon}, e^{\eta_1 t/\varepsilon}, e^{\eta_2 t/\varepsilon}\right) \cdot S^{-1}\mathbf{1}.$$

Then we can write it as

$$M(t) = \begin{pmatrix} c_{00} \\ c_{10} \\ c_{20} \end{pmatrix} e^{\eta_0 t/\varepsilon} + \begin{pmatrix} c_{01} \\ c_{11} \\ c_{21} \end{pmatrix} e^{\eta_1 t/\varepsilon} + \begin{pmatrix} c_{02} \\ c_{12} \\ c_{22} \end{pmatrix} e^{\eta_2 t/\varepsilon}$$

for some constants c_{ij}. For each $i = 0, 1, 2$, let $j_0(i) = \min\{j : c_{ij} \neq 0\}$. Then it is easy to see

$$H_i(y,t) = \lim_{\delta \to 0} \lim_{\varepsilon \to 0} \varepsilon \log M(\delta, i) = \eta_{j_0(i)}.$$

Note that $H(y,t)$ depends only on y when $\mathcal{G}$ is a constant matrix. When $\mathcal{G}$ changes with t, so does $H(y,t)$.

In particular, we let

$$\mathcal{G} = \begin{pmatrix} -2 & 2 & 0 \\ 2 & -3 & 1 \\ 0 & 3 & -3 \end{pmatrix}.$$

Then, the matrix satisfies the eigenvalue condition. The limit $H(y,t)$ is given in Figure 10.1.

To proceed, we come back to the three problems discussed briefly in the last section.

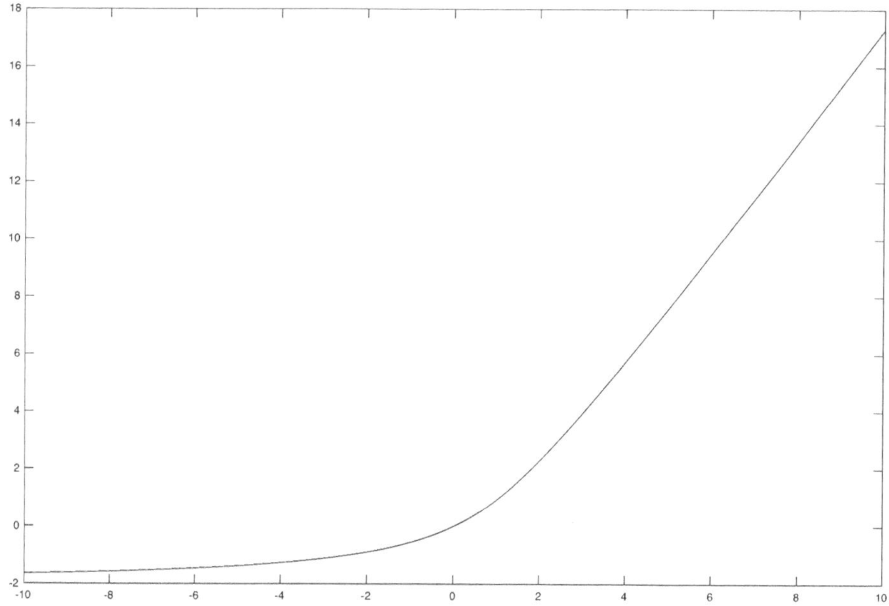

Figure 10.1. *Function* $H(y) = H(y,t)$

10.3.3. *Applications to queueing systems*

In many-server queues, the customer arrival rate is usually assumed to be very high, and the staffing level is also high regardless of which regime it takes among efficiency-driven, quality-driven, and both quality- and efficiency-driven regimes. Consider $M_t/M_t/s_t + M_t/N$ given by [10.1]. At time t, if the system is under the efficiency-driven regime, we have

$$s_t \approx \frac{\lambda(t)}{\mu(t)}(1-\theta) \quad \text{with } \theta \in (0,1), \tag{10.16}$$

while under the quality-driven regime,

$$s_t \approx \frac{\lambda(t)}{\mu(t)}(1+\eta) \quad \text{with } \eta \in (0,1), \tag{10.17}$$

and under the quality- and efficiency-driven regime,

$$s_t \approx \frac{\lambda(t)}{\mu(t)} + \vartheta\sqrt{\frac{\lambda(t)}{\mu(t)}} \quad \text{with } \vartheta \in (-\infty,\infty). \tag{10.18}$$

For a given time interval $[0, T]$, let

$$\bar{\lambda} = \sup_{0 \le t \le T} \lambda(t).$$

Under the setting of many-server queues, we assume that

$$\bar{\lambda} \gg 1.$$

This assumption makes us consider $1/\bar{\lambda}$ to be ε in theorem 10.1. The generator given by [10.1] can then be written as

$$\frac{1}{1/\bar{\lambda}} \begin{pmatrix} -\frac{q_0(t)}{\bar{\lambda}} & \frac{\lambda(t)}{\bar{\lambda}} & & & \\ \frac{d_1(t)}{\bar{\lambda}} & -\frac{q_1(t)}{\bar{\lambda}} & \frac{\lambda(t)}{\bar{\lambda}} & & \\ & \ddots & \ddots & \ddots & \\ & & \frac{d_{N-1}(t)}{\bar{\lambda}} & -\frac{q_{N-1}(t)}{\bar{\lambda}} & \frac{\lambda(t)}{\bar{\lambda}} \\ & & & \frac{d_N(t)}{\bar{\lambda}} & -\frac{q_N(t)}{\bar{\lambda}} \end{pmatrix}. \qquad [10.19]$$

Now we look at the cumulative time over $[0, T]$ in which the arriving customers get balking for problems 10.1–10.3. From [10.4], this is given by $L(T)$. In view of [10.13] and theorem 10.1, we have that for any $B \subset \mathbb{R}$ and large enough $\bar{\lambda}$,

$$\exp\Big(-\bar{\lambda} \cdot \inf_{v \in B^o} H_b(v)\Big) \le \mathsf{Pr}\Big(L(T) \in B\Big) \le \exp\Big(-\bar{\lambda} \cdot \inf_{v \in \bar{B}} H_b(v)\Big), \qquad [10.20]$$

where

$$H_b(x) = \sup_{\gamma \in \mathbb{R}} \Big\{x\gamma - \int_0^T H_b(\gamma, t) \mathrm{d}t\Big\},$$

$$H_b(\gamma, t) = \lim_{\delta \to 0} \frac{1}{\delta}\Big(\lim_{\bar{\lambda} \to \infty} \frac{1}{\bar{\lambda}} \log \mathsf{E}\Big[\exp\Big(\bar{\lambda} \int_t^{t+\delta} \gamma \cdot \mathcal{I}_{\{Q(v)=N\}} \mathrm{d}v\Big) \Big| Q(0) = i\Big]\Big).$$

Now for problem 10.1, we have that for any $B \subset \mathbb{R}$ and large enough $\bar{\lambda}$,

$$\exp\Big(-\bar{\lambda} \cdot \inf_{v \in B^o} H_{s_i}(v)\Big) \le \mathsf{Pr}\Big(n_{s_i}(T) \in B\Big) \le \exp\Big(-\bar{\lambda} \cdot \inf_{v \in \bar{B}} H_{s_i}(v)\Big), \quad [10.21]$$

where

$$H_{s_i}(x) = \sup_{\gamma\in\mathbb{R}} \Big\{x\gamma - \int_0^T H_{s_i}(\gamma, t)\mathrm{d}t\Big\},$$

$$H_{s_i}(\gamma, t) = \lim_{\delta\to 0}\frac{1}{\delta}\Big(\lim_{\bar{\lambda}\to\infty}\frac{1}{\bar{\lambda}}\log \mathsf{E}\Big[\exp\Big(\bar{\lambda}\int_t^{t+\delta}\gamma\cdot f_1(Q(v))\mathrm{d}v\Big)\Big|Q(0)=i\Big]\Big).$$

Here, $f_1(\cdot)$ is defined in [10.9] and [10.11] for $i = 1$ and $i = 2, \ldots, k$, respectively. Now consider its queue length given by

$$\int_0^T \sum_{i=1}^{k}\sum_{j=n_{i-1}+1}^{n_i}(j - s_i)^+ \cdot \mathcal{I}_{\{Q(v)=j\}}\mathrm{d}v. \qquad [10.22]$$

For any $B \subset \mathbb{R}$ and large enough $\bar{\lambda}$,

$$\begin{aligned}\exp\Big(-\bar{\lambda}\cdot\inf_{v\in B^o} H_q(v)\Big) &\le \Pr\Big(\int_0^T\sum_{i=1}^{k}\sum_{j=n_{i-1}+1}^{n_i}(j - s_i)^+\cdot\mathcal{I}_{\{Q(v)=j\}}\mathrm{d}v \in B\Big)\\ &\le \exp\Big(-\bar{\lambda}\cdot\inf_{v\in\bar{B}} H_q(v)\Big),\end{aligned} \qquad [10.23]$$

where

$$H_q(x) = \sup_{\gamma\in\mathbb{R}}\Big\{x\gamma - \int_0^T H_q(\gamma, t)\mathrm{d}t\Big\},$$

$$H_q(\gamma, t) = \lim_{\delta\to 0}\frac{1}{\delta}\Big(\lim_{\bar{\lambda}\to\infty}\frac{1}{\bar{\lambda}}\log \mathsf{E}\Big[\exp\Big(\bar{\lambda}\int_t^{t+\delta}\gamma\cdot \\ \cdot\sum_{i=1}^{k}\sum_{j=n_{i-1}}^{n_i}(j - s_i)^+\cdot\mathcal{I}_{\{Q(v)=j\}}\mathrm{d}v\Big)\Big|Q(0)=i\Big]\Big).$$

Note that equations [10.20]–[10.21] and [10.23] are concerned with a fixed time T. Now we look at the situation in which T is not fixed but time varying. In view of [10.8], consider the k-dimensional process $\{\beta^\varepsilon(t) : t \ge 0\}$ given by

$$\beta^\varepsilon(t) = \int_0^t f(X^\varepsilon(v))\mathrm{d}v,$$

where $f(\cdot)$ is defined by [10.7].

For $T > 0$, let $\mathcal{C}^k[0,T]$ denote the set of all the continuous functions $g(\cdot)$: $[0,T] \to \mathbb{R}^k$, and $\mathcal{C}_x^k[0,T] = \{h(\cdot) : h(\cdot) \in \mathcal{C}^k[0,T]$ and $h(0) = x\}$. From theorem 5.1 in He *et al.* (2011), we have the following:

THEOREM 10.2.– For each $B \subset \mathcal{C}_0^k[0,T]$,

$$- \inf_{x(\cdot)\in B^o} G_T(x(\cdot)) \leq \liminf_{\varepsilon\to 0} \varepsilon \log \Pr\Big(\beta^\varepsilon(\cdot) \in B\Big),$$

$$- \inf_{x(\cdot)\in \bar{B}} G_T(x(\cdot)) \geq \limsup_{\varepsilon\to 0} \varepsilon \log \Pr\Big(\beta^\varepsilon(\cdot) \in B\Big),$$

where B^o and $\bar{B}$ are the interior and closure of B in $\mathcal{C}_0^k[0,T]$ with the uniform metric, and

$$G_T(x(\cdot)) = \begin{cases} \int_0^T L\Big(\frac{\mathrm{d}x(v)}{\mathrm{d}v}, v\Big)\mathrm{d}v, & \text{if } x(\cdot) \in \mathcal{C}_0^k[0,T] \text{ is absolutely continuous,} \\ \infty, & \text{otherwise} \end{cases}$$

with $y \in \mathbb{R}^k$ and $t \geq 0$,

$$L(y,t) = \sup_{\gamma\in\mathbb{R}^k} \Big\{\sum_{i=1}^k \gamma_i y_i - H(\gamma,t)\Big\},$$

where $H(y,t)$ is defined in theorem 10.1.

Consider the process generated by the cumulative time during which the arriving customers get balking for problems 10.1–10.3. Consider $L(t)$ given by [10.4]. In more detail, the process $\{L(t) : t \geq 0\}$ characterizes the periods over $[0,\infty)$ that indicate when the customer reneging occurs. Then theorem 10.2 shows that for any subset B of $\mathcal{C}_0^k[0,T]$ and large enough $\bar{\lambda}$,

$$\begin{aligned} \exp\Big(-\bar{\lambda}\cdot \inf_{x(\cdot)\in B^o} G_T(x(\cdot))\Big) &\leq \Pr\Big(L(\cdot) \in B\Big) \\ &\leq \exp\Big(-\bar{\lambda}\cdot \inf_{x(\cdot)\in \bar{B}} G_T(x(\cdot))\Big). \end{aligned} \qquad [10.24]$$

Going along the same line of [10.24], by [10.21] and [10.23], we can give the upper and lower bounds for

$$\Pr\Big(n_{s_i}(\cdot) \in B\Big) \text{ and } \Pr\Big(\int_0^{\cdot} \sum_{i=1}^k \sum_{j=n_{i-1}}^{n_i} (j-s_i)^+ \cdot \mathcal{I}_{\{Q(v)=j\}}\mathrm{d}v \in B\Big).$$

Finally, we look at $M^{\varepsilon}(T)$ given in [10.5]. To this end, in view of [10.15], we consider the general Markov chain $\{X^{\varepsilon}(t) : t \geq 0\}$ given by [10.6].

$$\widehat{m}^{\varepsilon}(t) = \sum_{i=0}^{N} \int_0^t [\mathcal{I}_{\{X^{\varepsilon}(y)=i\}} - \nu_i(y)]\beta_i(y)\mathrm{d}y. \qquad [10.25]$$

Here, $\beta(t)$ so that $\beta(t) = (\beta_1(t), \ldots, \beta_k(t))' \in \mathbb{R}^k$ is a step function on $[0, T]$, and

$$(\nu_0(y), \ldots, \nu_N(y))$$

is the quasi-steady-state distribution of the queue with the corresponding $\mathcal{G}(y)$ in [10.6]. From theorems 3.1 and 3.2 of He and Yin (2014), we have the following result of moderate deviations. The moderate deviations are reflected from the choice of the parameter $\kappa \in (0, 0.5)$. The moderate deviations together with the large deviations give us a complete characterization of the approximating errors. In what follows, we first state a result for $\widehat{m}^{\varepsilon}(t)$.

THEOREM 10.3.– There exists a functional $\widehat{H}(\cdot,\cdot) : \mathbb{R}^k \times [0, \infty] \mapsto \mathbb{R}$ such that

$$\lim_{\varepsilon \to 0} \varepsilon \log \mathsf{E}\Big(\exp(\varepsilon^{-1}\widehat{m}^{\varepsilon}(T))\Big| X^{\varepsilon}(0) = \ell\Big) = \int_0^T \widehat{H}(\beta(t), t)\mathrm{d}t. \qquad [10.26]$$

Moreover, for each $\ell = 1, \ldots, k$ and $\kappa \in (0, 0.5)$,

$$\lim_{\varepsilon \to 0} \varepsilon^{1-2\kappa} \log \mathsf{E}\Big(\exp(\varepsilon^{\kappa-1}\widehat{m}^{\varepsilon}(T))\Big| X^{\varepsilon}(0) = \ell\Big) = \frac{1}{2}\int_0^T \langle C(t)\beta(t), \beta(t)\rangle \mathrm{d}t, \qquad [10.27]$$

where $C(t)$ is the Hessian matrix of $\widehat{H}(y, t)$ at $y = 0$, i.e.

$$C_{ij}(t) = \left.\frac{\partial^2 \widehat{H}(y,t)}{\partial y_i \partial y_j}\right|_{y=0}.$$

Let $(\nu_0^q(t), \ldots, \nu_N^q(t))$ be the quasi-steady-state distribution of the queue length with the generator

$$\begin{pmatrix} -\frac{q_0(t)}{\bar\lambda} & \frac{\lambda(t)}{\bar\lambda} & & & \\ \frac{d_1(t)}{\bar\lambda} & -\frac{q_1(t)}{\bar\lambda} & \frac{\lambda(t)}{\bar\lambda} & & \\ & \ddots & \ddots & \ddots & \\ & & \frac{d_{N-1}(t)}{\bar\lambda} & -\frac{q_{N-1}(t)}{\bar\lambda} & \frac{\lambda(t)}{\bar\lambda} \\ & & & \frac{d_N(t)}{\bar\lambda} & -\frac{q_N(t)}{\bar\lambda} \end{pmatrix}.$$

Now we give the probability estimate for $M^\varepsilon(T)$ given in [10.5]. Define

$$\beta(t) = (\beta_0(t), \ldots, \beta_N(t)) = (1, \ldots, 1) \cdot I(t),$$

where $I(t)$ is defined in [10.14]. Then it follows from [10.15] that

$$M(T) = \sum_{i=0}^{N} \int_0^T \mathcal{I}_{\{Q^\varepsilon(y)=i\}} \beta_i(y) \mathrm{d}y.$$

From theorem 10.3, there exists a functional $\widehat{H}(\cdot,\cdot) : \mathbb{R}^k \times [0,\infty] \mapsto \mathbb{R}$ such that

$$\begin{aligned} &\mathsf{E}\Big[\exp\Big(\frac{1}{\bar\lambda}\sum_{i=0}^{N}\int_0^T \Big(\mathcal{I}_{\{Q^\varepsilon(y)=i\}} - \nu_i^q(y)\Big)\beta_i(y)\mathrm{d}y\Big)\Big|Q^\varepsilon(0)=\ell\Big] \\ &\approx \exp\Big(\bar\lambda\int_0^T \widehat{H}(\beta(t),t)\mathrm{d}t\Big). \end{aligned} \qquad [10.28]$$

Consider the cumulative time in which the customer abandonment can occur for problem 10.2. That is,

$$A(t) = \int_0^t \mathcal{I}_{\{Q(v)>s_v\}} \mathrm{d}v. \qquad [10.29]$$

Recall that the problem is related to the time-dependent staffing. Formally, the staffing level at interval $[t_\ell, t_{\ell+1})$ is given by $s = s_{\ell+1}$ with

$$[t_0, t_1), [t_1, t_2), \ldots, [t_{k-1}, t_k), \;\; t_0 = 0 \text{ and } t_k = \infty.$$

Now define $\beta(t) = (\beta_1(t), \ldots, \beta_N(t))$ by

$$\beta_i(t) = \begin{cases} 0, & \text{if } t \in [t_0, t_1) \text{ and } i \leq s_1, \\ 1, & \text{if } t \in [t_0, t_1) \text{ and } i > s_1, \\ 0, & \text{if } t \in [t_1, t_2) \text{ and } i \leq s_2, \\ 1, & \text{if } t \in [t_1, t_2) \text{ and } i > s_2, \\ \cdots\cdots & \\ 0, & \text{if } t \in [t_\ell, t_{\ell+1}) \text{ and } i \leq s_{\ell+1}, \\ 1, & \text{if } t \in [t_\ell, t_{\ell+1}) \text{ and } i > s_{\ell+1}, \\ \cdots\cdots & \\ 0, & \text{if } t \in [t_{k-1}, t_k) \text{ and } i \leq s_k, \\ 1, & \text{if } t \in [t_{k-1}, t_k) \text{ and } i > s_k, \end{cases}$$

$i = 1, \ldots, N$. Then we have

$$A(t) = \int_0^t \sum_{i=1}^N \mathcal{I}_{\{Q(v)=i\}} \beta_i(v) \mathrm{d}v.$$

Similar to the discussion for problem 10.3 (see [10.28]), we have that there exists a functional $\widehat{H}(\cdot, \cdot) : \mathbb{R}^k \times [0, \infty] \mapsto \mathbb{R}$ such that

$$\begin{aligned} &\mathsf{E}\Big[\exp\Big(\frac{1}{\bar{\lambda}} \sum_{i=1}^N \int_0^T \Big(\mathcal{I}_{\{Q^\varepsilon(y)=i\}} - \nu_i^q(y)\Big)\beta_i(y)\mathrm{d}y\Big)\Big|Q^\varepsilon(0) = \ell\Big] \\ &\approx \exp\Big(\bar{\lambda} \int_0^T \widehat{H}(\beta(t), t)\mathrm{d}t\Big). \end{aligned} \qquad [10.30]$$

10.4. Concluding remarks

Time-varying queues are much more difficult to handle compared to the time-invariant systems. To face the challenges, we used the large deviations principle developed for the two-time-scale Markov chains to analyze the probabilities for some performance measures in time-varying queues. We also provided moderate deviations analysis for the queueing systems. These performance measures include the abandonment process, the customer balking probability and facility use. The significance of the large deviations and moderate deviations results indicates that tail probabilities of interest are exponentially small, which gives us tight error bounds. It is clear that the quantity of crucial importance is the rate function. Noting [10.20]–[10.21], [10.23]–[10.24], [10.28] and [10.30], however, we only give theoretically the upper and lower bounds or approximation for these probabilities. It is challenging to find the exact analytic or closed-form expressions for the functions such as $H_b(\cdot)$, $H_{s_i}(\cdot)$, $G_T(\cdot)$ and $\widehat{H}(\cdot,\cdot)$ for the general generator given by [10.19]. Hence, it is also rather difficult to find the precise expression of the rate function. In our future work, we will develop numerical methods to systematically and numerically compute these values and the rate functions. From another perspective, Koroliuk and Limnios (2001) considered semi-Markov processes and various applications. It will be a worthwhile effort to see how the approach we presented in this chapter can be applied to semi-Markov processes and applications in queueing systems with the renewal arrival or the general service time (see Çinlar 1975). In Anisimov (1992, 2002, 2008), Anisimov studied averaging principles for switching processes. In particular, an idea of quasi-ergodicity was brought in. It seems that this idea may have some connections to the quasi-steady-state distribution used in our approach. To further look into this and study properties, such a large deviation is another direction of worthwhile effort.

10.5. References

Anisimov, V. (1992). Averaging principle for switching processes. *Theory Probab. Math. Statist.*, 46, 1–10.

Anisimov, V. (1995). Switching processes: Averaging principle, diffusion approximation and applications. *Acta Appl. Math.*, 40, 95–141.

Anisimov, V. (2002). Diffusion approximation in overloaded switching queueing models. *Queueing Syst.*, 40, 141–180.

Anisimov, V. (2008). *Switching Processes in Queueing Models*. ISTE Ltd, London, and Wiley, New York.

Armony, M., Israelit, S., Mandelbaum, A., Marmor, Y.N., Tseytlin, Y., Yom-Tov, G. (2015). On patient flow in hospitals: A data-based queueing-science perspective. *Stochastic Syst.*, 5, 146–194.

Batt, R.J., Terwiesch, C. (2015). Waiting patiently: An empirical study of queue abandonment in an emergency department. *Management Sci.*, 61, 39–59.

Çinlar, E., (1975). *Introduction to Stochastic Processes*. Prentice-Hall, Englewood Cliffs.

Green, L.V., Kolesar, P.J. (1991). The pointwise stationary approximation for queues with nonstationary arrivals. *Management Sci.*, 37, 84–97.

He, Q., Yin, G. (2014). Moderate deviations for time-varying dynamic systems driven by non-homogeneous Markov chains with two-time scales. *Stochastics*, 86, 527–550.

He, Q., Yin, G., Zhang, Q. (2011). Large deviations for two-time scale systems driven by non-homogeneous Markov chains and associated optimal control problems. *SIAM J. Control Optim.*, 49, 1737–1765.

Koroliuk, V.S., Limnios, N. (2005). *Stochastic Systems in Merging Phase Space*. World Scientific Publishing Co. Pte. Ltd., Hackensack.

Liu, Y. (2018). Staffing to stabilize the tail probability of delay in service systems with time-varying demand. *Oper. Res.*, 66, 514–534.

Liu, Y., Whitt, W. (2014). Many-server heavy-traffic limit for queues with time-varying parameters. *Ann. Appl. Probab.*, 24, 378–421.

Whitt, W. (1991). The pointwise stationary approximation for $M_t/M_t/s$ queues is asymptotically correct as the rates increase. *Management Sci.*, 37, 307–314.

Whitt, W. (2018). Time-varying queues. *Queueing Models Service Management*, 1, 79–164.

Yin, G., Zhang, Q. (2013). *Continuous-Time Markov Chains and Applications: A Two-Time-Scale Approach*, 2nd edition. Springer, New York.

Yin, G., Zhang, H.Q., Zhang, Q. (2013). *Applications of Two-Time-Scale Markovian Systems*. Science Press, Beijing.

Yom-Tov, G., Mandelbaum, A. (2014). Erlang-R: A time-varying queue with reentrant customers, in support of healthcare staffing. *Manufacturing Service Oper. Management*, 16, 283–299.

List of Authors

Attahiru Sule ALFA
University of Manitoba
Canada
and
University of Pretoria
South Africa

Djamil AÏSSANI
Research Unit LaMOS (Modeling and Optimization of Systems)
University of Béjaïa
Algeria

Vladimir ANISIMOV
Centre for Design & Analysis
Amgen Inc.
London
United Kingdom

Onno BOXMA
Eindhoven University of Technology
The Netherlands

Fabrice GUILLEMIN
Orange Labs
Lannion
France

Dimitri KOROLIOUK
Institute of Telecommunications and Global Information Space
National Academy of Sciences of Ukraine
Kiev
Ukraine

Vladimir S. KOROLIUK
Institute of Mathematics
National Academy of Sciences of Ukraine
Kiev
Ukraine

Igor Nikolaevich KOVALENKO†
V.M. Glushkov Institute of Cybernetics
National Academy of Sciences of Ukraine
Kiev
Ukraine

Igor KUZNETSOV
National Technical University of Ukraine "Igor Sikorsky Kyiv Polytechnical Institute"
Kiev
Ukraine

Nickolay KUZNETSOV
V.M. Glushkov Institute of Cybernetics
National Academy of Sciences of Ukraine
Kiev
Ukraine

Anna KVACH
National Research Tomsk State University
Russia

Eugene LEBEDEV
National Taras Shevchenko University
Kiev
Ukraine

Ouiza LEKADIR
Research Unit LaMOS (Modeling and Optimization of Systems)
University of Béjaïa
Algeria

Nikolaos LIMNIOS
L.M.A.C. (Laboratoire de Mathématiques Appliquées de Compiègne)
University of Technology of Compiègne
France

Hanna LIVINSKA
National Taras Shevchenko University
Kiev
Ukraine

Anatoly NAZAROV
National Research Tomsk State University
Russia

Boualem RABTA
Research Unit LaMOS (Modeling and Optimization of Systems)
University of Béjaïa
Algeria

Marie-Ange REMICHE
University of Namur
Belgium

Bruno SERICOLA
Inria Rennes – Bretagne Atlantique
Rennes
France

János SZTRIK
University of Debrecen
Hungary

Lucas VAN KREVELD
Korteweg-de Vries Institute
University of Amsterdam
The Netherlands

George YIN
Wayne State University
Detroit
USA

Hanqin ZHANG
School of Business
National University of Singapore
Singapore

Qing ZHANG
University of Georgia
Athens
USA

Index

M, P

Q, R

S, U, V

Printed and bound by CPI Group (UK) Ltd, Croydon, CR0 4YY
13/09/2026
14961043-0005